Birkhäuser

Nečas Center Series

More information about this series at https://link.springer.com/bookseries/16005

Eduard Feireisl • Antonin Novotný

Mathematics of Open Fluid Systems

Birkhäuser

Eduard Feireisl
Institute of Mathematics
Czech Academy of Sciences
Praha, Czech Republic

Antonin Novotný (deceased)
Carqueiranne, France

ISSN 2523-3343 ISSN 2523-3351 (electronic)
Nečas Center Series
ISBN 978-3-030-94792-7 ISBN 978-3-030-94793-4 (eBook)
https://doi.org/10.1007/978-3-030-94793-4

This book is published under the imprint Birkhäuser, www.birkhauser-science.com by the registered company Springer Nature Switzerland AG
The registered company address is: Gewerbestrasse 11, 6330 Cham, Switzerland

Preface

The goal of the book is to develop a mathematical theory of open fluid systems in the framework of continuum thermodynamics. By *open* we mean actively interacting with the outer world through the boundary of its physical domain. This should be contrasted with the *closed* system characterized by the celebrated statement of Clausius:

> **Die Energie der Welt ist konstant. Die Entropie der Welt strebt einem Maximum zu.**
>
> **The energy of the world (closed system) is constant; its entropy tends to a maximum.**
>
> *Rudolf Clausius*, Poggendorff's Annals of Physics 1865 (125), 400

Turbulence can be identified with the behaviour of open fluid systems in the long run. In this context, one of the major still largely open questions can be formulated as

> **Ergodic hypothesis:**
>
> **Time averages along trajectories of the flow converge, for large enough times, to an ensemble average given by a certain probability measure.**

To attack or at least to discuss these problems at the level of a rigorous mathematical theory we need

- A **mathematical model** of a fluid in motion that takes into account the basic physical principles, including the first and second law of thermodynamics

- A proper concept of **solution** existing on large, ideally infinite time intervals to capture the long-time behaviour of the system
- **Theoretical results** concerning the long-time behaviour of solutions to the underlying mathematical model

The present book can be seen as a first step in the quest for attaining these rather complex mathematical problems. It is definitely not for beginners. We suppose the reader is fairly conversant with the functional analytic framework, in particular the concept of weak solution and the related function spaces used in the modern theory of partial differential equations. The knowledge of the relevant theory for *closed* systems developed in the introductory part of the monograph [43] is highly recommend although not really necessary for reading the present material.

At the level of modelling, we simply adopt the well-established *Navier–Stokes–Fourier* system describing the motion of a general compressible, *linearly* viscous, and heat conducting fluid. Special attention is paid to a proper choice of the (inhomogeneous) boundary conditions pertinent to open fluid systems.

As our aim is to address the behaviour of the fluid in the long run, we develop a concept of *weak solution* based on the entropy rather than the energy balance equation. Given the present state of the art of the mathematical theory, this seems to be the only way to avoid problems with (hypothetical) singularities the classical solutions may develop in a finite time. The main obstacles are the rather poor *a priori* bounds available for the Navier–Stokes–Fourier system, not strong enough to control the energy transfer via convection. Fortunately, they are sufficient to control the entropy convection, so a suitable weak formulation is based on the field equations describing the mass conservation (equation of continuity), the momentum balance (momentum equation), and the entropy balance (entropy inequality). The lack of information on the exact entropy production rate is compensated, similarly to [43], by augmenting system by the total energy balance. The main difference between the closed system studied in [43] and the present setting is that now neither the total mass nor the total energy is conserved quantities.

We show that weak solutions exist globally in time for any physically relevant initial/boundary data. Moreover, the satisfaction of the basic laws of thermodynamics imposed through the system of field equations has an important impact on the long-time behaviour. In particular, we observe that the resulting dynamical system is dissipative in the sense of Levinson, meaning there exists a bounded absorbing set in the energy space. This fact will be used to shed some light on the validity of *ergodic hypothesis*.

The material is divided into three main parts devoted to (**i**) modelling, (**ii**) analysis, and (**iii**) qualitative properties. Each part consists of several chapters. Chapter 1 introduces the general concepts of continuum fluid mechanics based on balance laws and constitutive equations. Chapter 2 discusses the difference between open and closed fluid systems. Chapter 3 is devoted to generalized solutions to systems of partial differential equations of fluid dynamics. In particular, the concept of relative energy is introduced and the property of weak–strong uniqueness is shown. The existence theory is split into several steps described in Chaps. 5–9. Here we introduce a multilevel approximation scheme, the solutions of which generate the weak solutions of the Navier–Stokes–Fourier system in the asymptotic limit. Chapter 10 is devoted to the long-time behaviour of solutions. In particular, we address the questions of asymptotic compactness and global boundedness of trajectories. In Chap. 11, we briefly address the statistical theory of turbulence, prove the existence of statistical stationary solutions, and discuss validity of the ergodic hypothesis. Finally, some related problems, in particular the Dirichlet boundary conditions for the temperature, are discussed in Chap. 12.

In June 2021, when the book was almost finished, Antonín Novotný unexpectedly passed away. I was really shocked to hear this sad news. He was an exceptional collaborator and my true friend for decades.

Prague, Czech Republic Eduard Feireisl

Acknowledgements

The work of Eduard Feireisl was supported by the Czech Sciences Foundation (GAČR), Grant Agreement 21–02411S.

We thank our colleagues Danica Basarić, Hana Mizerová, and Mădălina Petcu for careful reading of the manuscript.

Contents

Preliminaries, Notation

0.1 Vectors, Tensors, Sets

R^d	d-dimensional Euclidean space
$x = (x_1, \dots x_d)$	Spatial variable
$B(x, r)$	Ball in R^d of radius r centred at x
t	Time
$\Omega \subset R^d$	Domain (connected open set) in R^d
$\partial\Omega$	Boundary of Ω in R^d
$\overline{\Omega}$	Closure of Ω in R^d
$\Omega^c, \Omega^c = R^d \setminus \Omega$	Complement of Ω in R^d
$\mathbf{v} = (v_1, \dots, v_d)$	d-dimensional vector in R^d
$\mathbf{v} \cdot \mathbf{w} = \sum_{i=1}^d v_i w_i$	Scalar product in R^d
$\mathbf{v} \times \mathbf{w},\ [\mathbf{v} \times \mathbf{w}]_{i,j} = v_i w_j - w_i v_j$	Vector product
$\mathbf{v} \otimes \mathbf{w}, [\mathbf{v} \otimes \mathbf{w}]_{i,j} = v_i w_j$	Tensor product of vectors
$R^{d\times d}$	Space of $d \times d$ matrices (tensors)
$R^{d\times d}_{\text{sym}}$	Space of symmetric $d \times d$ matrices (tensors)
$\mathbb{A} = (A_{i,j})_{i,j=1}^d$	Tensor in $R^{d\times d}$
$\mathbb{A}^t,\ A^t_{i,j} = A_{j,i}$	Adjoint to $\mathbb{A}$
$\mathbb{I}, I_{i,j} = 1$ if $i = j, I_{i,j} = 0$ if $i \neq j$	Identity tensor
$\mathbb{A} : \mathbb{B}, \mathbb{A} : \mathbb{B} = \sum_{i,j=1}^d A_{i,j} B_{j,i}$	Scalar product of tensors

0.2 Relations

$a \lesssim b$	$a \leq cb$ for a positive constant c
$a \gtrsim b$	$a \geq cb$ for a positive constant c
$a \approx b$	$a \lesssim b$ and $b \lesssim a$

0.3 Differential Operators

$\partial_y, \frac{\partial}{\partial y}$	Partial derivative with respect to y
$\mathrm{div}_x \mathbf{v}, \mathrm{div}_x \mathbf{v} = \sum_{i=1}^d \partial_{x_i} v_i$	Spatial divergence of $\mathbf{v}$
$\mathrm{div}_x \mathbb{A}, [\mathrm{div}_x \mathbb{A}]_i = \sum_{j=1}^d \partial_{x_j} A_{i,j}$	Spatial divergence of $\mathbb{A}$
$\nabla_x v, \nabla_x v = (\partial_{x_1} v, \dots, \partial_{x_d} v)$	Spatial gradient of a scalar function v
$\nabla_x \mathbf{v}, [\nabla_x \mathbf{v}]_{i,j} = \partial_{x_i} v_j$	Spatial gradient of a vector function $\mathbf{v}$
$\mathbb{D}_x \mathbf{v}, [\mathbb{D}_x \mathbf{v}]_{i,j} = \frac{1}{2}\left(\partial_{x_i} v_j + \partial_{x_j} v_i\right)$	Symmetric gradient
Δ_x	Laplace operator in the x variable
Δ_x^{-1}	Inverse Laplace operator on R^d

$$\Delta_x^{-1}[f](x) = -\frac{1}{4\pi} \int_{R^3} \frac{f(\xi)}{|x-\xi|} \, \mathrm{d}\xi \text{ if } d = 3,$$

$$\Delta_x^{-1}[f](x) = \frac{1}{2\pi} \int_{R^2} \log(|x-\xi|) f(\xi) \, \mathrm{d}\xi \text{ if } d = 2.$$

0.4 Special Functions

$\mathbb{1}_B, \mathbb{1}_B(y) = \begin{cases} 1 \text{ if } y \in B \\ 0 \text{ if } y \notin B \end{cases}$	Indicator function of a set B
$[v]^+, [v]^+ = \max\{v, 0\}$	Positive part of a real function v
$[v]^-, [v]^- = \min\{v, 0\}$	Negative part of a real function v
$\|Q\|$	d-dimensional Lebesgue measure of a set $Q \subset R^d$
$\mathrm{sgn}^+[v] = \begin{cases} 0 \text{ if } v \leq 0, \\ 1 \text{ if } v > 0 \end{cases}$	
$\|Q\|_{\mathrm{d}\sigma_x}$	$(d-1)$-dimensional Hausdorff measure of a set $Q \subset R^d$
$\|Q\|_{\mathrm{d}t \otimes \mathrm{d}\sigma_x}$	d-dimensional Hausdorff measure of a set $Q \subset (0, T) \times R^d$
$\mathcal{F}_{x \to \xi}$	Fourier transform on R^d
$\mathcal{F}_{\xi \to x}^{-1}$	Inverse Fourier transform on R^d

0.5 State Variables in Fluid Mechanics

$\varrho = \varrho(t, x)$	Mass density
$\mathbf{u} = \mathbf{u}(t, x)$	Fluid (bulk) velocity
$\vartheta = \vartheta(t, x)$	(Absolute) temperature
p	Pressure
e	Internal energy
s	Entropy

$S = \varrho s$	Total entropy
$\mathbf{q}$	Internal energy (heat) flux
$\mathbb{S}$	Viscous stress tensor

0.6 Geometry of Spatial Domains

$\mathcal{H}^d$	d-dimensional Hausdorff measure

A *domain* $\Omega \subset R^d$ is an open connected subset of R^d.

Definition 1 (Domain of Class $\mathcal{C}$).
Let $\Omega \subset R^d$ be a bounded domain. We say the Ω *is of class* $\mathcal{C}$ if for each $x \in \partial\Omega$ there exists $r > 0$ and a mapping $b : R^{d-1} \to R$ belonging to the class $\mathcal{C}$ such that, rotating and relabelling the coordinate axes if necessary, we have

$$\Omega \cap B(x,r) = \left\{ y \;\middle|\; b(y') < y_d \right\} \cap B(x,r),$$

$$\partial\Omega \cap B(x,r) = \left\{ y \;\middle|\; b(y') = y_d \right\},$$

where $y' = (y_1, \ldots, y_{d-1})$. In particular, we say that Ω is Lipschitz if b is Lipschitz.

We also consider the domains that are *piecewise* of class $\mathcal{C}$.

Definition 2 (Domain of Class Piecewise $\mathcal{C}$).
Let $\Omega \subset R^d$ be a bounded domain. We say the Ω *is of class piecewise* $\mathcal{C}$ if there exists a sequence $\{\Omega_n\}_{n=1}^{\infty}$ of domains of class $\mathcal{C}$ such that

- $$\Omega \subset \Omega_n \text{ for any } n = 1, 2 \ldots$$
- $$|\partial\Omega_n \setminus \partial\Omega|_{d-1} + |\partial\Omega \setminus \partial\Omega_n|_{d-1} \to 0 \text{ as } n \to \infty$$

where $|\cdot|_{d-1}$ denotes the $(d-1)$-Hausdorff measure.

We mostly consider domains that are piecewise C^k, $k \geq 1$, the boundary of which is a union of a finite number of C^k-manifolds with boundaries. It is easy to check that they fit in the class of piecewise C^k domains in the sense of Definition 2.

0.6.1 Weakly Lipschitz Domains

In certain applications, and, in particular in numerical analysis, the spatial domain may not be Lipschitz in the sense of Definition 1.

Definition 3 (Weakly Lipschitz Domain).
A bounded domain $\Omega \subset R^d$ is weakly Lipschitz if for any $x \in \Omega$ there exists a closed neighbourhood $U(x)$ and a bi-Lipschitz (Lipschitz with Lipschitz inverse) mapping

$$\Phi_x : U(x) \to [-1,1]^d$$

such that

- $$\Phi_x(x) = 0,$$
- $$\Phi_x(\Omega \cap U(x)) = [-1,1]^{d-1} \times [-1,0),$$
- $$\Phi_x(\partial\Omega \cap U(x)) = [-1,1]^{d-1} \times \{0\},$$
- $$\Phi_x((R^d \setminus \overline{\Omega}) \cap U(x)) = [-1,1]^{d-1} \times (0,1].$$

The following result was proved by Licht [66, Theorem 2.3].

Theorem 1 (Two-Sided Lipschitz Collar).
Let $\Omega \subset R^d$ be a weakly Lipschitz domain in the sense of Definition 3. Then there exists a bi-Lipschitz homeomorphism

$$\Phi : \partial\Omega \times [-1,1] \to R^d$$

such that

- $$\Phi(x,0) = x \text{ for all } x \in \partial\Omega,$$
  $$\Phi(\partial\Omega, [-1,0)) \subset \Omega,$$
  $$\Phi(\partial\Omega, (0,1]) \subset R^d \setminus \overline{\Omega}.$$

Note that the existence of (one-sided) Lipschitz collar is required by Chen, Torres, and Ziemer [18] in their analysis of normal traces of fields with measure divergence. In this book, we focus on Lipschitz domains to use the well-developed theory of Sobolev spaces, and, in particular, the embedding theorems. As our priority is to handle the open fluid systems with inhomogoneous boundary conditions, certain regularity of the boundary is necessary.

0.6.2 Distance Function, Nearest Point

If $\partial\Omega$ is a compact set, then the distance function

$$d = \mathrm{dist}[x, \partial\Omega] \text{ is Lipschitz continuous}$$

and

$$\nabla_x d(x) = \frac{x - \Pi(x)}{d(x)} \text{ for a.a. } x \in \partial\Omega,$$

where $\Pi(x)$ is the nearest point to x on $\partial\Omega$,

$$|x - \Pi(x)| = d(x),$$

see Ziemer [92, Chapter 1]. If, in addition, the boundary is of class C^k, $k \geq 2$, then d is of class C^k in a neighbourhood of $\partial\Omega$, see Foote [52]. In particular the nearest point mapping Π is of class C^{k-1} in the same neighbourhood.

0.7 Function Spaces

0.7.1 Spaces of Continuous Functions

$C(Q, X)$ Space of continuous functions ranging in X

$C_{\mathrm{weak}}(Q; X)$ Space of continuous functions with respect to the weak–topology on X

$C_c(Q; X)$ Space of continuous functions ranging in a Banach space X with compact support in Q

$BC(Q; X)$ Space of bounded continuous functions ranging in a Banach space X

$C^k(Q; X)$ Space of k-times continuously differentiable functions ranging in a Banach space X

$C^\alpha(Q; X)$ Space of α-Hölder continuous functions ranging in a Banach space X

$C^{\alpha,\beta}(I \times Q; X)$ Space of functions of $t \in I$, $x \in Q$, α-Hölder continuous in t and β-Hölder continuous in x ranging in a Banach space X

0.7.2 Spaces of Integrable Vector-Valued Functions

$L^p(Q; X)$ Lebesgue space of p-integrable functions ranging in X

If not otherwise specified, the integral is understood in Bochner's sense. Basic properties of the vector-valued Lebesgue functions can be found in Gajewski et al. [53, Chapter 2].

0.7.3 Sobolev Spaces

$W^{k,p}(Q; R^d)$ Sobolev space of functions on Q ranging in R^d with distributional derivatives up to order k in the Lebesgue space $L^p(\Omega)$, $1 \leq p \leq \infty$

$W_0^{k,p}(Q; R^d)$ Closure of $C_c^\infty(Q; R^d)$ in $W^{k,p}(\Omega; R^d)$

$W^{-k,p}(Q; R^d)$ Dual to $W^{k,p'}(Q; R^d)$, $\frac{1}{p} + \frac{1}{p'} = 1$

The basic properties of Sobolev functions, in particular the embedding theorems used in the present monograph, can be found in the classical textbooks Adams [1] or Ziemer [92].

0.7.4 Measures

$\mathcal{M}(Q)$ Set of regular Borel measures on Q

$\mathfrak{P}(Q)$ Set of probability measures on Q

0.8 Inverse of Div-Operator

Consider the problem

$$\text{div}_x \mathbf{v} = f \text{ in } \Omega, \ \mathbf{v}|_{\partial\Omega} = 0. \tag{1}$$

Denote $\mathbf{v} = \mathcal{B}[f]$. Bogovskii [10] proposed an explicit formula for $\mathcal{B}$. The specific properties of the operator $\mathcal{B}$ were elaborated by Galdi [54] and Geißert et al. [56]. The available results are summarized in [43, Chapter 11, Section 11.6].

Theorem 2 (Bogovskii Operator).
Let $\Omega \subset R^d$ be a bounded Lipschitz domain.
There exists a linear mapping

$$\mathcal{B} : \left\{ f \;\middle|\; f \in C_c^\infty(\Omega),\ \int_\Omega f \ \mathrm{d}x = 0 \right\} \to C_c^\infty(\Omega; R^d)$$

such that $\mathrm{div}_x \mathcal{B}[f] = f$, *meaning* $\mathbf{v} = \mathcal{B}[f]$ *solves* (1). *In addition, the following holds:*

-
$$\|\mathcal{B}[f]\|_{W_0^{m+1,p}(\Omega;R^d)} \le c(p, m, \Omega)\|f\|_{W_0^{m,p}(\Omega)}$$
for any $m = 0, 1, \ldots,\ 1 < p < \infty$.
- *If* $f \in L^p(\Omega)$, $f = \mathrm{div}_x \mathbf{h}$, *where*
$$\mathbf{h} \in L^q(\Omega; R^d),\ \mathbf{h} \cdot \mathbf{n}|_{\partial\Omega} = 0,$$
$1 < p < \infty$, $1 < q < \infty$, *then*
$$\|\mathcal{B}[f]\|_{L^q(\Omega;R^d)} \le c(p, q, \Omega)\|\mathbf{h}\|_{L^q(\Omega;R^d)}.$$

0.9 By Parts Integration for Vector-Valued Functions

Let V be a separable reflexive Banach space and H a Hilbert space such that

$$V \hookrightarrow H,\ V \text{ dense in } H.$$

Identifying H with its dual via the Riesz isometry we obtain the so-called *Gelfand triple*,

$$V \hookrightarrow H \approx H^* \hookrightarrow V^*.$$

Finally, consider

$$X = L^{p_1}(0, T; V) \cap L^{p_2}(0, T; H),\ 1 < p_1 \le p_2 < \infty,$$

with its dual

$$X^* = L^{q_1}(0, T; V^*) + L^{q_2}(0, T; H),\ \frac{1}{p_i} + \frac{1}{q_i} = 1,\ i = 1, 2.$$

Set

$$W = \left\{ v \;\middle|\; v \in X,\ \frac{\mathrm{d}}{\mathrm{d}t} v \in X^* \right\}.$$

The following by-parts integration formula was proved in Gajewski et al. [53, Chapter 4, & 1, Theorem 1.17].

Proposition 1 (By Parts Integration).

- $$W \hookrightarrow C([0,T]; H);$$
- *For any $v, w \in W$, we have*

$$[\langle v; w\rangle_H]_{t=\tau_1}^{t=\tau_2} = \int_{\tau_1}^{\tau_2} \left\langle \frac{\mathrm{d}}{\mathrm{d}t} v; w \right\rangle_{V^*;V} \mathrm{d}t + \left\langle \frac{\mathrm{d}}{\mathrm{d}t} w; v \right\rangle_{V^*;V} \mathrm{d}t$$

$0 \le \tau_1 \le \tau_2 \le T.$

0.10 Compensated Compactness

0.10.1 Div–Curl Lemma

There are several forms of this celebrated result that is considered to be a cornerstone of the theory of compensated compactness. We will be using the following version.

Lemma 1 (Div–Curl Lemma).
Let $Q \subset R^d$ be open and $1 < p, q < \infty$. Suppose

$$\mathbf{U}_n \to \mathbf{U} \text{ weakly in } L^p(Q; R^d),$$
$$\mathbf{V}_n \to \mathbf{V} \text{ weakly in } L^q(Q; R^d),$$
$$\frac{1}{p} + \frac{1}{q} = \frac{1}{r} < 1.$$

In addition, assume

$$\mathrm{div}_x \mathbf{U}_n \equiv \nabla \cdot \mathbf{U}_n \text{ is precompact in } W^{-1,s}(Q)$$
$$\mathrm{curl} \mathbf{V}_n \equiv \nabla \mathbf{V}_n - \nabla^t \cdot \mathbf{V}_n \text{ is precompact in } W^{-1,s}(Q; R^{d\times d})$$

for some $s > 1$.

Then

$$\mathbf{U}_n \cdot \mathbf{V}_n \to \mathbf{U} \cdot \mathbf{V} \text{ weakly in } L^r(W).$$

For the proof see Tartar [85], Murat [76], and [43, Chapter 11, Section 11.14].

0.10.2 Commutators Involving Riesz Operator

Let

$$\mathcal{R} = \mathcal{R}_{i,j=1}^{d}, \ \mathcal{R}_{i,j}[v](x) = \mathcal{F}^{-1}_{\xi\to x}\left[\frac{\xi_i\xi_j}{|\xi|^2}\mathcal{F}_{x\to\xi}[v]\right]$$

where $\mathcal{F} : R^d \to R^d$ denotes the Fourier transform. The following result is a straightforward consequence of Div–Curl Lemma.

Lemma 2 (Compactness Lemma).
Let

$$\mathbf{U}_n \to \mathbf{U} \textit{ weakly in } L^p(R^d; R^d),$$
$$\mathbf{V}_n \to \mathbf{V} \textit{ weakly in } L^q(R^d; R^d),$$
$$\frac{1}{p} + \frac{1}{q} = \frac{1}{r} < 1.$$

Then

$$\mathbf{U}_n\mathcal{R}\cdot[\mathbf{V}_n] - \mathcal{R}[\mathbf{U}_n]\cdot\mathbf{V}_n \to \mathbf{U}\mathcal{R}\cdot[\mathbf{V}] - \mathcal{R}[\mathbf{U}]\cdot\mathbf{V} \textit{ weakly in } L^r(R^d).$$

See [43, Chapter 11, Section 11.18, Theorem 11.34].

0.10.3 Commutator Lemma

Lemma 3 (Commutator Lemma).
Let $w \in W^{1,r}(R^d)$, $v \in L^p(R^d; R^d)$,

$$1 < r < d,\ 1 < p < \infty,\ \frac{1}{r} + \frac{1}{p} - \frac{1}{d} < 1$$

be given.
Then for any s *satisfying*

$$\frac{1}{r} + \frac{1}{p} - \frac{1}{d} < \frac{1}{s} < 1,$$

there exists

$$0 < \beta < 1,\ \frac{\beta}{d} = \frac{1}{s} + \frac{1}{d} - \frac{1}{p} - \frac{1}{r}$$

such that

$$\left\| \mathcal{R}[w\mathbf{v}] - w\mathcal{R}[\mathbf{V}] \right\|_{W^{\beta,s}(R^d;R^d)} \le c \|w\|_{W^{1,r}(R^d)} \|\mathbf{V}\|_{L^p(R^d;R^d)},$$

with a positive constant $c = c(s, p, r)$.

For the proof inspired by Coifman and Meyer [21] see [43, Chapter 11, Section 11.18, Theorem 11.35].

0.11 Measures on Infinite-Dimensional Spaces

Let $(\mathcal{T}, d_{\mathcal{T}})$ be a separable metric space, let $\mathfrak{P}[\mathcal{T}]$ denote the set of all Borel probability measures on $\mathcal{T}$. we say the

$$M_n \in \mathfrak{P}[\mathcal{T}] \to M \in \mathfrak{P}[\mathcal{T}] \text{ narrowly}$$

if

$$\int_{\mathcal{T}} \mathcal{F} \mathrm{d}M_n \equiv \langle M_n; \mathcal{F} \rangle \to \langle M; \mathcal{F} \rangle \equiv \int_{\mathcal{T}} \mathcal{F} \mathrm{d}M \text{ for any } \mathcal{F} \in BC(\mathcal{T}).$$

We say that a family $\mathcal{K} \subset \mathfrak{P}[\mathcal{T}]$ is tight if for any $\varepsilon > 0$ there exists a compact $C_\varepsilon \subset \mathcal{T}$ such that

$$M[\mathcal{T} \setminus C_\varepsilon] < \varepsilon \text{ for any } M \in \mathcal{K}.$$

Theorem 3 (Prokhorov Theorem).
Let $(\mathcal{T}, d_{\mathcal{T}})$ be a separable metric space. A collection $\mathcal{K} \subset \mathfrak{P}[\mathcal{T}]$ of probability measures on $\mathcal{T}$ is tight if and only if $\mathcal{K}$ is sequentially pre-compact in the topology of narrow convergence on $\mathfrak{P}[\mathcal{T}]$.

0.12 Miscellaneous Results Used in the Text

0.12.1 Korn–Poincaré Inequality

$$\|\mathbf{v}\|^p_{W^{1,p}(\Omega;R^d)} \leq c_{K,P}(p,\Omega) \int_\Omega \left| \nabla_x \mathbf{v} + \nabla_x^t \mathbf{v} - \frac{2}{d} \mathrm{div}_x \mathbf{v} \mathbb{I} \right|^p \, \mathrm{d}x, \ 1 < p < \infty,$$

whenever $\Omega \subset R^d$, $d = 2,3$ is a bounded Lipschitz domain and $\mathbf{v} \in W_0^{1,p}(\Omega; R^d)$, see e.g. [43, Chapter 11, Section 11.10].

0.12.2 Lions–Aubin Compactness Argument

There are several versions of this result that asserts compactness of a sequence of Lebesgue function provided some very mild information on the time derivative is available. We report the following result proved in [36, Chapter 6, Lemma 6.3].

Lemma 4 (Lions–Aubin Lemma).
Let $\{v_n\}_{n=1}^\infty$ be a sequence of functions,

$$\{v_n\}_{n=1}^\infty \text{ bounded in } L^2(0,T; L^q(\Omega)) \cap L^\infty(0,T; L^1(\Omega)), \ q > \frac{2d}{d+2}.$$

In addition, suppose

$$\partial_t v_n \geq g_n \text{ in } \mathcal{D}'((0,T) \times \Omega),$$

where

$$\{g_n\}_{n=1}^\infty \text{ is bounded in } L^1(0,T; W^{-m,r}(\Omega))$$

for some $m \geq 1$, $r > 1$.
Then, up to a suitable subsequence,

$$v_n \to v \text{ in } L^2(0,T; W^{-1,2}(\Omega)).$$

[illegible]

0.12 Miscellaneous Results Used in the Text

0.12.1 [illegible]

[illegible]

0.12.2 [illegible]

[illegible]

[illegible]

Part I
Modelling

Chapter 1
Mathematical Models of Fluids in Continuum Mechanics

Fluids in continuum mechanics are described through a family of observable physical quantities called *fields*. Examples of fields are the mass density ϱ, the macroscopic fluid velocity $\mathbf{u}$, the (absolute) temperature ϑ, among others. To describe the time evolution of fields, we use exclusively the *Eulerian reference system*. This means a particular value of a field is evaluated at a given time t and a spatial position $x \in \Omega$, where $\Omega \subset R^d$ is the cavity occupied by the fluid. This should be contrasted with the Lagrangian description, where the fluid paths—streamlines—are used to determine the reference coordinate system, see e.g. Chorin and Marsden [20]. Although both descriptions are equivalent as soon as the fluid motion is smooth, the Eulerian frame is more convenient to accommodate possible singularities in the weak formulation used in this book.

1.1 Conservation/Balance Laws

Basic physical principles can be expressed in terms of *balance laws*. Given a field of *volume density* d, we identify its *flux* $\mathbf{F}$ through any given surface, and its volume source s. The associated *balance law* is then written in the form:

$$\left[\int_B d(t,x) \mathrm{d}x \right]_{t=t_1}^{t=t_2} = - \int_{t_1}^{t_2} \int_{\partial B} \mathbf{F}(t,x) \cdot \mathbf{n} \ \mathrm{d}\sigma_x \mathrm{d}t + \int_{t_1}^{t_2} \int_B s(t,x) \ \mathrm{d}x \mathrm{d}t$$

for any volume element $B \subset \Omega$, any $t_1 < t_2$,

where $\mathbf{n}$ denotes the outer normal vector to ∂B. (1.1)

E. Feireisl, A. Novotný, *Mathematics of Open Fluid Systems*, Nečas Center Series, https://doi.org/10.1007/978-3-030-94793-4_1

At this stage, we assume that all quantities in question admit the regularity properties required and that B is smooth enough for the outer normal vector $\mathbf{n}$ to exist at any $x \in \partial B$.

As the above relation should hold for *any* $t_1 < t_2$ and *any* (Borel) subset $B \subset \Omega$, a routine application of Gauss–Green theorem converts the balance law (1.1) to its *differential form*:

$$\partial_t d(t,x) + \mathrm{div}_x \mathbf{F}(t,x) = s(t,x) \text{ for any } t, x \tag{1.2}$$

as long as all quantities are smooth enough. Although (1.1) represents a natural form of the balance law compatible with the laws of physics, it is the differential form (1.2) that is mostly used in textbooks.

1.1.1 Balance Laws of Continuum Fluid Dynamics

We review briefly the balance laws of continuum mechanics yielding the system of equations studied in this monograph. The first one is the principle of *mass conservation* that can be written as

equation of continuity:

$$\partial_t \varrho + \mathrm{div}_x(\varrho \mathbf{u}) = 0,$$

where $\varrho = \varrho(t,x)$ is the mass density, while $\mathbf{u} = \mathbf{u}(t,x)$ is the (bulk) velocity.

Next, *Newton's second law* or the balance of (linear) momentum can be written as

momentum balance:

$$\partial_t(\varrho \mathbf{u}) + \mathrm{div}_x(\varrho \mathbf{u} \otimes \mathbf{u}) = \mathrm{div}_x \mathbb{T} + \varrho \mathbf{g}, \tag{1.3}$$

where $\mathbb{T}$ is the *Cauchy stress tensor* and $\mathbf{g}$ is the volume density of external forces. In addition, fluids are characterized by *Stokes' law*

$$\mathbb{T} = \mathbb{S} - p\mathbb{I},$$

where $\mathbb{S}$ is the *viscous stress tensor* and p is a scalar quantity called *pressure*. The Cauchy stress $\mathbb{T}$, and, accordingly, the viscous stress tensor $\mathbb{S}$ are symmetric.

The next goal is to formulate the *First law of thermodynamics* or the energy balance. The scalar product of the momentum equation (1.3) with $\mathbf{u}$ yields the kinetic energy balance

$$\partial_t \left(\frac{1}{2} \varrho |\mathbf{u}|^2 \right) + \operatorname{div}_x \left(\frac{1}{2} \varrho |\mathbf{u}|^2 \mathbf{u} + p\mathbf{u} \right)$$
$$= \operatorname{div}_x (\mathbb{S} \cdot \mathbf{u}) - \mathbb{S} : \nabla_x \mathbf{u} + p \operatorname{div}_x \mathbf{u} + \varrho \mathbf{g} \cdot \mathbf{u}. \tag{1.4}$$

As $\mathbb{S}$ is symmetric, we can replace $\mathbb{S} : \nabla_x \mathbf{u}$ by $\mathbb{S} : \mathbb{D}_x \mathbf{u}$, where

$$\mathbb{D}_x \mathbf{u} = \frac{1}{2} \left(\nabla_x \mathbf{u} + \nabla_x^t \mathbf{u} \right)$$

is the symmetric part of the velocity gradient.

If viscous effects are active, Eq. (1.4) cannot be written as a conservation law; whence the kinetic energy must be augmented by the *internal energy*, denoted by e, to achieve the total energy conservation. Setting

$$\text{the total energy: } E = \left(\frac{1}{2} \varrho |\mathbf{u}|^2 + \varrho e \right)$$

we obtain

$$\partial_t E + \operatorname{div}_x (E\mathbf{u} + p\mathbf{u}) = \operatorname{div}_x \Big(\mathbb{S} \cdot \mathbf{u} - \mathbf{q} \Big) + \varrho \mathbf{g} \cdot \mathbf{u},$$

where $\mathbf{q}$ denotes the *internal energy flux*. Accordingly, we may formulate

internal energy balance:

$$\partial_t (\varrho e) + \operatorname{div}_x (\varrho e \mathbf{u}) + \operatorname{div}_x \mathbf{q} = \mathbb{S} : \mathbb{D}_x \mathbf{u} - p \operatorname{div}_x \mathbf{u}. \tag{1.5}$$

Obviously, the system of equations just derived is not closed; there are more unknowns than equations. Constitutive relations characterizing the *material properties* of the fluid must be added to obtain a (formally) closed system.

1.1.2 Constitutive Relations

The constitutive relations characterize the material properties of the fluid in question. We focus on linearly viscous (Newtonian) fluids, where the viscous stress is a linear function of $\mathbb{D}_x \mathbf{u}$. Accordingly, we consider the internal energy (heat) flux to be a linear function of $\nabla_x \vartheta$.

1.1.2.1 Thermodynamics

We start with *Gibbs' law*

$$\vartheta Ds = De + pD\left(\frac{1}{\varrho}\right), \tag{1.6}$$

where ϑ denotes the (absolute) temperature and s is a new thermodynamic quantity called *entropy*. The symbol D denotes differential and (1.6) should be interpreted in the following way: The quantity

$$\frac{1}{\vartheta}\left[De + pD\left(\frac{1}{\varrho}\right)\right]$$

is a perfect differential of a new field s.

Going back to (1.5) and using (1.6) we can derive a balance law for the entropy. Indeed we have

$$\begin{aligned}\frac{1}{\vartheta}\varrho\partial_t e &= \varrho\partial_t s + \frac{1}{\vartheta\varrho}p\partial_t\varrho = \varrho\partial_t s - \frac{1}{\vartheta\varrho}p\mathrm{div}_x(\varrho\mathbf{u})\\ &= \varrho\partial_t s - \frac{1}{\vartheta\varrho}p\nabla_x\varrho\cdot\mathbf{u} - \frac{1}{\vartheta}p\mathrm{div}_x\mathbf{u},\end{aligned}$$

and

$$\frac{1}{\vartheta}\varrho\mathbf{u}\cdot\nabla_x e = \varrho\mathbf{u}\cdot\nabla_x s + \frac{1}{\vartheta\varrho}p\mathbf{u}\cdot\nabla_x\varrho.$$

Thus dividing the internal energy equation (1.5) on ϑ and using the equation of continuity we derive

entropy balance equation:

$$\partial_t(\varrho s) + \mathrm{div}_x(\varrho s\mathbf{u}) + \mathrm{div}_x\left(\frac{\mathbf{q}}{\vartheta}\right) = \frac{1}{\vartheta}\left(\mathbb{S}:\mathbb{D}_x\mathbf{u} - \frac{\mathbf{q}\cdot\nabla_x\vartheta}{\vartheta}\right). \tag{1.7}$$

The quantity on the right-hand side of (1.7) is the *entropy production rate* and, in accordance with the *Second law of thermodynamics*,

$$\frac{1}{\vartheta}\left(\mathbb{S}:\mathbb{D}_x\mathbf{u} - \frac{\mathbf{q}\cdot\nabla_x\vartheta}{\vartheta}\right) \geq 0. \tag{1.8}$$

As the above inequality must hold for *any* physically admissible process, it imposes certain restrictions on the specific form of $\mathbb{S}$ and $\mathbf{q}$ discussed in the forthcoming section.

1.1.2.2 Transport Coefficients

To close the set of constitutive relations, the viscous stress as well as the internal energy flux $\mathbf{q}$ must be expressed in terms of the remaining thermodynamic quantities. We suppose the simplest possible relation between the velocity gradient and the viscous stress—the linear one. This can be formulated in terms of

Newton's rheological law:

$$\mathbb{S}(\mathbb{D}_x\mathbf{u}) = \mu\left(\nabla_x\mathbf{u} + \nabla_x\mathbf{u}^t - \frac{2}{d}\mathrm{div}_x\mathbf{u}\mathbb{I}\right) + \eta\mathrm{div}_x\mathbf{u}\mathbb{I}, \tag{1.9}$$

where $\mu \geq 0$ denotes the *shear viscosity coefficient* and $\eta \geq 0$ the *bulk viscosity coefficient*. Note that formula (1.9) looks rather awkward if $d = 1$, where the "shear" viscosity in fact vanishes. The present book, however, focuses on the physically relevant higher dimensional cases, in particular $d = 3$.

Similarly, we consider the internal energy flux $\mathbf{q}$ linearly proportional to the temperature gradient via

Fourier's law:

$$\mathbf{q} = -\kappa\nabla_x\vartheta, \tag{1.10}$$

where $\kappa \geq 0$ is the heat conductivity coefficient.

1.2 Navier–Stokes–Fourier System

Summarizing the material collected in the previous part we obtain the basic equations describing a general compressible, (linearly) viscous or Newtonian, and heat conducting fluid in motion.

- **Equation of continuity**

$$\partial_t\varrho + \mathrm{div}_x(\varrho\mathbf{u}) = 0. \tag{1.11}$$

- **Momentum equation**

$$\partial_t(\varrho\mathbf{u}) + \mathrm{div}_x(\varrho\mathbf{u}\otimes\mathbf{u}) + \nabla_x p = \mathrm{div}_x\mathbb{S}(\mathbb{D}_x\mathbf{u}) + \varrho\mathbf{g}. \tag{1.12}$$

- **Total energy balance**

$$\partial_t E + \mathrm{div}_x(E\mathbf{u} + p\mathbf{u}) = \mathrm{div}_x\Big(\mathbb{S}(\mathbb{D}_x\mathbf{u})\cdot\mathbf{u} - \mathbf{q}(\nabla_x\vartheta)\Big) + \varrho\mathbf{g}\cdot\mathbf{u}, \tag{1.13}$$

or, equivalently,

- **Internal energy balance**

$$\partial_t(\varrho e) + \mathrm{div}_x(\varrho e\mathbf{u}) + \mathrm{div}_x\mathbf{q}(\nabla_x\vartheta) = \mathbb{S}(\mathbb{D}_x\mathbf{u}) : \mathbb{D}_x\mathbf{u} - p\mathrm{div}_x\mathbf{u}, \tag{1.14}$$

or, equivalently,

- **Entropy balance**

$$\partial_t(\varrho s) + \mathrm{div}_x(\varrho s\mathbf{u}) + \mathrm{div}_x\left(\frac{\mathbf{q}(\nabla_x\vartheta)}{\vartheta}\right) = \frac{1}{\vartheta}\left(\mathbb{S}(\mathbb{D}_x\mathbf{u}) : \mathbb{D}_x\mathbf{u} - \frac{\mathbf{q}(\nabla_x\vartheta)\cdot\nabla_x\vartheta}{\vartheta}\right). \tag{1.15}$$

- **Gibbs' law**

$$\vartheta Ds = De + pD\left(\frac{1}{\varrho}\right). \tag{1.16}$$

- **Newton's rheological law**

$$\mathbb{S}(\mathbb{D}_x\mathbf{u}) = \mu\left(\nabla_x\mathbf{u} + \nabla_x\mathbf{u}^t - \frac{2}{d}\mathrm{div}_x\mathbf{u}\mathbb{I}\right) + \eta\mathrm{div}_x\mathbf{u}\mathbb{I}. \tag{1.17}$$

- **Fourier's law**

$$\mathbf{q} = -\kappa \nabla_x \vartheta. \tag{1.18}$$

The field equations (1.11)–(1.13), with the constitutive relations (1.17), (1.18), form the *Navier–Stokes–Fourier system.* Strictly speaking, the Eqs. (1.13), (1.14), (1.15) are equivalent only if all quantities in question are smooth (differentiable). This is no longer the case in the framework of weak solutions considered in this book. As we shall see below, a convenient weak formulation is based on a judicious combination of the entropy balance (inequality) combined with the total energy balance.

1.3 Thermodynamic Stability

The concept of *thermodynamic stability* will play a crucial role in the analysis of the Navier–Stokes–Fourier system. It concerns the constitutive equations relating the pressure p, the internal energy e, and the entropy s. If all these quantities are expressed as functions of the *standard thermodynamic variables* ϱ and ϑ,

the **hypothesis of thermodynamic stability** reads

$$\frac{\partial p(\varrho, \vartheta)}{\partial \varrho} > 0, \ \frac{\partial e(\varrho, \vartheta)}{\partial \vartheta} > 0 \text{ for any } \varrho > 0, \ \vartheta > 0. \tag{1.19}$$

The former condition says that compressibility of the fluid is positive, while the latter expresses positivity of the specific heat at constant volume.

Later, it will be useful to consider different sets of thermodynamic variables, in particular the *conservative entropy variables* —the density ϱ, the momentum $\mathbf{m} = \varrho\mathbf{u}$, and the total entropy $S = \varrho s$. To avoid confusion, we shall use the common notation

$$\frac{\partial e|_S}{\partial \varrho} = \frac{\partial e(\varrho, S)}{\partial \varrho}, \ \frac{\partial e|_\varrho}{\partial S} = \frac{\partial e(\varrho, S)}{\partial S},$$

and similarly for other thermodynamic functions.

Hypothesis of thermodynamic stability (1.19) can be equivalently stated as *convexity* of the internal energy $(\varrho, S) \mapsto \varrho e(\varrho, S)$ with respect to the conservative entropy variables (ϱ, S). Thanks to Gibbs' relation (1.16) and positivity of the absolute temperature, the mapping

$$(\varrho, \vartheta) \in (0, \infty)^2 \mapsto (\varrho, \varrho s(\varrho, \vartheta)) \in (0, \infty) \times R$$

is a diffeomorphism, the range of which depends on the specific form of the entropy s as a function of ϑ. We consider two situations conformal with the behaviour of real fluids:

$$\lim_{\vartheta \to 0+} s(\varrho, \vartheta) = -\infty, \ \lim_{\vartheta \to \infty} s(\varrho, \vartheta) = \infty \text{ for any } \varrho > 0, \tag{1.20}$$

or

$$\lim_{\vartheta \to 0+} s(\varrho, \vartheta) = 0, \ \lim_{\vartheta \to \infty} s(\varrho, \vartheta) = \infty \text{ for any } \varrho > 0. \tag{1.21}$$

The second alternative expresses the *Third law of thermodynamics* —the entropy vanishes for the temperature approaching the absolute zero, see Belgiorno [8, 9]. Accordingly, the total entropy $S = \varrho s$ ranges in the interval $(-\infty, \infty)$ if (1.20) holds, or in $(0, \infty)$ in the case (1.21).

Consider the internal energy

$$E_{\text{int}} = \varrho e : (\varrho, S) \mapsto \varrho e(\varrho, S)$$

as a function of (ϱ, S) defined on $(0, \infty) \times R$, $(0, \infty)^2$, respectively. In view of Gibbs' relation (1.16), it is easy to check that

$$\frac{\partial(\varrho e)|_S}{\partial \varrho} = e - \vartheta s + \frac{p}{\varrho}, \ \frac{\partial(\varrho e)|_\varrho}{\partial S} = \vartheta, \tag{1.22}$$

where the latter identity can be seen as a definition of the absolute temperature in terms of the conservative entropy variables.

Next, we introduce the *ballistic free energy*

$$H_{\tilde{\vartheta}}(\varrho, \vartheta) = \varrho\Big(e(\varrho, \vartheta) - \tilde{\vartheta} s(\varrho, \vartheta)\Big).$$

By means of a simple manipulation, we check the identity

$$\begin{aligned} E_{\text{int}}(\varrho, S) - \partial_\varrho E_{\text{int}}(\tilde{\varrho}, \widetilde{S})(\varrho - \tilde{\varrho}) - \partial_S E_{\text{int}}(\tilde{\varrho}, \widetilde{S})(S - \widetilde{S}) - E_{\text{int}}(\tilde{\varrho}, \widetilde{S}) \\ = H_{\tilde{\vartheta}}(\varrho, \vartheta) - \frac{\partial H_{\tilde{\vartheta}}(\tilde{\varrho}, \tilde{\vartheta})}{\partial \varrho}(\varrho - \tilde{\varrho}) - H_{\tilde{\vartheta}}(\tilde{\varrho}, \tilde{\vartheta}). \end{aligned} \tag{1.23}$$

Our goal is to show that the right-hand side of (1.23) is always non-negative and vanishes only if $\varrho = \tilde{\varrho}$, $\vartheta = \tilde{\vartheta}$. This implies (strict) convexity of E_{int} in the interior of its domain of definition.

First write

$$\begin{aligned} &H_{\tilde{\vartheta}}(\varrho, \vartheta) - \frac{\partial H_{\tilde{\vartheta}}(\tilde{\varrho}, \tilde{\vartheta})}{\partial \varrho}(\varrho - \tilde{\varrho}) - H_{\tilde{\vartheta}}(\tilde{\varrho}, \tilde{\vartheta}) \\ &\quad = H_{\tilde{\vartheta}}(\varrho, \tilde{\vartheta}) - \frac{\partial H_{\tilde{\vartheta}}(\tilde{\varrho}, \tilde{\vartheta})}{\partial \varrho}(\varrho - \tilde{\varrho}) - H_{\tilde{\vartheta}}(\tilde{\varrho}, \tilde{\vartheta}) + H_{\tilde{\vartheta}}(\varrho, \vartheta) - H_{\tilde{\vartheta}}(\varrho, \tilde{\vartheta}). \end{aligned}$$

Next, with the help of Gibbs' relation, compute

$$\frac{\partial^2 H_{\tilde{\vartheta}}(\varrho,\tilde{\vartheta})}{\partial \varrho^2} = \frac{1}{\varrho}\frac{\partial p(\varrho,\tilde{\vartheta})}{\partial \varrho}, \text{ and } \frac{\partial H_{\tilde{\vartheta}}(\varrho,\vartheta)}{\partial \vartheta} = \frac{\varrho}{\vartheta}(\vartheta - \tilde{\vartheta})\frac{\partial e(\varrho,\vartheta)}{\partial \vartheta}.$$

As a consequence of hypothesis of thermodynamic stability (1.19), we may infer that

- $$\varrho \mapsto H_{\tilde{\vartheta}}(\varrho,\tilde{\vartheta}) \text{ is strictly convex for any } \tilde{\vartheta} > 0,$$

- $$\vartheta \mapsto H_{\tilde{\vartheta}}(\varrho,\vartheta) \geq 0 \text{ and admits its strict local minimum (zero) at } \vartheta = \tilde{\vartheta}.$$

This yields the desired positivity of the right-hand side in (1.23).

Finally, we extend E_{int} to the whole space R^2. If (1.20) holds, we set

$$E_{\text{int}}(\varrho,S) = \begin{cases} \varrho e(\varrho,S) \text{ if } \varrho > 0, \\ \liminf_{r\to 0+, z\to S} re(r,z) \text{ if } \varrho = 0, \\ \infty \text{ otherwise.} \end{cases} \tag{1.24}$$

If (1.21) holds, we set

$$E_{\text{int}}(\varrho,S) = \begin{cases} \varrho e(\varrho,S) \text{ if } \varrho > 0,\ S > 0, \\ \liminf_{r\to 0+, z\to S} re(r,S) \text{ if } \varrho = 0,\ S > 0, \\ \liminf_{r\to \varrho, z\to 0+} \varrho e(\varrho,z) \text{ if } \varrho > 0,\ S = 0, \\ 0 \text{ if } \varrho = S = 0, \\ \infty \text{ otherwise.} \end{cases} \tag{1.25}$$

In both cases, E_{int} is a convex l.s.c. function of $(\varrho,S) \in R^2$, and, in accordance with (1.23), strictly convex in its domain.

1.4 Concluding Remarks

The material presented in this chapter is entirely standard except, possibly, some parts of Sect. 1.3. The interested reader may consult the existing monographs concerning the derivation of models in fluid mechanics: Batchelor [4], Becker [7], Chorin and Marsden [20], Gallavotti [55] to name only a few. The hypothesis of thermodynamic stability and its relation to convexity/concavity of thermodynamic functions was studied by Bechtel et al. [6]. A very elegant and illuminating treatment of convexity in thermodynamics was elaborated by Evans [31]. For general properties of convex functions see the monograph by Ekeland and Temam [29].

Chapter 2
Open vs. Closed Systems

By a *closed* system we mean a system that does not exchange any matter with its surroundings. In addition, an *isolated* system does not allow the transfer of energy either. This can be contrasted to *open* systems, where both matter and energy can be exchanged with the outer world. Although most systems of interest in applications are open, a significant amount of mathematics literature focuses, "for the sake of simplicity", on closed systems. The aim of this monograph is to fill up the gap.

From the mathematical point of view, the main difference between closed and open systems is the choice of boundary conditions. The boundary of a closed system is impermeable, meaning there is no mass flux through it. Similarly, for isolated systems, the normal component of the internal energy (heat) flux vanishes on the boundary. The open systems are characterized by more complex boundary conditions. In particular, the fluid velocity as well as the internal energy flux can be prescribed on the boundary.

We suppose the fluid occupies a bounded domain $\Omega \subset R^d$, $d = 1, 2, 3$. The mono-dimensional case $d = 1$ is rather special with a limited range of applications. We therefore mostly focus on the physically relevant case $d = 3$ whereas all results apply, with obvious modifications, to the cases $d = 1, 2$.

2.1 Closed and Isolated Systems

For a closed system, the physical boundary $\partial\Omega$ is *impermeable* . Mathematically speaking,

$$\mathbf{u} \cdot \mathbf{n}|_{\partial\Omega} = 0, \tag{2.1}$$

where $\mathbf{u}$ is the fluid bulk velocity and $\mathbf{n}$ denotes the outer normal vector to $\partial\Omega$. In addition, viscous fluids are expected to adhere completely to the boundary. If the latter is at rest, this gives rise to *no-slip* boundary condition

$$\mathbf{u} \times \mathbf{n}|_{\partial\Omega} = 0. \tag{2.2}$$

E. Feireisl, A. Novotný, *Mathematics of Open Fluid Systems*, Nečas Center Series, https://doi.org/10.1007/978-3-030-94793-4_2

Putting (2.1), (2.2) together yields

$$\mathbf{u}|_{\partial\Omega} = 0. \tag{2.3}$$

Very often, it is (2.3) that is referred to as no-slip.

Isolated systems do not allow for the transfer of energy through the boundary. In the present context, this yields, besides (2.3),

$$\mathbf{q} \cdot \mathbf{n}|_{\partial\Omega} = 0, \tag{2.4}$$

where $\mathbf{q}$ is the internal energy (heat) flux. Integrating the total energy balance (1.13) over the physical space Ω, we therefore obtain

$$\frac{\mathrm{d}}{\mathrm{d}t} \int_\Omega E \, \mathrm{d}x = \int_\Omega \varrho \mathbf{g} \cdot \mathbf{u} \, \mathrm{d}x.$$

Thus the total energy of the fluid is conserved unless stirred up by the external force $\mathbf{g}$.

If, moreover, the driving force is potential,

$$\mathbf{g}(x) = \nabla_x G(x),$$

then we can use the equation of continuity (1.11) to rewrite the forcing term as

$$\int_\Omega \varrho \mathbf{g} \cdot \mathbf{u} \, \mathrm{d}x = - \int_\Omega G \mathrm{div}_x(\varrho \mathbf{u}) \, \mathrm{d}x = \frac{\mathrm{d}}{\mathrm{d}t} \int_\Omega \varrho G \, \mathrm{d}x.$$

Consequently,

$$\frac{\mathrm{d}}{\mathrm{d}t} \int_\Omega (E - \varrho G) \, \mathrm{d}x = 0; \tag{2.5}$$

whence this modified energy is a constant of motion. By the same token, the total mass of a closed system is conserved,

$$\frac{\mathrm{d}}{\mathrm{d}t} \int_\Omega \varrho \, \mathrm{d}x = 0. \tag{2.6}$$

Finally, the entropy balance (1.15) integrated over Ω yields

$$\frac{\mathrm{d}}{\mathrm{d}t} \int_\Omega \varrho s \, \mathrm{d}x = \int_\Omega \frac{1}{\vartheta} \left(\mathbb{S} : \mathbb{D}_x \mathbf{u} - \frac{\mathbf{q} \cdot \mathbf{n}}{\vartheta} \right) \mathrm{d}x \geq 0; \tag{2.7}$$

whence the total entropy $- \int_\Omega \varrho s \, \mathrm{d}x$ plays a role of a Lyapunov function in accordance with the Second law of thermodynamics.

Under the conditions (2.5)–(2.7), the system will approach a time independent *equilibrium solution* for $t \to \infty$. We give a rigorous proof of this statement in Chap. 10, Sect. 10.2.1. Moreover, we also show that

$$\int_{\Omega} E(t, \cdot)\ \mathrm{d}x \to \infty \text{ as } t \to \infty$$

whenever $\mathbf{g}(x) \neq \nabla_x G(x)$.

In the light of these arguments, we may infer that

> the large time behaviour of closed energetically isolated fluid systems is apparently not compatible with the phenomena observed in a turbulent regime.

Turbulent fluid flows are excited by the outer sources interacting with the fluid through the physical boundary. Accordingly, relevant mathematical models must be based on open fluid systems. What is more, open systems typically operate far from the thermodynamic equilibrium. Mathematically speaking, a global non-perturbative theory is needed to describe their dynamics. Given the present state of the art, such a theory is possibly available in the framework of *weak solutions.*

2.2 Open Systems

For open systems, the impermeability/no slip boundary conditions are replaced by a general

> **Dirichlet type boundary condition for the velocity:**
>
> $$\mathbf{u}|_{\partial\Omega} = \mathbf{u}_B, \tag{2.8}$$

where $\mathbf{u}_B = \mathbf{u}_B(t, x)$ is a given field that may depend on time. Accordingly, given a time interval $[0, T]$, the boundary of the space–time cylinder $[0, T] \times \Omega$ can be written as

$$[0, T] \times \partial\Omega = \Gamma_{\text{in}} \cup \Gamma_{\text{out}} \cup \Gamma_{\text{wall}}, \tag{2.9}$$

where

$$\begin{aligned}
\Gamma_{\text{in}} &= \left\{ (t, x) \in [0, T] \times \partial\Omega \;\middle|\; \mathbf{u}_B(t, x) \cdot \mathbf{n}(x) < 0 \right\}, \\
\Gamma_{\text{out}} &= \left\{ (t, x) \in [0, T] \times \partial\Omega \;\middle|\; \mathbf{u}_B(t, x) \cdot \mathbf{n}(x) > 0 \right\}, \\
\Gamma_{\text{wall}} &= \left\{ (t, x) \in [0, T] \times \partial\Omega \;\middle|\; \mathbf{u}_B(t, x) \cdot \mathbf{n}(x) = 0 \right\}.
\end{aligned} \tag{2.10}$$

Recall our convention that $\mathbf{n}$ denotes the *outer* normal vector to $\partial\Omega$. We also admit that the fluid can flow along Γ_{wall}, a bit at odds with our statement that viscous fluids adhere to a rigid boundary.

Here, we have tacitly assumed that the outer normal vector exists at any boundary point $x \in \Omega$. One can easily imagine the situation when the in/out flow parts are separated by an edge on which $\mathbf{n}$ does not exist, in general. This type of domains will be included in the mathematical theory developed later in this book.

The density satisfies the equation of continuity that can be written as a transport equation

$$\partial_t \varrho + \mathbf{u} \cdot \nabla_x \varrho = -\varrho \mathrm{div}_x \mathbf{u}.$$

If the velocity $\mathbf{u}$ is smooth, we may define the characteristics (streamlines) $\mathbf{X} = \mathbf{X}(t, x)$ as the (unique) solution of the system of ordinary differential equations

$$\frac{\mathrm{d}}{\mathrm{d}t}\mathbf{X}(t,x) = \mathbf{u}(t, \mathbf{X}(t,x)), \ \mathbf{X}(0,x) = x.$$

Given a time $t \geq 0$ and a point $x \in \Omega$ there is exactly one characteristic curve that passes through x and the time t,

$$\mathbf{X}(t, x_0) = x.$$

Let

$$\tau = \inf \left\{ s \in [0,t] \ \middle| \ \mathbf{X}(z, x_0) \subset \Omega \text{ for all } z \in (s,t) \right\}.$$

If $\tau = 0$, the value of $\varrho(t,x)$ is determined by the initial density distribution $\varrho_0(x) = \varrho(0, \cdot)$,

$$\varrho(t,x) = \varrho_0(x_0) \exp \left(- \int_0^t \mathrm{div}_x \mathbf{u}(s, \mathbf{X}(s, x_0)) \mathrm{d}s \right).$$

If $\tau > 0$, then, necessarily, $\mathbf{X}(\tau, x_0) \in \partial\Omega$. One is tempted to say that in this case $(\tau, \mathbf{X}(\tau, x_0)) \in \Gamma_{\mathrm{in}}$ discarding automatically $(\tau, \mathbf{X}(\tau, x_0)) \in \Gamma_{\mathrm{out}}$ as well as the interior of Γ_{wall}. Indeed it can be shown that the complement of points, for which the ingoing characteristics do not emanate from Γ_{in} is of zero measure. Accordingly, we set

$$\varrho(t,x) = \varrho_B(\tau, \mathbf{X}(\tau, x_0)) \exp \left(- \int_\tau^t \mathrm{div}_x \mathbf{u}(s, \mathbf{X}(s, \mathbf{X}(\tau, x_0)) \mathrm{d}s \right),$$

where ϱ_B is the density prescribed on Γ_{in}, through

the **mass inflow boundary condition**

$$\varrho|_{\Gamma_{\mathrm{in}}} = \varrho_B. \tag{2.11}$$

Unfortunately, the weak solutions considered in this book do not enjoy sufficient regularity for the characteristics to be well defined. In particular, the velocity field fails (is not known) to satisfy $\mathrm{div}_x \mathbf{u} \in L^1(0,T; L^\infty(\Omega; R^d))$—

the weakest condition for the characteristics to be unique in some sense. Instead, our approach to the equation of continuity is based on the concept of renormalized solutions in the spirit of the seminal work of DiPerna and Lions [28].

Our choice of boundary conditions for the internal energy (heat) flux is inspired by Norman [77]. We consider a general

internal energy flux condition:

$$\left(\varrho e \mathbf{u}_B + \mathbf{q}\right) \cdot \mathbf{n}|_{\Gamma_{\rm in}} = F_{i,B} \leq 0, \ \ \mathbf{q} \cdot \mathbf{n}|_{\Gamma_{\rm out} \cup \Gamma_{\rm wall}} = 0. \tag{2.12}$$

We prescribe positive (remember $\mathbf{n}$ is the outer normal) input of heat $F_{i,B}$ on the inflow boundary while keeping the outflow part thermally isolated. The positivity of $-F_{i,B}$ is crucial and produces a stabilization effect used in the existence theory.

Needless to say there is a large variety of other possibilities, among which

the Dirichlet boundary conditions for the temperature:

$$\vartheta|_{\partial\Omega} = \vartheta_B \tag{2.13}$$

discussed in Chap. 12.

2.3 Global Form of Conservation Laws

The basic physical principles of conservation of mass, momentum, and energy are enforced *locally* in the fluid domain through the corresponding field equations. The *total amount* of these quantities, however, may vary in time for open fluid systems. Below, we assume that the physical space Ω is a *bounded* domain with sufficiently smooth boundary.

2.3.1 Total Mass Balance

Integrating the equation of continuity (1.11) over Ω and keeping the boundary conditions (2.8), (2.11) in mind, we obtain

total mass balance equation:

$$
\begin{aligned}
\frac{\mathrm{d}}{\mathrm{d}t}\int_{\Omega}\varrho(t,\cdot)\,\mathrm{d}x &= -\int_{\partial\Omega}\varrho\mathbf{u}(t,\cdot)\cdot\mathbf{n}\,\mathrm{d}\sigma_x \\
&= -\int_{\Gamma_{\mathrm{in}}\cap\{t\}}\varrho_B\mathbf{u}_B\cdot\mathbf{n}\,\mathrm{d}\sigma_x - \int_{\Gamma_{\mathrm{out}}\cap\{t\}}\varrho\mathbf{u}_B\cdot\mathbf{n}\,\mathrm{d}\sigma_x \\
&= -\int_{\partial\Omega}\varrho_B(t,\cdot)[\mathbf{u}_B(t,\cdot)\cdot\mathbf{n}]^-\mathrm{d}\sigma_x - \int_{\partial\Omega}\varrho(t,\cdot)[\mathbf{u}_B(t,\cdot)\cdot\mathbf{n}]^+\mathrm{d}\sigma_x.
\end{aligned}
\tag{2.14}
$$

We recall that here and always hereafter,

$$
[v]^- \equiv \min\{v,0\},\ [v]^+ \equiv \max\{v,0\}
$$

for any real valued function v. Note that if $\Gamma_{\mathrm{in}} = \emptyset$, and, simultaneously, $\Gamma_{\mathrm{out}} \neq \emptyset$, then the fluid may entirely disappear in a finite time $\tau > 0$, meaning $\varrho(\tau,\cdot) \equiv 0$. The system of field equations then becomes completely degenerate, and, in fact, physically irrelevant. To avoid such a singular situation, we shall always assume

$$
\Gamma_{\mathrm{in}} = \emptyset \ \Rightarrow \ \Gamma_{\mathrm{out}} = \emptyset. \tag{2.15}
$$

2.3.2 Total Energy Balance

Similarly to the above, we may integrate the energy balance (1.13) to obtain

$$
\begin{aligned}
&\frac{\mathrm{d}}{\mathrm{d}t}\int_{\Omega}\left(\frac{1}{2}\varrho|\mathbf{u}|^2 + \varrho e\right)\,\mathrm{d}x \\
&= -\int_{\partial\Omega}\frac{1}{2}\varrho_B|\mathbf{u}_B|^2\,[\mathbf{u}_B\cdot\mathbf{n}]^-\mathrm{d}\sigma_x - \int_{\partial\Omega}\frac{1}{2}\varrho|\mathbf{u}_B|^2\,[\mathbf{u}_B\cdot\mathbf{n}]^+\mathrm{d}\sigma_x \\
&\quad + \int_{\partial\Omega}F_{i,B}\frac{[\mathbf{u}_B\cdot\mathbf{n}]^-}{|\mathbf{u}_B\cdot\mathbf{n}|}\mathrm{d}\sigma_x - \int_{\partial\Omega}\varrho e\,[\mathbf{u}_B\cdot\mathbf{n}]^+\mathrm{d}\sigma_x \\
&\quad - \int_{\partial\Omega}p\,[\mathbf{u}_B\cdot\mathbf{n}]^-\mathrm{d}\sigma_x - \int_{\partial\Omega}p\,[\mathbf{u}_B\cdot\mathbf{n}]^+\mathrm{d}\sigma_x \\
&\quad + \int_{\Omega}\varrho\mathbf{g}\cdot\mathbf{u}\,\mathrm{d}x + \int_{\partial\Omega}\mathbb{S}(\mathbb{D}_x\mathbf{u}):(\mathbf{u}_B\otimes\mathbf{n})\mathrm{d}\sigma_x.
\end{aligned}
\tag{2.16}
$$

Here, the rightmost integral is a priori not well defined as the normal component of $\mathbb{D}_x\mathbf{u}$ is not prescribed on the boundary. Fortunately, the corresponding integral can be computed multiplying the momentum balance (1.12) on $\mathbf{u}_B$ and integrating the resulting expression over Ω. Of course, this step requires a suitable *extension* of the boundary velocity $\mathbf{u}_B$ inside Ω. A routine manipulation yields

$$\begin{aligned}
&\frac{\mathrm{d}}{\mathrm{d}t}\int_\Omega \varrho\mathbf{u}\cdot\mathbf{u}_B \ \mathrm{d}x \\
&= -\int_{\partial\Omega} \varrho_B|\mathbf{u}_B|^2 \ [\mathbf{u}_B\cdot\mathbf{n}]^- \mathrm{d}\sigma_x - \int_{\partial\Omega} \varrho|\mathbf{u}_B|^2 \ [\mathbf{u}_B\cdot\mathbf{n}]^+ \mathrm{d}\sigma_x \\
&\quad + \int_\Omega \Big[\varrho(\mathbf{u}\otimes\mathbf{u}) + p\mathbb{I}\Big] : \mathbb{D}_x\mathbf{u}_B \ \mathrm{d}x \\
&\quad - \int_{\partial\Omega} p \ [\mathbf{u}_B\cdot\mathbf{n}]^- \mathrm{d}\sigma_x - \int_{\partial\Omega} p \ [\mathbf{u}_B\cdot\mathbf{n}]^+ \mathrm{d}\sigma_x \\
&\quad + \int_\Omega \Big(\varrho\mathbf{u}\cdot\partial_t\mathbf{u}_B + \varrho\mathbf{g}\cdot\mathbf{u}_B\Big) \ \mathrm{d}x \\
&\quad - \int_\Omega \mathbb{S} : \mathbb{D}_x\mathbf{u}_B \ \mathrm{d}x + \int_{\partial\Omega} \mathbb{S} : (\mathbf{u}_B\otimes\mathbf{n})\mathrm{d}\sigma_x
\end{aligned} \tag{2.17}$$

In addition, we use the equation of continuity (1.11) to compute

$$\begin{aligned}
&\frac{\mathrm{d}}{\mathrm{d}t}\int_\Omega \frac{1}{2}\varrho|\mathbf{u}_B|^2 \ \mathrm{d}x \\
&= \int_\Omega \varrho\mathbf{u}_B\cdot\partial_t\mathbf{u}_B \ \mathrm{d}x + \frac{1}{2}\int_\Omega \varrho\mathbf{u}\cdot\nabla_x|\mathbf{u}_B|^2 \ \mathrm{d}x \\
&\quad - \frac{1}{2}\int_{\partial\Omega} \varrho_B|\mathbf{u}_B|^2 \ [\mathbf{u}_B\cdot\mathbf{n}]^- \mathrm{d}\sigma_x - \frac{1}{2}\int_{\partial\Omega} \varrho|\mathbf{u}_B|^2 \ [\mathbf{u}_B\cdot\mathbf{n}]^+ \mathrm{d}\sigma_x.
\end{aligned} \tag{2.18}$$

Thus summing up (2.16) with (2.18) and subtracting (2.17) we get

$$\begin{aligned}
&\frac{\mathrm{d}}{\mathrm{d}t}\int_\Omega \left[\frac{1}{2}\varrho|\mathbf{u}-\mathbf{u}_B|^2 + \varrho e\right] \mathrm{d}x \\
&= -\int_\Omega \Big[\varrho(\mathbf{u}\otimes\mathbf{u}) + p\mathbb{I} - \mathbb{S}(\mathbb{D}_x\mathbf{u})\Big] : \mathbb{D}_x\mathbf{u}_B \ \mathrm{d}x \\
&\quad + \int_{\partial\Omega} F_{i,B}\frac{[\mathbf{u}_B\cdot\mathbf{n}]^-}{|\mathbf{u}_B\cdot\mathbf{n}|}\mathrm{d}\sigma_x - \int_{\partial\Omega} \varrho e \ [\mathbf{u}_B\cdot\mathbf{n}]^+ \mathrm{d}\sigma_x \\
&\quad + \frac{1}{2}\int_\Omega \varrho\mathbf{u}\cdot\nabla_x|\mathbf{u}_B|^2 \ \mathrm{d}x + \int_\Omega \varrho(\mathbf{u}-\mathbf{u}_B)\cdot(\mathbf{g}-\partial_t\mathbf{u}_B) \ \mathrm{d}x.
\end{aligned} \tag{2.19}$$

Finally, we use the identity

$$\begin{aligned}
&-(\mathbf{u}\otimes\mathbf{u}) : \mathbb{D}_x\mathbf{u}_B + \frac{1}{2}\mathbf{u}\cdot\nabla_x|\mathbf{u}_B|^2 \\
&\qquad = -(\mathbf{u}-\mathbf{u}_B)\otimes(\mathbf{u}-\mathbf{u}_B) : \mathbb{D}_x\mathbf{u}_B + (\mathbf{u}_B-\mathbf{u})\cdot\mathbf{u}_B\cdot\nabla_x\mathbf{u}_B
\end{aligned} \tag{2.20}$$

to rewrite (2.19) as

the **total energy balance:**

$$\begin{aligned}
&\frac{\mathrm{d}}{\mathrm{d}t}\int_{\Omega}\left[\frac{1}{2}\varrho|\mathbf{u}-\mathbf{u}_B|^2+\varrho e\right]\,\mathrm{d}x+\int_{\partial\Omega}\varrho e\,[\mathbf{u}_B\cdot\mathbf{n}]^+\mathrm{d}\sigma_x\\
&=-\int_{\Omega}\Big[\varrho(\mathbf{u}-\mathbf{u}_B)\otimes(\mathbf{u}-\mathbf{u}_B)+p\mathbb{I}-\mathbb{S}(\mathbb{D}_x\mathbf{u})\Big]:\mathbb{D}_x\mathbf{u}_B\,\mathrm{d}x\\
&\quad+\int_{\Omega}\varrho(\mathbf{u}-\mathbf{u}_B)\cdot(\mathbf{g}-\partial_t\mathbf{u}_B-\mathbf{u}_B\cdot\nabla_x\mathbf{u}_B)\,\mathrm{d}x\\
&\quad+\int_{\partial\Omega}F_{i,B}\frac{[\mathbf{u}_B\cdot\mathbf{n}]^-}{|\mathbf{u}_B\cdot\mathbf{n}|}\mathrm{d}\sigma_x
\end{aligned}\tag{2.21}$$

It is easy to see that (2.21) is independent of the specific form of the extension $\mathbf{u}_B$, meaning the same equality holds with $\mathbf{u}_B$ replaced by $\widetilde{\mathbf{u}}_B$ as long as

$$\mathbf{u}_B=\widetilde{\mathbf{u}}_B \text{ in } [0,T]\times\partial\Omega.$$

It is worth noting the general structure of the outflow boundary integral that can be formally written as

$$\int_{\partial\Omega}\varrho e\,[\mathbf{u}_B\cdot\mathbf{n}]^+\mathrm{d}\sigma_x=\int_{\partial\Omega}\left[\frac{1}{2}\varrho|\mathbf{u}-\mathbf{u}_B|^2+\varrho e\right]\,[\mathbf{u}_B\cdot\mathbf{n}]^+\mathrm{d}\sigma_x;$$

whence the left-hand side of (2.21) reads

$$\frac{\mathrm{d}}{\mathrm{d}t}\int_{\Omega}\left[\frac{1}{2}\varrho|\mathbf{u}-\mathbf{u}_B|^2+\varrho e\right]\,\mathrm{d}x+\int_{\partial\Omega}\left[\frac{1}{2}\varrho|\mathbf{u}-\mathbf{u}_B|^2+\varrho e\right]\,[\mathbf{u}_B\cdot\mathbf{n}]^+\mathrm{d}\sigma_x.$$

In accordance with the First law of thermodynamics, the total energy of a *closed* system ($\mathbf{u}_B=0$, $\mathbf{g}=0$) is conserved.

In a similar way, we could derive the total entropy balance, however, we postpone this step to Sect. 3.1.

2.4 Concluding Remarks

Obviously, we have omitted a large number of physically relevant and mathematically challenging boundary conditions relevant to open systems. In particular, the Navier slip type boundary conditions

$$(\mathbb{S}\cdot\mathbf{n}+\lambda\mathbf{u})\times\mathbf{n}|_{\partial\Omega}=0$$

investigated in detail by Bulíček, Málek and Rajagopal [15], or the controlled heat flux conditions

$$\mathbf{q} \cdot \mathbf{n}|_{\partial\Omega} = d(\vartheta - \overline{\vartheta})|_{\partial\Omega}$$

used in [40]. As in both cases the flow of mechanical/internal energy is controlled on the boundary, they can be easily handled in the framework of the theory presented in this book.

Part II
Analysis

The main goal of this part is to introduce a class of generalized solutions to the Navier–Stokes–Fourier system, to show existence of global in time solutions for any finite energy initial data, and to establish the weak–strong uniqueness principle. Our approach is based on a suitable weak formulation of the problem, where the entropy balance and the total energy balance are replaced by inequalities. Although this may seem like enlarging the set of admissible weak solutions, the natural compatibility as well as the weak–strong uniqueness principle will be retained for this weak formulation.

Chapter 3
Generalized Solutions

Our concept of weak solutions is based on the *entropy formulation* of the *Navier–Stokes–Fourier system* considered in $(0,T)\times\Omega$:

- **Equation of continuity**

$$\partial_t \varrho + \mathrm{div}_x(\varrho \mathbf{u}) = 0. \tag{3.1}$$

- **Momentum balance**

$$\partial_t(\varrho \mathbf{u}) + \mathrm{div}_x(\varrho \mathbf{u}\otimes\mathbf{u}) + \nabla_x p = \mathrm{div}_x \mathbb{S} + \varrho \mathbf{g}. \tag{3.2}$$

- **Entropy balance**

$$\partial_t(\varrho s) + \mathrm{div}_x(\varrho s \mathbf{u}) + \mathrm{div}_x\left(\frac{\mathbf{q}}{\vartheta}\right) = \frac{1}{\vartheta}\left(\mathbb{S} : \mathbb{D}_x \mathbf{u} - \frac{\mathbf{q}\cdot\nabla_x\vartheta}{\vartheta}\right). \tag{3.3}$$

- **Dirichlet boundary condition for the velocity**

$$\mathbf{u}|_{\partial\Omega} = \mathbf{u}_B. \tag{3.4}$$

E. Feireisl, A. Novotný, *Mathematics of Open Fluid Systems*, Nečas Center Series, https://doi.org/10.1007/978-3-030-94793-4_3

- **Inflow boundary conditions for the density**

$$\varrho|_{\Gamma_{\text{in}}} = \varrho_B. \tag{3.5}$$

- **Boundary conditions for the internal energy flux**

$$\Big(\varrho e \mathbf{u}_B + \mathbf{q}\Big) \cdot \mathbf{n}|_{\Gamma_{\text{in}}} = F_{i,B} \leq 0, \ \mathbf{q} \cdot \mathbf{n}|_{\Gamma_{\text{out}}} = 0. \tag{3.6}$$

- **Total energy balance**

$$\begin{aligned}
&\frac{\mathrm{d}}{\mathrm{d}t} \int_\Omega \left[\frac{1}{2} \varrho |\mathbf{u} - \mathbf{u}_B|^2 + \varrho e \right] \, \mathrm{d}x + \int_{\partial\Omega} \varrho e \, [\mathbf{u}_B \cdot \mathbf{n}]^+ \mathrm{d}\sigma_x \\
&= - \int_\Omega \Big[\varrho (\mathbf{u} - \mathbf{u}_B) \otimes (\mathbf{u} - \mathbf{u}_B) + p\mathbb{I} - \mathbb{S} \Big] : \mathbb{D}_x \mathbf{u}_B \, \mathrm{d}x \\
&+ \int_\Omega \varrho (\mathbf{u} - \mathbf{u}_B) \cdot (\mathbf{g} - \partial_t \mathbf{u}_B - \mathbf{u}_B \cdot \nabla_x \mathbf{u}_B) \, \mathrm{d}x \\
&+ \int_{\partial\Omega} F_{i,B} \frac{[\mathbf{u}_B \cdot \mathbf{n}]^-}{|\mathbf{u}_B \cdot \mathbf{n}|} \mathrm{d}\sigma_x
\end{aligned} \tag{3.7}$$

Here and hereafter we suppose that $\Omega \subset R^d$, $d = 2, 3$ is a bounded domain with Lipschitz boundary. In particular, by Rademacher's theorem, the outer normal vector $\mathbf{n}$ exists at a.a. $x \in \partial\Omega$, where a.a. refers to the $\mathcal{H}^{d-1}$ Hausdorff measure on $\partial\Omega$. We will impose more restrictions on $\partial\Omega$ in the existence theory.

The boundary data $\mathbf{u}_B = \mathbf{u}_B(t, x)$, $\varrho_B = \varrho_B(t, x)$, and $F_{i,B} = F_{i,B}(t, x)$ are restrictions of smooth functions defined for $(t, x) \in R^{d+1}$, the required smoothness will be specified when necessary. Although we mostly discuss solutions defined on a bounded time interval $(0, T)$, the existence theory developed in Chap. 5 will provide global-in-time solutions defined for $t \in [0, \infty)$.

Finally, to avoid the awkward notation in the last boundary integral in (3.7), we write

$$F_{i,B} = f_{i,B} [\mathbf{u}_B \cdot \mathbf{n}]^-, \ f_{i,B} \geq 0; \tag{3.8}$$

whence

$$\int_{\partial\Omega} F_{i,B} \frac{[\mathbf{u}_B \cdot \mathbf{n}]^-}{|\mathbf{u}_B \cdot \mathbf{n}|} \mathrm{d}\sigma_x = - \int_{\partial\Omega} f_{i,B} \, [\mathbf{u}_B \cdot \mathbf{n}]^- \mathrm{d}\sigma_x.$$

3.1 Relative Energy

The *relative energy* plays a crucial role in the analysis of stability of weak solutions, in particular, it represents an indispensable tool in the proof of the *weak–strong uniqueness principle.* Expressed in the standard variables $(\varrho, \vartheta, \mathbf{u})$,

the **relative energy** reads

$$E\left(\varrho, \vartheta, \mathbf{u} \middle| \tilde{\varrho}, \tilde{\vartheta}, \tilde{\mathbf{u}}\right) = \frac{1}{2}\varrho|\mathbf{u}-\tilde{\mathbf{u}}|^2 + H_{\tilde{\vartheta}}(\varrho, \vartheta) - \frac{\partial H_{\tilde{\vartheta}}(\tilde{\varrho}, \tilde{\vartheta})}{\partial \varrho}(\varrho - \tilde{\varrho}) - H_{\tilde{\vartheta}}(\tilde{\varrho}, \tilde{\vartheta}), \tag{3.9}$$

where

$$H_{\tilde{\vartheta}}(\varrho, \vartheta) = \varrho\Big(e(\varrho, \vartheta) - \tilde{\vartheta} s(\varrho, \vartheta)\Big)$$

is the ballistic free energy introduced in Sect. 1.3. The same quantity written in the entropy conservative variables $(\varrho, S, \mathbf{m})$ reads

$$E\left(\varrho, S, \mathbf{m} \middle| \tilde{\varrho}, \widetilde{S}, \widetilde{\mathbf{m}}\right) = \frac{1}{2}\varrho\left|\frac{\mathbf{m}}{\varrho} - \frac{\widetilde{\mathbf{m}}}{\tilde{\varrho}}\right|^2$$
$$+ E_{\rm int}(\varrho, S) - \partial_\varrho E_{\rm int}(\tilde{\varrho}, \widetilde{S})(\varrho - \tilde{\varrho}) - \partial_S E_{\rm int}(\tilde{\varrho}, \widetilde{S})(S - \widetilde{S}) - E_{\rm int}(\tilde{\varrho}, \widetilde{S}), \tag{3.10}$$

where the total internal energy $E_{\rm int}$ considered as a function of (ϱ, S) has been specified in (1.24) and (1.25).

Similarly to (1.24) and (1.25), we introduce the *kinetic energy*

$$E_{\rm kin}(\varrho, \mathbf{m}) = \begin{cases} \frac{1}{2}\frac{|\mathbf{m}|^2}{\varrho} \text{ if } \varrho > 0, \\ 0 \text{ if } \varrho = 0, \ \mathbf{m} = 0, \\ \infty \text{ otherwise.} \end{cases} \tag{3.11}$$

Note that $E_{\rm kin} : R^{d+1} \to [0, \infty]$ is a convex lower semi-continuous function.

Seeing that

$$\frac{\partial E_{\rm kin}(\tilde{\varrho}, \widetilde{\mathbf{m}})}{\partial \varrho} = -\frac{1}{2}\frac{|\widetilde{\mathbf{m}}|^2}{\tilde{\varrho}^2}, \ \frac{\partial E_{\rm kin}(\tilde{\varrho}, \widetilde{\mathbf{m}})}{\partial m_i} = \frac{\widetilde{m}_i}{\tilde{\varrho}}, \ i = 1, \ldots, d,$$

we may rewrite (3.10) in a concise form

$$
\begin{aligned}
&E\left(\varrho, S, \mathbf{m} \middle| \tilde{\varrho}, \widetilde{S}, \widetilde{\mathbf{m}}\right) \\
&\quad = E(\varrho, S, \mathbf{m}) - \frac{\partial E(\tilde{\varrho}, \widetilde{S}, \widetilde{\mathbf{m}})}{\partial \varrho}(\varrho - \tilde{\varrho}) - \frac{\partial E(\tilde{\varrho}, \widetilde{S}, \widetilde{\mathbf{m}})}{\partial S}(S - \widetilde{S}) \\
&\quad - \frac{\partial E(\tilde{\varrho}, \widetilde{S}, \widetilde{\mathbf{m}})}{\partial \mathbf{m}} \cdot (\mathbf{m} - \widetilde{\mathbf{m}}) - E(\tilde{\varrho}, \widetilde{S}, \widetilde{\mathbf{m}}),
\end{aligned}
\tag{3.12}
$$

where we have set

$$
E(\varrho, S, \mathbf{m}) = E_{\mathrm{kin}}(\varrho, \mathbf{m}) + E_{\mathrm{int}}(\varrho, S).
$$

Under the hypothesis of thermodynamic stability introduced in Sect. 1.3, the relations (1.24), (1.25), and (3.11) imply that the (total) energy E is a convex lower semi-continuous function of the conservative entropy variables $(\varrho, S, \mathbf{m})$. The relative energy (3.12) is therefore well defined as long as $(\tilde{\varrho}, \widetilde{S}, \widetilde{\mathbf{m}})$ belong to the interior of $\mathrm{Dom}(E)$.

3.1.1 Relative Energy as Bregman Distance

As already pointed out, the relative energy is a simple but very useful tool in the analysis of the Navier–Stokes–Fourier system. It may be seen as a *Bregman distance (divergence)* associated to the total energy. Given a convex function F on a Banach space X, the Bregman distance associated to F is defined as

$$
d_B(f; g) = F(f) - \langle \partial F(g); (f - g) \rangle - F(g),
$$

see e.g. Sprung [84]. In general, d_B is not symmetric, however, it shares an important property with a distance, namely

$$
d_B(f; g) \geq 0 \text{ and } d_B(f; g) = 0 \ \Leftrightarrow \ f = g
$$

as soon as F is strictly convex. Consequently, as we have seen in Sect. 1.3, the relative energy expressed in terms of the conservative entropy variables is a Bregman distance as long as the hypothesis of thermodynamic stability holds. In the future, we mostly consider the integrated form of the relative energy,

$$
\mathcal{E}\left(\varrho, S, \mathbf{m} \middle| \tilde{\varrho}, \widetilde{S}, \widetilde{\mathbf{m}}\right) \equiv \int_{\Omega} E\left(\varrho, S, \mathbf{m} \middle| \tilde{\varrho}, \widetilde{S}, \widetilde{\mathbf{m}}\right) \, \mathrm{d}x
$$

3.2 Energy Balance Equations

We recall the balance laws satisfied by the kinetic, internal, and total energy. Some of these relations will be later retained as an integral part of the definition of weak solutions. The low regularity expected in the class of weak solutions does not allow, in general, to derive them directly from the field equations. Some steps in this direction have already been done in Sect. 2.3, where the total energy balance (3.7) was derived.

3.2.1 Kinetic Energy Balance

We start by a formal derivation of the kinetic energy balance. We suppose that all quantities in question are smooth, and, consequently, the momentum equation can be written as

$$\varrho \partial_t \mathbf{u} + \varrho \mathbf{u} \cdot \nabla_x \mathbf{u} + \nabla_x p = \mathrm{div}_x \mathbb{S} + \varrho \mathbf{g}.$$

Moreover, extending the boundary velocity $\mathbf{u}_B$ to the whole space we can write

$$\varrho \partial_t (\mathbf{u} - \mathbf{u}_B) + \varrho \mathbf{u} \cdot \nabla_x (\mathbf{u} - \mathbf{u}_B) + \nabla_x p = \mathrm{div}_x \mathbb{S} + \varrho \mathbf{g} - \varrho \partial_t \mathbf{u}_B - \varrho \mathbf{u} \cdot \nabla_x \mathbf{u}_B. \tag{3.13}$$

Next, we consider the scalar product of (3.13) with $\mathbf{u} - \mathbf{u}_B$ and use the equation of continuity

$$\partial_t \varrho + \mathrm{div}_x (\varrho \mathbf{u}) = 0$$

to obtain

$$\partial_t \left(\frac{1}{2} \varrho |\mathbf{u} - \mathbf{u}_B|^2 \right) + \mathrm{div}_x \left[\left(\frac{1}{2} \varrho |\mathbf{u} - \mathbf{u}_B|^2 \right) \mathbf{u} \right] + \nabla_x p \cdot (\mathbf{u} - \mathbf{u}_B)$$
$$= \mathrm{div}_x \mathbb{S} \cdot (\mathbf{u} - \mathbf{u}_B) + \left[\varrho \mathbf{g} - \varrho \partial_t \mathbf{u}_B - \varrho \mathbf{u} \cdot \nabla_x \mathbf{u}_B \right] \cdot (\mathbf{u} - \mathbf{u}_B)$$

Furthermore,

$$\nabla_x p \cdot (\mathbf{u} - \mathbf{u}_B) = \mathrm{div}_x (p(\mathbf{u} - \mathbf{u}_B)) - p \mathrm{div}_x (\mathbf{u} - \mathbf{u}_B),$$

and, similarly,

$$\mathrm{div}_x \mathbb{S} \cdot (\mathbf{u} - \mathbf{u}_B) = \mathrm{div}_x \left(\mathbb{S} \cdot (\mathbf{u} - \mathbf{u}_B) \right) - \mathbb{S} : \nabla_x (\mathbf{u} - \mathbf{u}_B).$$

Summing up the previous computation we obtain

kinetic energy balance:

$$\partial_t \left[\frac{1}{2}\varrho|\mathbf{u}-\mathbf{u}_B|^2\right] + \mathrm{div}_x\left(\left[\frac{1}{2}\varrho|\mathbf{u}-\mathbf{u}_B|^2\right]\mathbf{u}\right)$$
$$+\,\mathrm{div}_x(p(\mathbf{u}-\mathbf{u}_B)) + \mathrm{div}_x\left(\mathbb{S}\cdot(\mathbf{u}_B-\mathbf{u})\right)$$
$$= -\left[\varrho(\mathbf{u}-\mathbf{u}_B)\otimes(\mathbf{u}-\mathbf{u}_B) + p\mathbb{I} - \mathbb{S}\right] : \mathbb{D}_x\mathbf{u}_B$$
$$+\,\varrho\left[\mathbf{g} - \partial_t\mathbf{u}_B - (\mathbf{u}_B\cdot\nabla_x)\mathbf{u}_B\right]\cdot(\mathbf{u}-\mathbf{u}_B)$$
$$+\,p\mathrm{div}_x\mathbf{u} - \mathbb{S} : \mathbb{D}_x\mathbf{u}. \tag{3.14}$$

All terms depending on $\mathbf{u}_B$ or $\mathbf{g}$ can be associated with the effect of external forces. The remaining quantity

$$p\mathrm{div}_x\mathbf{u} - \mathbb{S} : \mathbb{D}_x\mathbf{u}$$

coincides with the forcing term in the internal energy balance discussed in the next section.

3.2.2 Internal and Total Energy Balance

The *internal energy balance* equation reads

$$\partial_t(\varrho e) + \mathrm{div}_x(\varrho e\mathbf{u}) + \mathrm{div}_x\mathbf{q} = \mathbb{S} : \mathbb{D}_x\mathbf{u} - p\mathrm{div}_x\mathbf{u}. \tag{3.15}$$

The sum of (3.14) and (3.15) gives rise to

total energy balance

$$\partial_t \left[\frac{1}{2}\varrho|\mathbf{u}-\mathbf{u}_B|^2 + \varrho e\right] + \mathrm{div}_x\left(\left[\frac{1}{2}\varrho|\mathbf{u}-\mathbf{u}_B|^2 + \varrho e\right]\mathbf{u}\right) + \mathrm{div}_x\mathbf{q}$$
$$+\,\mathrm{div}_x(p(\mathbf{u}-\mathbf{u}_B)) + \mathrm{div}_x\left(\mathbb{S}\cdot(\mathbf{u}_B-\mathbf{u})\right)$$
$$= -\left[\varrho(\mathbf{u}-\mathbf{u}_B)\otimes(\mathbf{u}-\mathbf{u}_B) + p\mathbb{I} - \mathbb{S}\right] : \mathbb{D}_x\mathbf{u}_B$$
$$+\,\varrho\left[\mathbf{g} - \partial_t\mathbf{u}_B - (\mathbf{u}_B\cdot\nabla_x)\mathbf{u}_B\right]\cdot(\mathbf{u}-\mathbf{u}_B). \tag{3.16}$$

Note carefully that (3.16) reduces to a conservation law for the total energy

$$\frac{1}{2}\varrho|\mathbf{u}|^2 + \varrho e$$

in the absence of external forcing. Relation (3.16) is a mathematical formulation of the *First law of thermodynamics.* Unfortunately, the total energy balance in its differential form (3.16) is not suitable for the *weak formulation* of the problem because of the lack of sufficiently strong a priori bounds to control the fluxes. Instead the integrated version of (3.16) stated in (3.7) will be used.

3.2.2.1 Barotropic Fluids

We say that a fluid is *barotropic* if the pressure $p = p(\varrho)$ depends only on the mass density. Leaving apart any discussion concerning the physical relevance of such restriction, we may compute $p\mathrm{div}_x\mathbf{u}$ directly from the equation of continuity. Indeed a straightforward manipulation yields

$$\partial_t P(\varrho) + \mathrm{div}_x(P(\varrho)\mathbf{u}) + p(\varrho)\mathrm{div}_x\mathbf{u} = 0, \tag{3.17}$$

where $P = P(\varrho)$ is termed *pressure potential* related to the pressure through the equation

$$P'(\varrho)\varrho - P(\varrho) = p(\varrho).$$

Note that P is determined by p modulo a linear function of ϱ.

Accordingly, the internal energy can be written in the form

$$\varrho e = \varrho Q(\vartheta) + P(\varrho),$$

where $Q(\vartheta)$ is determined by the thermal energy balance equation

$$\partial_t(\varrho Q(\vartheta)) + \mathrm{div}_x(\varrho Q(\vartheta)\mathbf{u}) + \mathrm{div}_x\mathbf{q} = \mathbb{S} : \mathbb{D}_x\mathbf{u}. \tag{3.18}$$

Moreover, the kinetic energy balance (3.14) can be rewritten as

mechanical energy balance:

$$\begin{aligned}
&\partial_t \left[\frac{1}{2}\varrho|\mathbf{u}-\mathbf{u}_B|^2 + P(\varrho)\right] + \mathrm{div}_x\left(\left[\frac{1}{2}\varrho|\mathbf{u}-\mathbf{u}_B|^2 + P(\varrho)\right]\mathbf{u}\right)\\
&+ \mathrm{div}_x(p(\mathbf{u}-\mathbf{u}_B)) + \mathrm{div}_x(\mathbb{S}\cdot(\mathbf{u}_B-\mathbf{u})) + \mathbb{S} : \mathbb{D}_x\mathbf{u}\\
&= -\Big[\varrho(\mathbf{u}-\mathbf{u}_B)\otimes(\mathbf{u}-\mathbf{u}_B) + p\mathbb{I} - \mathbb{S}\Big] : \mathbb{D}_x\mathbf{u}_B\\
&+ \Big[\varrho\mathbf{g} - \varrho\partial_t\mathbf{u}_B - \varrho(\mathbf{u}_B\cdot\nabla_x)\mathbf{u}_B\Big]\cdot(\mathbf{u}-\mathbf{u}_B).
\end{aligned} \tag{3.19}$$

In view of (3.18), the quantity $Q(\vartheta)$ can be seen as an "entropy" for the barotropic system. If neither $\mathbb{S}$ nor the data $\mathbf{g}$, $\mathbf{u}_B$ depend on ϑ, the system of field equations completely decouples and its mechanical part determines uniquely ϱ and $\mathbf{u}$ and can be handled as a separate problem. If this is the case, the term

$$\mathbb{S} : \mathbb{D}_x \mathbf{u}$$

on the left-hand side of (3.19) represents the irreversible dissipation of the mechanical energy into heat. Rather incorrectly from the point of view of physics but commonly in mathematics, the mechanical energy

$$\frac{1}{2}\varrho|\mathbf{u}|^2 + P(\varrho)$$

is then termed "entropy" of the barotropic system.

3.3 Weak Formulation

The theory developed in this book is based on the concept of generalized or weak solutions. There are several reasons among which the most prominent is possibly the absence of a priori bounds that would allow to control higher order derivatives of local (strong) solutions globally in time. Global in time solutions, in turn, are necessary to perform the asymptotic analysis in Part III.

Although there are many different ways how to introduce generalized (weak) solutions to the Navier–Stokes–Fourier system the following properties are mandatory:

- **Existence.** The weak solution exists globally in time for any physically admissible choice of initial/boundary data.
- **Compatibility.** A weak solution that possesses the regularity required for classical solutions is a classical solution.
- **Weak–strong uniqueness.** A weak solution coincides with the strong solution generated by the same initial/boundary data as long as the strong solution exists.

Even in the framework of the above general principles, there is a large space for introducing various degrees of generalizations. In numerical studies, it is convenient to deal with "very weak" solutions—measure-valued or weaker—that can be easily identified as limits of approximate schemes. If the main interest are the qualitative properties, the class of weak solutions is usually the closest to classical ones, for which the global in time existence can still be established. We pursue the latter philosophy in this book.

We consider the Navier–Stokes–Fourier system (3.1)–(3.7) in the time interval $[0, T)$, $T \leq \infty$ and in a bounded Lipschitz domain $\Omega \subset R^d$, $d = 2, 3$.

3.3.1 Equation of Continuity

The weak formulation of the equation of continuity (3.1) consists of a family of integral identities

$$\int_0^\tau \int_\Omega \left[\varrho \partial_t \varphi + \varrho \mathbf{u} \cdot \nabla_x \varphi \right] \mathrm{d}x\mathrm{d}t = \left[\int_\Omega \varrho \varphi \ \mathrm{d}x \right]_{t=0}^{t=\tau}$$
$$+ \int_0^\tau \int_{\partial\Omega} \varphi \varrho_B \left[\mathbf{u}_B \cdot \mathbf{n} \right]^- \mathrm{d}\sigma_x \mathrm{d}t + \int_0^\tau \int_{\partial\Omega} \varphi \varrho \left[\mathbf{u}_B \cdot \mathbf{n} \right]^+ \mathrm{d}\sigma_x \mathrm{d}t,$$
$$\varrho(0, \cdot) = \varrho_0 \tag{3.20}$$

for any $0 \leq \tau < T$, and any test function $\varphi \in C^1_c([0,T) \times \overline{\Omega})$.

In addition, we impose the natural restriction

$$\varrho \geq 0 \text{ a.a. in } (0,T) \times \Omega.$$

Remark 1. Some comments are in order.

- By imposing (3.20) we tacitly assume that all integrals are well defined and finite.
- Saying that $\varphi \in C^1_c([0,T) \times \overline{\Omega})$ we mean that φ is a restriction of a function $\varphi \in C^1_c((-\infty,T) \times R^d)$ on $[0,T) \times \overline{\Omega}$.
- Unlike the velocity $\mathbf{u}$, the density ϱ is merely an integrable function without a well defined *trace* on the boundary of the space–time cylinder. Still the $(d+1)$–dimensional quantity $[\varrho, \varrho\mathbf{u}]$ fits in the category of vector fields with spatio–temporal divergence

  $$\mathrm{DIV}_{t,x}[\varrho, \varrho\mathbf{u}] = \partial_t \varrho + \mathrm{div}_x(\varrho\mathbf{u})$$

  being a measure on $[0,T) \times \overline{\Omega}$, see Chen et al. [18]. In particular, we may define the normal trace of $[\varrho, \varrho\mathbf{u}]$ on the space–time cylinder $[0,\tau] \times \partial\Omega$ formally via Gauss–Green theorem

  $$\left[\int_\Omega \varrho\varphi \ \mathrm{d}x \right]_{t=0}^{t=\tau} + \int_0^\tau \int_{\partial\Omega} \varphi \varrho \mathbf{u} \cdot \mathbf{n} \ \mathrm{d}\sigma_x \mathrm{d}t \equiv \int_0^\tau \int_\Omega [\varrho, \varrho\mathbf{u}] \cdot \nabla_{t,x} \varphi \ \mathrm{d}x\mathrm{d}t$$

  for any Lipschitz $\varphi \in W^{1,\infty}(\{t=0, t=\tau\} \times \Omega \cup [0,\tau] \times \partial\Omega)$ extended as a Lipschitz function inside $[0,\tau] \times \overline{\Omega}$. Accordingly, equality (3.20) can interpreted as

$$t \mapsto \int_\Omega \varrho(t,\cdot)\phi \, \mathrm{d}x \in C_{\mathrm{loc}}[0,T) \text{ for any } \phi \in C_c^1(\Omega), \ \varrho(0,\cdot) = \varrho_0,$$
$$\varrho \mathbf{u} \cdot \mathbf{n}|_{[0,T)\times\partial\Omega} = \varrho_B[\mathbf{u}_B \cdot \mathbf{n}]^- + \varrho[\mathbf{u}_B \cdot \mathbf{n}]^+. \tag{3.21}$$

- The property (3.21) is frequently called *weak continuity*. If, in addition, $\varrho \in L^\infty(0,T;L^\gamma(\Omega))$, $\gamma > 1$, then (3.21) yields

$$\varrho \in C_{\mathrm{weak}}([0,\tau];L^\gamma(\Omega)).$$

Similar convention is tacitly adopted in the weak formulation of the remaining field equations presented below.

Alternatively, we may rewrite (3.20) in the form

$$\frac{\mathrm{d}}{\mathrm{d}t}\int_\Omega \varrho\phi \, \mathrm{d}x + \int_{\partial\Omega} \varrho\phi \, [\mathbf{u}_B \cdot \mathbf{n}]^+ \mathrm{d}\sigma_x = \int_\Omega \varrho\mathbf{u} \cdot \nabla_x\phi \, \mathrm{d}x - \int_{\partial\Omega} \phi\varrho_B \, [\mathbf{u}_B \cdot \mathbf{n}]^- \mathrm{d}\sigma_x$$

for any $\phi \in C^1(\overline{\Omega})$, reminiscent of the total energy balance (3.7).

3.3.2 Momentum Equation

In view of the viscous effects, we anticipate the velocity field $\mathbf{u}$ to enjoy at least higher spatial regularity,

$$\mathbf{u} \in L^q(0,T;W^{1,q}(\Omega;R^d)) \text{ for some } q > 1. \tag{3.22}$$

Accordingly, the boundary condition (3.4) can be interpreted as

$$(\mathbf{u} - \mathbf{u}_B) \in L^q(0,T;W_0^{1,q}(\Omega;R^d)). \tag{3.23}$$

The weak formulation of the momentum equation (3.2) reads

$$\int_0^\tau \int_\Omega \Big[\varrho\mathbf{u} \cdot \partial_t\boldsymbol{\varphi} + \varrho(\mathbf{u}\otimes\mathbf{u}) : \nabla_x\boldsymbol{\varphi} + p\mathrm{div}_x\boldsymbol{\varphi}\Big] \, \mathrm{d}x\mathrm{d}t$$
$$= \int_0^\tau \int_\Omega \Big[\mathbb{S} : \nabla_x\boldsymbol{\varphi} - \varrho\mathbf{g}\cdot\boldsymbol{\varphi}\Big] \, \mathrm{d}x\mathrm{d}t + \left[\int_\Omega \varrho\mathbf{u}\cdot\boldsymbol{\varphi} \, \mathrm{d}x\right]_{t=0}^{t=\tau},$$
$$\varrho\mathbf{u}(0,\cdot) = \mathbf{m}_0 \tag{3.24}$$

for any $0 \le \tau < T$, and any $\boldsymbol{\varphi} \in C_c^1([0,T)\times\Omega;R^d)$.

Similarly to the density, the momentum $\mathbf{m}$ is a weakly continuous function of t, therefore we are allowed to prescribe its initial value $\mathbf{m}_0$, cf. Remark 1.

Remark 2. Validity of (3.24) can be extended to the class of test functions $\varphi \in C^1_c([0,T) \times \overline{\Omega}; R^d)$, $\varphi|_{\partial\Omega} = 0$.

3.3.3 Entropy Balance

The concept of weak solution promoted in this monograph is based on the *Second law of thermodynamics* enforced through the entropy balance (3.3). Note that the boundary conditions (3.6) expressed in terms of the entropy yield

$$\begin{aligned} \varrho s \mathbf{u}\cdot\mathbf{n} + \frac{\mathbf{q}}{\vartheta}\cdot\mathbf{n} &= \varrho_B\left(s(\varrho_B,\vartheta) - \frac{e(\varrho_B,\vartheta)}{\vartheta}\right)[\mathbf{u}_B\cdot\mathbf{n}]^- \\ &\quad + \frac{f_{i,b}}{\vartheta}[\mathbf{u}_B\cdot\mathbf{n}]^- \text{ on } \Gamma_{\rm in}, \\ \varrho s \mathbf{u}\cdot\mathbf{n} + \frac{\mathbf{q}}{\vartheta}\cdot\mathbf{n} &= \varrho s[\mathbf{u}_B\cdot\mathbf{n}]^+ \text{ on } \Gamma_{\rm out},\ \varrho s\mathbf{u}\cdot\mathbf{n} + \frac{\mathbf{q}}{\vartheta}\cdot\mathbf{n} = 0 \text{ on } \Gamma_{\rm wall}. \end{aligned} \tag{3.25}$$

Being aware of the fact that we are not able to control higher order derivatives of $\mathbf{u}$ and ϑ we replace the entropy balance (3.3) by inequality

$$\partial_t(\varrho s) + {\rm div}_x(\varrho s\mathbf{u}) + {\rm div}_x\left(\frac{\mathbf{q}}{\vartheta}\right) \geq \frac{1}{\vartheta}\left(\mathbb{S} : \mathbb{D}_x\mathbf{u} - \frac{\mathbf{q}\cdot\nabla_x\vartheta}{\vartheta}\right). \tag{3.26}$$

Accordingly, the weak formulation of (3.26) and (3.25) reads

$$\begin{aligned} &\int_0^T\int_\Omega\left[\varrho s\partial_t\varphi + \varrho s\mathbf{u}\cdot\nabla_x\varphi + \left(\frac{\mathbf{q}}{\vartheta}\right)\cdot\nabla_x\varphi\right]{\rm d}x{\rm d}t \\ &\leq -\int_0^T\int_\Omega\frac{\varphi}{\vartheta}\left(\mathbb{S}:\mathbb{D}_x\mathbf{u} - \frac{\mathbf{q}\cdot\nabla_x\vartheta}{\vartheta}\right){\rm d}x{\rm d}t \\ &+ \int_0^T\int_{\partial\Omega}\varphi\left(\frac{f_{i,B}}{\vartheta} + \varrho_B\left[s(\varrho_B,\vartheta) - \frac{e(\varrho_B,\vartheta)}{\vartheta}\right]\right)[\mathbf{u}_B\cdot\mathbf{n}]^-{\rm d}\sigma_x{\rm d}t \\ &+ \left[\int_\Omega\varphi\varrho s\,{\rm d}x\right]_{t=0}^{t=\tau} + \int_0^T\int_{\partial\Omega}\varphi\varrho s\,[\mathbf{u}_B\cdot\mathbf{n}]^+{\rm d}\sigma_x{\rm d}t, \\ &\varrho s(0,\cdot) = S_0, \end{aligned} \tag{3.27}$$

for any $\varphi \in C^1_c([0,T) \times \overline{\Omega})$, $\varphi \geq 0$.

Here again some comments are in order. First, formula (3.27) contains the "trace" of $S = \varrho s$ on $\Gamma_{\rm out}$. On the one hand, this quantity is a priori not well defined, not even in the sense of normal traces as the vector field that admits the normal trace is

$$\left[\varrho s, \varrho s \mathbf{u} + \frac{\mathbf{q}}{\vartheta}\right].$$

On the other hand, however, the value of the density ϱ on $\Gamma_{\rm out}$ is uniquely determined by (3.20), while the trace of ϑ exists in the Sobolev sense since that dissipation rate

$$\int_0^T \int_\Omega \frac{1}{\vartheta} \left(\mathbb{S} : \mathbb{D}_x \mathbf{u} - \frac{\mathbf{q} \cdot \nabla_x \vartheta}{\vartheta} \right) \ {\rm d}x$$

is bounded, which yields bounds on $\nabla_x \vartheta$. Consistently with (3.20), we may therefore define ϱs on $\Gamma_{\rm out}$ as $\varrho s(\varrho, \vartheta)$, where ϱ is the same as in (3.20), while $\vartheta|_{\partial\Omega}$ is the Sobolev trace of ϑ.

Second, unlike ϱ and $\mathbf{m}$, the total entropy S is not weakly continuous in time. Still using (3.27) we deduce

$$\begin{aligned} &\int_0^T \partial_t \psi \int_\Omega \varrho s \phi {\rm d}x + \int_0^T \psi \int_\Omega \left[\varrho s \mathbf{u} \cdot \nabla_x \phi + \left(\frac{\mathbf{q}}{\vartheta} \right) \cdot \nabla_x \phi \right] \ {\rm d}x{\rm d}t \\ &\quad - \int_0^T \psi \int_{\partial\Omega} \phi \varrho s \ [\mathbf{u}_B \cdot \mathbf{n}]^+ {\rm d}\sigma_x {\rm d}t \\ &\quad - \int_0^T \psi \int_{\partial\Omega} \phi \varrho_B \left[s(\varrho_B, \vartheta) - \frac{e(\varrho_B, \vartheta)}{\vartheta} \right] \ [\mathbf{u}_B \cdot \mathbf{n}]^- {\rm d}\sigma_x {\rm d}t \leq 0 \end{aligned}$$

for any $\psi \in C^\infty_c(0,T)$, $\phi \in C^1(\overline{\Omega})$, $\psi \geq 0$, $\phi \geq 0$. This implies that for any $\phi \in C^1(\overline{\Omega})$, $\phi \geq 0$, the function of time

$$t \in [0,T) \mapsto \int_\Omega (\varrho s)(t, \cdot) \phi \ {\rm d}x$$

can be written as a sum of a non-decreasing and a continuous function. Consequently, the one-sided limits

$$\begin{aligned} \int_\Omega \varrho s(\tau-, \cdot) \phi \ {\rm d}x &\equiv \lim_{\delta \searrow 0} \frac{1}{\delta} \int_{\tau-\delta}^\tau \int_\Omega (\varrho s)(t, \cdot) \phi \ {\rm d}x \ {\rm d}t, \\ \int_\Omega \varrho s(\tau+, \cdot) \phi \ {\rm d}x &\equiv \lim_{\delta \searrow 0} \frac{1}{\delta} \int_\tau^{\tau+\delta} \int_\Omega (\varrho s)(t, \cdot) \phi \ {\rm d}x{\rm d}t \end{aligned} \tag{3.28}$$

exist for any $0 < \tau < T$ and any $\phi \in C^1(\overline{\Omega})$, $\phi \geq 0$. Anticipating that ϱs is integrable, the above relation can be extended to $\phi \in C(\overline{\Omega})$, $\phi \geq 0$, and, then, writing $\phi = [\phi]^+ + [\phi]^-$, to any $\phi \in C(\overline{\Omega})$. Thus we can identify the *instantaneous value* of the total entropy S as

$$\int_\Omega S(\tau, \cdot) \phi \ {\rm d}x \equiv \int_\Omega \varrho s(\tau-, \cdot) \phi \ {\rm d}x \ \text{for any } \phi \in C(\overline{\Omega}),$$

meaning the integral mean

$$t \in [0,T) \mapsto \int_{\Omega} S(\tau,\cdot)\phi \, \mathrm{d}x$$

is interpreted as a càglàd function of time (continuous from the left, admitting a limit from the right) for any $\phi \in C(\overline{\Omega})$. In particular, the instantaneous value $S(\tau,\cdot)$ can be identified with a regular Borel measure on $\overline{\Omega}$. Accordingly, the rightmost term in the entropy inequality (3.27) should be interpreted in the same way.

3.3.4 Total Energy Balance

Replacing the entropy balance (3.3) by inequality (3.27) obviously extends the class of admissible weak solutions to the Navier–Stokes–Fourier system. To recover uniqueness at least in the class of strong solutions (weak–strong uniqueness), some form of the total energy balance must be imposed. Note that our strategy is somehow complementary to the theory of conservation laws, where the energy equation is imposed while the entropy balance/inequality is considered as an admissibility criterion, cf. Dafermos [24].

Ideally, we should consider the weak formulation of the total energy balance (3.16) as an integral part of the definition of weak solutions. Unfortunately, this is impossible because of missing a priori bounds to control the convective terms. Instead we postulate its integrated version (3.7). Here we experience a problem similar to the entropy balance, namely how to define the boundary integral on Γ_{out}, specifically,

$$\int_{\partial\Omega} \varrho e \, [\mathbf{u}_B \cdot \mathbf{n}]^+ \mathrm{d}\sigma_x.$$

Adopting the same convention as in the entropy inequality (3.27) we may define $\varrho e = \varrho e(\varrho, \vartheta)$ in terms of the Sobolev trace of ϑ and the boundary density ϱ appearing in the weak formulation of the equation of continuity (3.20). Alternatively, we may consider

$$\varrho e = E_{\mathrm{int}}(\varrho, S)$$

as a (convex) function of the density ϱ and the total entropy $S = \varrho s$. In this case, the boundary value of S should match its counterpart in the boundary integral in the entropy balance (3.27).

Keeping the above agreement in mind, we write down the weak form of the *energy balance* (inequality),

$$
\begin{aligned}
&- \int_0^T \partial_t \psi \int_\Omega \left(\frac{1}{2} \varrho |\mathbf{u} - \mathbf{u}_B|^2 + \varrho e \right) \, \mathrm{d}x\mathrm{d}t + \int_0^T \int_{\partial\Omega} \psi \varrho e \, [\mathbf{u}_B \cdot \mathbf{n}]^+ \mathrm{d}\sigma_x \mathrm{d}t \\
&+ \int_0^T \int_{\partial\Omega} \psi f_{i,B} \, [\mathbf{u}_B \cdot \mathbf{n}]^- \mathrm{d}\sigma_x \mathrm{d}t \\
&\leq \psi(0) \int_\Omega \left(\frac{1}{2} \varrho_0 \left| \frac{\mathbf{m}_0}{\varrho_0} - \mathbf{u}_B \right|^2 + E_{\mathrm{int}}(\varrho_0, S_0) \right) \, \mathrm{d}x \\
&- \int_0^T \psi \int_\Omega \Big[\varrho (\mathbf{u} - \mathbf{u}_B) \otimes (\mathbf{u} - \mathbf{u}_B) + p \mathbb{I} - \mathbb{S} \Big] : \mathbb{D}_x \mathbf{u}_B \, \mathrm{d}x\mathrm{d}t \\
&+ \int_0^T \psi \int_\Omega \varrho \Big[\mathbf{g} - \partial_t \mathbf{u}_B - (\mathbf{u}_B \cdot \nabla_x) \mathbf{u}_B \Big] \cdot (\mathbf{u} - \mathbf{u}_B) \, \mathrm{d}x\mathrm{d}t
\end{aligned}
\tag{3.29}
$$

for any $\psi \in C^1_c[0,T)$, $\psi \geq 0$.

Similarly to the entropy balance, we have replaced (3.7) by an inequality. The reason is that boundary values of the internal energy appearing in the second integral in (3.29) are only controlled in the non-reflexive space $L^1(\Gamma_{\mathrm{out}})$.

It is important to notice that the specific form of (3.29) depends solely on the boundary values of $\mathbf{u}_B$ and not on its extension inside Ω. Indeed suppose that $\tilde{\mathbf{u}} \in C^1([0,T] \times \overline{\Omega}; R^d)$ satisfies

$$
\tilde{\mathbf{u}} = \mathbf{u}_B \text{ in } [0,T] \times \overline{\Omega}.
$$

Thus we may consider $\boldsymbol{\varphi} = \psi(\mathbf{u}_B - \tilde{\mathbf{u}})$, $\psi \in C^1_c[0,T)$, as a test function in the momentum balance (3.24) obtaining

$$
\begin{aligned}
&\int_0^T \int_\Omega \varrho \mathbf{u} \cdot (\mathbf{u}_B - \tilde{\mathbf{u}}) \partial_t \psi \, \mathrm{d}x\mathrm{d}t \\
&\quad + \int_0^T \psi \int_\Omega \Big[\varrho \mathbf{u} \cdot \partial_t (\mathbf{u}_B - \tilde{\mathbf{u}}) + \varrho (\mathbf{u} \otimes \mathbf{u}) : \nabla_x (\mathbf{u}_B - \tilde{\mathbf{u}}) \Big] \, \mathrm{d}x\mathrm{d}t \\
&\quad + \int_0^T \psi \int_\Omega p \mathrm{div}_x (\mathbf{u}_B - \tilde{\mathbf{u}}) \, \mathrm{d}x\mathrm{d}t \\
&\quad = \int_0^T \psi \int_\Omega \Big[\mathbb{S} : \nabla_x (\mathbf{u}_B - \tilde{\mathbf{u}}) - \varrho \mathbf{g} \cdot (\mathbf{u}_B - \tilde{\mathbf{u}}) \Big] \, \mathrm{d}x\mathrm{d}t \\
&\quad - \psi(0) \int_\Omega \mathbf{m}_0 \cdot (\mathbf{u}_B - \tilde{\mathbf{u}}) \, \mathrm{d}x.
\end{aligned}
\tag{3.30}
$$

Subtracting (3.30) from (3.29) we get the desired conclusion.

We have shown the following result.

Lemma 5. *If $(\varrho, \vartheta, \mathbf{u})$ satisfy the weak formulation of the momentum balance (3.24), together with the energy inequality (3.29) for some $\mathbf{u}_B \in C^1([0,T] \times \overline{\Omega}; R^d)$, then (3.29) holds with $\mathbf{u}_B$ replaced by any $\tilde{\mathbf{u}} \in C^1([0,T] \times \overline{\Omega}; R^d)$ satisfying*

$$\tilde{\mathbf{u}} = \mathbf{u}_B \text{ on } [0,T] \times \partial\Omega.$$

Remark 3 (Instantaneous values of total energy).

Unlike the entropy inequality (3.27), the total energy balance does not contain the term

$$\left[\psi \int_\Omega \left(\frac{1}{2}\varrho|\mathbf{u} - \mathbf{u}_B|^2 + \varrho e \right) \, \mathrm{d}x \right]_{t=0}^{t=\tau}$$

depending on the instantaneous value of the energy at the time τ. The reason is that we would need to write the energy in terms of "weakly continuous" conservative entropy variables $(\varrho, S, \mathbf{m})$. Unfortunately, the instantaneous values of the càglàd total entropy S can be interpreted only as a measure on $\overline{\Omega}$ at least for certain zero measure set of times. As it is a bit problematic to define the internal energy $\varrho e = E_{\rm int}(\varrho, S)$ for measure-valued arguments, we prefer to keep the integral form of (3.29). Let us point out, however, that (3.29) still yields the energy inequality at any time τ that is a Lebesgue point of the total energy, meaning

$$\begin{aligned}
&\int_\Omega \left(\frac{1}{2}\varrho|\mathbf{u} - \mathbf{u}_B|^2 + \varrho e \right) (\tau, \cdot) \, \mathrm{d}x \\
&+ \int_0^\tau \int_{\partial\Omega} f_{i,B} \, [\mathbf{u}_B \cdot \mathbf{n}]^- \mathrm{d}\sigma_x \mathrm{d}t + \int_0^\tau \int_{\partial\Omega} \varrho e \, [\mathbf{u}_B \cdot \mathbf{n}]^+ \mathrm{d}\sigma_x \mathrm{d}t \\
&\leq \int_\Omega \left(\frac{1}{2}\varrho_0 \left| \frac{\mathbf{m}_0}{\varrho_0} - \mathbf{u}_B \right|^2 + E_{\rm int}(\varrho_0, S_0) \right) \, \mathrm{d}x \\
&- \int_0^\tau \int_\Omega \Big[\varrho(\mathbf{u} - \mathbf{u}_B) \otimes (\mathbf{u} - \mathbf{u}_B) + p\mathbb{I} - \mathbb{S} \Big] : \mathbb{D}_x \mathbf{u}_B \, \mathrm{d}x\mathrm{d}t \\
&+ \int_0^\tau \int_\Omega \varrho \Big[\mathbf{g} - \partial_t \mathbf{u}_B - (\mathbf{u}_B \cdot \nabla_x)\mathbf{u}_B \Big] \cdot (\mathbf{u} - \mathbf{u}_B) \, \mathrm{d}x\mathrm{d}t
\end{aligned} \tag{3.31}$$

for a.a. $\tau \in (0,T)$.

3.3.4.1 Energy Balance for the Barotropic System

If $p = p(\varrho)$, the weak form of the mechanical energy balance (3.19) reads

$$
\begin{aligned}
& - \int_0^T \partial_t \psi \int_\Omega \left(\frac{1}{2} \varrho |\mathbf{u} - \mathbf{u}_B|^2 + P(\varrho) \right) \, \mathrm{d}x\mathrm{d}t \\
& + \int_0^T \int_{\partial\Omega} \psi P(\varrho_B) \, [\mathbf{u}_B \cdot \mathbf{n}]^- \mathrm{d}\sigma_x \mathrm{d}t + \int_0^T \int_{\partial\Omega} \psi P(\varrho) \, [\mathbf{u}_B \cdot \mathbf{n}]^+ \mathrm{d}\sigma_x \mathrm{d}t \\
& \leq \psi(0) \int_\Omega \left(\frac{1}{2} \varrho_0 \left| \frac{\mathbf{m}_0}{\varrho_0} - \mathbf{u}_B \right|^2 + P(\varrho_0) \right) \, \mathrm{d}x \\
& - \int_0^T \psi \int_\Omega \mathbb{S} : \mathbb{D}_x \mathbf{u} \, \mathrm{d}x\mathrm{d}t \\
& - \int_0^T \psi \int_\Omega \Big[\varrho (\mathbf{u} - \mathbf{u}_B) \otimes (\mathbf{u} - \mathbf{u}_B) + p\mathbb{I} - \mathbb{S} \Big] : \nabla_x \mathbf{u}_B \, \mathrm{d}x\mathrm{d}t \\
& + \int_0^T \psi \int_\Omega \Big[\varrho \mathbf{g} - \varrho \partial_t \mathbf{u}_B - \varrho (\mathbf{u}_B \cdot \nabla_x) \mathbf{u}_B \Big] \cdot (\mathbf{u} - \mathbf{u}_B) \, \mathrm{d}x\mathrm{d}t
\end{aligned}
\tag{3.32}
$$

for any $\psi \in C^1[0,T)$, $\psi \geq 0$.

3.3.5 Weak Solutions

Having collected the necessary material we are ready to introduce the concept of *weak solution* to the Navier–Stokes–Fourier system. We consider a bounded Lipschitz domain $\Omega \subset R^d$, $d = 2,3$ so that all surface integrals are well defined and the standard Gauss–Green integration formula applies, see e.g. Evan and Gariepy [32]. As we have already observed, the velocity $\mathbf{u}$ and the temperature ϑ admit the standard Sobolev traces on $\partial\Omega$ as their spatial gradients are integrable. In addition, the density ϱ or rather the momentum $\varrho\mathbf{u}$ admits a normal trace determined on Γ_{out} via (3.20). Consequently, we consider ϱ as a function defined a.a. in $(0,T) \times \Omega$ as well as on $(0,T) \times \partial\Omega$ integrable with respect to the weighted surface measure $[\mathbf{u}_B \cdot \mathbf{n}]^+ (\mathrm{d}t \otimes \mathrm{d}\sigma_x)$. We use the same symbol ϱ for both. For the sake of simplicity, we suppose that the boundary data ϱ_B, $\mathbf{u}_B$, $f_{i,B}$ are restrictions of continuously differentiable functions and that $\varrho_B|_{\Gamma_{\text{in}}} > 0$. Similarly, we suppose that the external driving force $\mathbf{g}$ is a bounded measurable function.

Definition 4 (Weak Solution of Navier–Stokes–Fourier System).
Let $\Omega \subset R^d$, $d = 2, 3$ be a bounded Lipschitz domain. We say that the trio $(\varrho, \vartheta, \mathbf{u})$ is a *weak solution* of the Navier–Stokes–Fourier system (3.1)–(3.7) in $[0, T) \times \Omega$, $0 < T \leq \infty$ with the initial data

$$\varrho(0, \cdot) = \varrho_0, \ \varrho\mathbf{u}(0, \cdot) = \mathbf{m}_0, \ S(0, \cdot) = S_0 \tag{3.33}$$

if the following holds:

- **Integrability, regularity.**

$$\begin{aligned} &\varrho \in L^\infty(0, \tau; L^\gamma(\Omega)) \cap L^\gamma((0, \tau) \times \partial\Omega; [\mathbf{u}_B \cdot \mathbf{n}]^+ (\mathrm{d}t \otimes \mathrm{d}\sigma_x)) \\ &\quad \text{for any } 0 \leq 0 < \tau < T, \text{for some } \gamma > 1, \\ &\varrho \geq 0 \text{ a.a. in } (0, T) \times \Omega \text{ and a.a. in } \Gamma_{\mathrm{out}}, \end{aligned} \tag{3.34}$$

$$\begin{aligned} \mathbf{u} &\in L^q(0, T; W^{1,q}(\Omega; R^d)) \text{ for some } q > 1, \\ \varrho\mathbf{u} &\in L^\infty(0, T; L^{\frac{2\gamma}{\gamma+1}}(\Omega; R^d)), \end{aligned} \tag{3.35}$$

$$\begin{aligned} &\vartheta, \ \log(\vartheta) \in L^2(0, T; W^{1,2}(\Omega)), \ \vartheta > 0 \text{ a.a. in } (0, T) \times \Omega, \\ &\frac{1}{\vartheta} \in L^1((0, T) \times \partial\Omega; [\mathbf{u}_B \cdot \mathbf{n}]^- (\mathrm{d}t \otimes \mathrm{d}\sigma_x)), \end{aligned} \tag{3.36}$$

$$\begin{aligned} &\varrho s \in L^\infty(0, T; L^1(\Omega)) \cap L^1((0, T) \times \partial\Omega, [\mathbf{u}_B \cdot \mathbf{n}]^+ (\mathrm{d}t \otimes \mathrm{d}\sigma_x)), \\ &\varrho e \in L^\infty(0, T; L^1(\Omega)) \cap L^1((0, T) \times \partial\Omega, [\mathbf{u}_B \cdot \mathbf{n}]^+ (\mathrm{d}t \otimes \mathrm{d}\sigma_x)). \end{aligned} \tag{3.37}$$

- **Equation of continuity.** The integral identity

$$\begin{aligned} &\int_0^T \int_\Omega \left[\varrho \partial_t \varphi + \varrho \mathbf{u} \cdot \nabla_x \varphi \right] \mathrm{d}x\mathrm{d}t \\ &\quad = \int_0^T \left[\int_{\partial\Omega} \varphi \varrho_B \left[\mathbf{u}_B \cdot \mathbf{n} \right]^- \mathrm{d}\sigma_x + \int_{\partial\Omega} \varphi \varrho \left[\mathbf{u}_B \cdot \mathbf{n} \right]^+ \mathrm{d}\sigma_x \right] \mathrm{d}t \\ &\quad - \int_\Omega \varrho_0 \varphi(0, \cdot) \, \mathrm{d}x \end{aligned} \tag{3.38}$$

holds for any test function $\varphi \in C^1_c([0, T) \times \overline{\Omega})$.

- **Momentum equation.**

$$(\mathbf{u} - \mathbf{u}_B) \in L^q(0,T; W_0^{1,q}(\Omega; R^d)), \tag{3.39}$$

and the integral identity

$$\int_0^T \int_\Omega \Big[\varrho \mathbf{u} \cdot \partial_t \boldsymbol{\varphi} + \varrho (\mathbf{u} \otimes \mathbf{u}) : \nabla_x \boldsymbol{\varphi} + p \mathrm{div}_x \boldsymbol{\varphi} \Big] \ \mathrm{d}x\mathrm{d}t$$
$$= \int_0^T \int_\Omega \Big[\mathbb{S} : \nabla_x \boldsymbol{\varphi} - \varrho \mathbf{g} \cdot \boldsymbol{\varphi} \Big] \ \mathrm{d}x\mathrm{d}t - \int_\Omega \mathbf{m}_0 \cdot \boldsymbol{\varphi}(0, \cdot) \ \mathrm{d}x \tag{3.40}$$

holds for any $\boldsymbol{\varphi} \in C_c^1([0,T) \times \overline{\Omega}; R^d)$, $\boldsymbol{\varphi}|_{\partial\Omega} = 0$.

- **Entropy inequality.** The inequality

$$\int_0^T \int_\Omega \Big[\varrho s \partial_t \varphi + \varrho s \mathbf{u} \cdot \nabla_x \varphi + \left(\frac{\mathbf{q}}{\vartheta} \right) \cdot \nabla_x \varphi \Big] \mathrm{d}x\mathrm{d}t$$
$$\leq - \int_0^T \int_\Omega \frac{\varphi}{\vartheta} \left(\mathbb{S} : \nabla_x \mathbf{u} - \frac{\mathbf{q} \cdot \nabla_x \vartheta}{\vartheta} \right) \ \mathrm{d}x\mathrm{d}t$$
$$+ \int_0^T \int_{\partial\Omega} \varphi \left(\frac{f_{i,B}}{\vartheta} + \varrho_B \Big[s(\varrho_B, \vartheta) - \frac{e(\varrho_B, \vartheta)}{\vartheta} \Big] \right) \ [\mathbf{u}_B \cdot \mathbf{n}]^- \mathrm{d}\sigma_x \mathrm{d}t$$
$$- \int_\Omega \varphi(0, \cdot) S_0 \ \mathrm{d}x + \int_0^T \int_{\partial\Omega} \varphi \varrho s \ [\mathbf{u}_B \cdot \mathbf{n}]^+ \mathrm{d}\sigma_x \mathrm{d}t \tag{3.41}$$

holds for any test function $\varphi \in C_c^1([0,T) \times \overline{\Omega})$, $\varphi \geq 0$.

- **Energy inequality.** The inequality

$$- \int_0^T \partial_t \psi \int_\Omega \left(\frac{1}{2} \varrho |\mathbf{u} - \mathbf{u}_B|^2 + \varrho e \right) \ \mathrm{d}x\mathrm{d}t$$
$$+ \int_0^T \left[\int_{\partial\Omega} \psi f_{i,B} \ [\mathbf{u}_B \cdot \mathbf{n}]^- \mathrm{d}\sigma_x + \int_{\partial\Omega} \psi \varrho e \ [\mathbf{u}_B \cdot \mathbf{n}]^+ \mathrm{d}\sigma_x \right] \mathrm{d}t$$
$$\leq \psi(0) \int_\Omega \left(\frac{1}{2} \varrho_0 \left| \frac{\mathbf{m}_0}{\varrho_0} - \mathbf{u}_B \right|^2 + E_{\mathrm{int}}(\varrho_0, S_0) \right) \ \mathrm{d}x$$
$$- \int_0^T \psi \int_\Omega \Big[\varrho (\mathbf{u} - \mathbf{u}_B) \otimes (\mathbf{u} - \mathbf{u}_B) + p \mathbb{I} - \mathbb{S} \Big] : \mathbb{D}_x \mathbf{u}_B \ \mathrm{d}x\mathrm{d}t$$
$$+ \int_0^T \psi \int_\Omega \varrho \Big[\mathbf{g} - \partial_t \mathbf{u}_B - (\mathbf{u}_B \cdot \nabla_x) \mathbf{u}_B \Big] \cdot (\mathbf{u} - \mathbf{u}_B) \ \mathrm{d}x\mathrm{d}t \tag{3.42}$$

is satisfied for any $\psi \in C^1_c[0,T)$, $\psi \geq 0$.

In view of the assumed regularity/integrability of the weak solutions, we easily deduce that

$$\varrho \in C_{\rm weak}([0,\tau]; L^\gamma(\Omega)),\ \varrho\mathbf{u} \in C_{\rm weak}([0,\tau]; L^{\frac{2\gamma}{\gamma+1}}(\Omega; R^d))$$

for any $0 < \tau < T$, and the refined form (3.20) of the equation of continuity and (3.24) of the momentum equation hold. Similarly, the total entropy $S = \varrho s$ can be interpreted as càglàd function of t ranging in $\mathcal{M}(\overline{\Omega})$ and (3.41) yields (3.27).

Our approach to weak solutions is based on imposing a weak form of the *First and Second law of thermodynamics* through the inequalities (3.42) and (3.41). As we have already pointed out, the a priori bounds available (known) for the Navier–Stokes–Fourier system are not sufficient to control the fluxes in the total energy balance (1.13); whence the latter is not convenient for the weak formulation. By the same token, the source term in the internal energy balance (1.14) cannot be controlled either. The present hybrid formulation—replacing one single equation by two inequalities—is therefore the only one that allows analytical treatment in the framework of available a priori bounds. As we shall see below, all (known) a priori bounds for the Navier–Stokes–Fourier system can be deduced from the weak formulation, specifically, from boundedness of the total energy and the entropy production rate.

3.4 Relative Energy Revisited

In Sect. 3.1 we have introduced the relative energy

$$E\left(\varrho, \vartheta, \mathbf{u} \middle| \tilde{\varrho}, \tilde{\vartheta}, \tilde{\mathbf{u}}\right) \ \text{or}\ E\left(\varrho, S, \mathbf{m} \middle| \tilde{\varrho}, \widetilde{S}, \widetilde{\mathbf{m}}\right),$$

interpreted in the latter case as a kind of Bregman distance. The advantage of the weak formulation specified in Definition 4 is the possibility to capture the time evolution of the integrated relative energy

$$\mathcal{E}\left(\varrho, S, \mathbf{m} \middle| \tilde{\varrho}, \widetilde{S}, \widetilde{\mathbf{m}}\right)(\tau) \equiv \int_\Omega E\left(\varrho, S, \mathbf{m} \middle| \tilde{\varrho}, \widetilde{S}, \widetilde{\mathbf{m}}\right)(\tau, \cdot)\ {\rm d}x$$

as soon as $(\varrho, \vartheta, \mathbf{u})$, or, equivalently $(\varrho, S, \mathbf{m})$, is a weak solution of the Navier–Stokes–Fourier system.

Consider a weak solution $(\varrho, \vartheta, \mathbf{u})$ of the Navier–Stokes–Fourier system in the sense of Definition 4 and an arbitrary trio of sufficiently smooth "test" functions $(\tilde{\varrho}, \tilde{\vartheta}, \tilde{\mathbf{u}})$. Fixing the constitutive laws (state equations), we may associate to $(\varrho, \vartheta, \mathbf{u})$ the trio $(\varrho, S = \varrho s(\varrho, \vartheta), \mathbf{m} = \varrho\mathbf{u})$ and to $(\tilde{\varrho}, \tilde{\vartheta}, \tilde{\mathbf{u}})$ the trio $(\tilde{\varrho}, \widetilde{S} = \tilde{\varrho} s(\tilde{\varrho}, \tilde{\vartheta}), \widetilde{\mathbf{m}} = \tilde{\varrho}\tilde{\mathbf{u}})$.

In view of Lemma 5, the energy inequality (3.42) already includes the velocity/momentum part of the relative energy. It is therefore enough to evaluate the perturbation of the internal energy.

Recalling

$$\frac{\partial(\varrho e)|_{\varrho}(\tilde{\varrho},\widetilde{S})}{\partial S}=\tilde{\vartheta}$$

and assuming $\tilde{\vartheta}>0$ is smooth enough, we may use $\varphi(t,x)=\psi(t)\tilde{\vartheta}(t,x)$, $\psi\in C^1_c[0,T)$, $\psi\geq 0$, as a test function in the entropy inequality (3.41) to obtain

$$\begin{aligned}
&\int_0^T\int_\Omega\Big[S\tilde{\vartheta}\partial_t\psi+\psi\Big(S\partial_t\tilde{\vartheta}+\varrho s\mathbf{u}\cdot\nabla_x\tilde{\vartheta}+\Big(\frac{\mathbf{q}}{\vartheta}\Big)\cdot\nabla_x\tilde{\vartheta}\Big)\Big]\,\mathrm{d}x\mathrm{d}t\\
&\quad-\int_0^T\int_{\partial\Omega}\psi\tilde{\vartheta}S[\mathbf{u}_B\cdot\mathbf{n}]^+\,\mathrm{d}\sigma_x\mathrm{d}t\\
&\quad\leq-\int_0^T\psi\int_\Omega\frac{\tilde{\vartheta}}{\vartheta}\Big(\mathbb{S}:\nabla_x\mathbf{u}-\frac{\mathbf{q}\cdot\nabla_x\vartheta}{\vartheta}\Big)\,\mathrm{d}x\mathrm{d}t\\
&\quad+\int_0^T\int_{\partial\Omega}\psi\tilde{\vartheta}\Big(\frac{f_{i,B}}{\vartheta}+\varrho_B\Big[s(\varrho_B,\vartheta)-\frac{e(\varrho_B,\vartheta)}{\vartheta}\Big]\Big)\,[\mathbf{u}_B\cdot\mathbf{n}]^-\mathrm{d}\sigma_x\mathrm{d}t\\
&\quad-\psi(0)\int_\Omega\tilde{\vartheta}(0,\cdot)S_0\,\mathrm{d}x.
\end{aligned}\tag{3.43}$$

Adding (3.43) to the energy balance (3.42) and regrouping several terms we get

$$\begin{aligned}
&-\int_0^T\partial_t\psi\int_\Omega\Big(\frac{1}{2}\varrho|\mathbf{u}-\tilde{\mathbf{u}}|^2+E_{\mathrm{int}}(\varrho,S)-\partial_S E_{\mathrm{int}}(\tilde{\varrho},\widetilde{S})S\Big)\,\mathrm{d}x\mathrm{d}t\\
&\quad+\int_0^T\int_{\partial\Omega}\psi\Big(E_{\mathrm{int}}(\varrho,S)-\partial_S E_{\mathrm{int}}(\tilde{\varrho},\widetilde{S})S\Big)\,[\mathbf{u}_B\cdot\mathbf{n}]^+\mathrm{d}\sigma_x\mathrm{d}t\\
&\quad+\int_0^T\psi\int_\Omega\frac{\tilde{\vartheta}}{\vartheta}\Big(\mathbb{S}:\nabla_x\mathbf{u}-\frac{\mathbf{q}\cdot\nabla_x\vartheta}{\vartheta}\Big)\,\mathrm{d}x\mathrm{d}t\\
&\quad\leq\psi(0)\int_\Omega\left(\frac{1}{2}\varrho_0\left|\frac{\mathbf{m}_0}{\varrho_0}-\tilde{\mathbf{u}}\right|^2+E_{\mathrm{int}}(\varrho_0,S_0)-\partial_S E_{\mathrm{int}}(\tilde{\varrho},\widetilde{S})(0,\cdot)S_0\right)\,\mathrm{d}x\\
&\quad-\int_0^T\int_\Omega\psi\Big(S\partial_t\tilde{\vartheta}+\varrho s\mathbf{u}\cdot\nabla_x\tilde{\vartheta}+\Big(\frac{\mathbf{q}}{\vartheta}\Big)\cdot\nabla_x\tilde{\vartheta}\Big)\,\mathrm{d}x\mathrm{d}t\\
&\quad+\int_0^T\int_{\partial\Omega}\psi\Big(\Big(\frac{\tilde{\vartheta}}{\vartheta}-1\Big)f_{i,B}+\tilde{\vartheta}\varrho_B\Big[s(\varrho_B,\vartheta)-\frac{e(\varrho_B,\vartheta)}{\vartheta}\Big]\Big)\,[\mathbf{u}_B\cdot\mathbf{n}]^-\mathrm{d}\sigma_x\mathrm{d}t\\
&\quad-\int_0^T\psi\int_\Omega\Big[\varrho(\mathbf{u}-\tilde{\mathbf{u}})\otimes(\mathbf{u}-\tilde{\mathbf{u}})+p\mathbb{I}-\mathbb{S}\Big]:\mathbb{D}_x\tilde{\mathbf{u}}\,\mathrm{d}x\mathrm{d}t\\
&\quad+\int_0^T\psi\int_\Omega\varrho\Big[\mathbf{g}-\partial_t\tilde{\mathbf{u}}-(\tilde{\mathbf{u}}\cdot\nabla_x)\tilde{\mathbf{u}}\Big]\cdot(\mathbf{u}-\tilde{\mathbf{u}})\,\mathrm{d}x\mathrm{d}t
\end{aligned}\tag{3.44}$$

for any $\psi \in C^1_c[0,T)$, $\psi \geq 0$, and $\tilde{\mathbf{u}} \in C^1_c([0,T) \times \overline{\Omega}; R^d)$ satisfying

$$\tilde{\mathbf{u}}|_{\partial\Omega} = \mathbf{u}_B.$$

Similarly, we use $\psi \partial_\varrho E_{\rm int}(\tilde{\varrho}, \widetilde{S})$ as a test function in the equation of continuity (3.38):

$$\begin{aligned}
&\int_0^T \int_\Omega \left[\varrho \partial_\varrho E_{\rm int}(\tilde{\varrho}, \widetilde{S}) \partial_t \psi + \psi \left(\varrho \partial_t \partial_\varrho E_{\rm int}(\tilde{\varrho}, \widetilde{S}) + \varrho \mathbf{u} \cdot \nabla_x \partial_\varrho E_{\rm int}(\tilde{\varrho}, \widetilde{S}) \right) \right] \, {\rm d}x \\
&\quad = -\psi(0) \int_\Omega \varrho_0 \partial_\varrho E_{\rm int}(\tilde{\varrho}, \widetilde{S})(0, \cdot) \, {\rm d}x \\
&\quad + \int_0^T \int_{\partial\Omega} \psi \varrho_B \partial_\varrho E_{\rm int}(\tilde{\varrho}, \widetilde{S}) \, [\mathbf{u}_B \cdot \mathbf{n}]^- {\rm d}\sigma_x {\rm d}t \\
&\quad + \int_0^T \int_{\partial\Omega} \psi \varrho \partial_\varrho E_{\rm int}(\tilde{\varrho}, \widetilde{S}) \, [\mathbf{u}_B \cdot \mathbf{n}]^+ {\rm d}\sigma_x {\rm d}t.
\end{aligned}$$

Adding this to (3.44) we obtain

$$\begin{aligned}
&- \int_0^T \partial_t \psi \int_\Omega \frac{1}{2} \varrho |\mathbf{u} - \tilde{\mathbf{u}}|^2 \, {\rm d}x{\rm d}t \\
&- \int_0^T \partial_t \psi \int_\Omega \left(E_{\rm int}(\varrho, S) - \partial_S E_{\rm int}(\tilde{\varrho}, \widetilde{S}) S - \partial_\varrho E_{\rm int}(\tilde{\varrho}, \widetilde{S}) \varrho \right) \, {\rm d}x{\rm d}t \\
&+ \int_0^T \int_{\partial\Omega} \psi \Big(E_{\rm int}(\varrho, S) - \partial_S E_{\rm int}(\tilde{\varrho}, \widetilde{S}) S - \partial_\varrho E_{\rm int}(\tilde{\varrho}, \widetilde{S}) \varrho \Big) \, [\mathbf{u}_B \cdot \mathbf{n}]^+ {\rm d}\sigma_x {\rm d}t \\
&+ \int_0^T \psi \int_\Omega \frac{\tilde{\vartheta}}{\vartheta} \left(\mathbb{S} : \mathbb{D}_x \mathbf{u} - \frac{\mathbf{q} \cdot \nabla_x \vartheta}{\vartheta} \right) \, {\rm d}x{\rm d}t \\
\leq & \psi(0) \int_\Omega \frac{1}{2} \varrho_0 \left| \frac{\mathbf{m}_0}{\varrho_0} - \tilde{\mathbf{u}} \right|^2 \, {\rm d}x \\
&+ \psi(0) \int_\Omega \left(E_{\rm int}(\varrho_0, S_0) - \partial_S E_{\rm int}(\tilde{\varrho}, \widetilde{S})(0, \cdot) S_0 - \partial_\varrho E_{\rm int}(\tilde{\varrho}, \widetilde{S})(0, \cdot) \varrho_0 \right) \, {\rm d}x \\
&- \int_0^T \int_\Omega \psi \left(S \partial_t \tilde{\vartheta} + \varrho s \mathbf{u} \cdot \nabla_x \tilde{\vartheta} + \left(\frac{\mathbf{q}}{\vartheta} \right) \cdot \nabla_x \tilde{\vartheta} \right) {\rm d}x{\rm d}t \\
&+ \int_0^T \int_{\partial\Omega} \psi \left(\left(\frac{\tilde{\vartheta}}{\vartheta} - 1 \right) f_{i,B} + \tilde{\vartheta} \varrho_B \left[s(\varrho_B, \vartheta) - \frac{e(\varrho_B, \vartheta)}{\vartheta} \right] \right) \, [\mathbf{u}_B \cdot \mathbf{n}]^- {\rm d}\sigma_x {\rm d}t \\
&+ \int_0^T \int_{\partial\Omega} \psi \varrho_B \partial_\varrho E_{\rm int}(\tilde{\varrho}, \widetilde{S}) \, [\mathbf{u}_B \cdot \mathbf{n}]^- {\rm d}\sigma_x {\rm d}t \\
&- \int_0^T \psi \int_\Omega \Big[\varrho(\mathbf{u} - \tilde{\mathbf{u}}) \otimes (\mathbf{u} - \tilde{\mathbf{u}}) + p\mathbb{I} - \mathbb{S} \Big] : \mathbb{D}_x \tilde{\mathbf{u}} \, {\rm d}x{\rm d}t \\
&+ \int_0^T \psi \int_\Omega \varrho \Big[\mathbf{g} - \partial_t \tilde{\mathbf{u}} - (\tilde{\mathbf{u}} \cdot \nabla_x) \tilde{\mathbf{u}} \Big] \cdot (\mathbf{u} - \tilde{\mathbf{u}}) \, {\rm d}x{\rm d}t
\end{aligned}$$

$$-\int_0^T \psi \int_\Omega \Big(\varrho \partial_t \partial_\varrho E_{\rm int}(\tilde\varrho, \widetilde S) + \varrho \mathbf{u} \cdot \nabla_x \partial_\varrho E_{\rm int}(\tilde\varrho, \widetilde S) \Big) \ {\rm d}x{\rm d}t. \tag{3.45}$$

Finally, we use the trivial identity

$$\begin{aligned}
&\int_0^T \int_\Omega \psi \partial_t \Big(\partial_S E_{\rm int}(\tilde\varrho, \widetilde S)\widetilde S + \partial_\varrho E_{\rm int}(\tilde\varrho, \widetilde S)\tilde\varrho - E_{\rm int}(\tilde\varrho, \widetilde S) \Big) \ {\rm d}x{\rm d}t \\
&= - \psi(0) \int_\Omega \Big(\partial_S E_{\rm int}(\tilde\varrho, \widetilde S)\widetilde S + \partial_\varrho E_{\rm int}(\tilde\varrho, \widetilde S)\tilde\varrho - E_{\rm int}(\tilde\varrho, \widetilde S) \Big)(0, \cdot) \ {\rm d}x \\
&\quad - \int_0^T \int_\Omega \partial_t \psi \Big(\partial_S E_{\rm int}(\tilde\varrho, \widetilde S)\widetilde S + \partial_\varrho E_{\rm int}(\tilde\varrho, \widetilde S)\tilde\varrho - E_{\rm int}(\tilde\varrho, \widetilde S) \Big) \ {\rm d}x{\rm d}t
\end{aligned}$$

to conclude

$$\begin{aligned}
&- \int_0^T \partial_t \psi \int_\Omega \frac{1}{2} \varrho \left| \frac{\mathbf{m}}{\varrho} - \tilde{\mathbf{u}} \right|^2 \ {\rm d}x{\rm d}t \\
&- \int_0^T \partial_t \psi \int_\Omega \Big(E_{\rm int}(\varrho, S) - \nabla_{\varrho, S} E_{\rm int}(\tilde\varrho, \widetilde S) \cdot (\varrho - \tilde\varrho, S - \widetilde S) - E_{\rm int}(\tilde\varrho, \widetilde S) \Big) \ {\rm d}x{\rm d}t \\
&+ \int_0^T \int_{\partial\Omega} \psi \Big(E_{\rm int}(\varrho, S) - \partial_S E_{\rm int}(\tilde\varrho, \widetilde S) S - \partial_\varrho E_{\rm int}(\tilde\varrho, \widetilde S)\varrho \Big) \ [\mathbf{u}_B \cdot \mathbf{n}]^+ {\rm d}\sigma_x {\rm d}t \\
&+ \int_0^T \psi \int_\Omega \frac{\tilde\vartheta}{\vartheta} \left(\mathbb{S} : \mathbb{D}_x \mathbf{u} - \frac{\mathbf{q} \cdot \nabla_x \vartheta}{\vartheta} \right) \ {\rm d}x{\rm d}t \\
\leq &\psi(0) \int_\Omega \frac{1}{2} \varrho_0 \left| \frac{\mathbf{m}_0}{\varrho_0} - \tilde{\mathbf{u}}(0, \cdot) \right|^2 \ {\rm d}x \\
&+ \psi(0) \int_\Omega \Big(E_{\rm int}(\varrho_0, S_0) - \nabla_{\varrho, S} E_{\rm int}(\tilde\varrho, \widetilde S)(S_0 - \widetilde S, \varrho_0 - \tilde\varrho) - E_{\rm int}(\tilde\varrho, \widetilde S) \Big) \ (0, \cdot) \ {\rm d}x \\
&- \int_0^T \int_\Omega \psi \left(S \partial_t \tilde\vartheta + \varrho s \mathbf{u} \cdot \nabla_x \tilde\vartheta + \left(\frac{\mathbf{q}}{\vartheta} \right) \cdot \nabla_x \tilde\vartheta \right) {\rm d}x{\rm d}t \\
&+ \int_0^T \int_{\partial\Omega} \psi \left(\left(\frac{\tilde\vartheta}{\vartheta} - 1 \right) f_{i,B} + \tilde\vartheta \varrho_B \Big[s(\varrho_B, \vartheta) - \frac{e(\varrho_B, \vartheta)}{\vartheta} \Big] \right) \ [\mathbf{u}_B \cdot \mathbf{n}]^- {\rm d}\sigma_x {\rm d}t \\
&+ \int_0^T \int_{\partial\Omega} \psi \varrho_B \partial_\varrho E_{\rm int}(\tilde\varrho, \widetilde S) \ [\mathbf{u}_B \cdot \mathbf{n}]^- {\rm d}\sigma_x {\rm d}t \\
&- \int_0^T \psi \int_\Omega \Big[\varrho (\mathbf{u} - \tilde{\mathbf{u}}) \otimes (\mathbf{u} - \tilde{\mathbf{u}}) + p \mathbb{I} - \mathbb{S} \Big] : \mathbb{D}_x \tilde{\mathbf{u}} \ {\rm d}x{\rm d}t \\
&+ \int_0^T \psi \int_\Omega \varrho \Big[\mathbf{g} - \partial_t \tilde{\mathbf{u}} - (\tilde{\mathbf{u}} \cdot \nabla_x) \tilde{\mathbf{u}} \Big] \cdot (\mathbf{u} - \tilde{\mathbf{u}}) \ {\rm d}x{\rm d}t \\
&- \int_0^T \psi \int_\Omega \Big(\varrho \partial_t \partial_\varrho E_{\rm int}(\tilde\varrho, \widetilde S) + \varrho \mathbf{u} \cdot \nabla_x \partial_\varrho E_{\rm int}(\tilde\varrho, \widetilde S) \Big) \ {\rm d}x{\rm d}t \\
&+ \int_0^T \int_\Omega \psi \partial_t \Big(\partial_S E_{\rm int}(\tilde\varrho, \widetilde S)\widetilde S + \partial_\varrho E_{\rm int}(\tilde\varrho, \widetilde S)\tilde\varrho - E_{\rm int}(\tilde\varrho, \widetilde S) \Big) \ {\rm d}x{\rm d}t
\end{aligned} \tag{3.46}$$

for any

$$\psi \in C_c^1[0,T),\ \psi \geq 0,\ \tilde{\mathbf{u}}|_{\partial\Omega} = \mathbf{u}_B,\ \tilde{\vartheta} > 0.$$

On the left-hand side of (3.46), we easily recognize the distributional time derivative of the relative energy. Relation (3.46) is termed *relative energy inequality* and plays a crucial role in the analysis of weak solutions to the Navier–Stokes–Fourier system. It can be seen as an extended version of the energy inequality (3.29). Indeed the relative energy inequality holds for any weak solution of the Navier–Stokes–Fourier system satisfying the energy inequality.

The present form (3.46) is rather awkward and not practical in applications. For future use, it is convenient to rewrite (3.46) in a more concise form with the help of Gibbs' relation (1.6). More specifically, using (1.22) we have

$$\partial_\varrho E_{\rm int}(\tilde{\varrho}, \widetilde{S}) = \widetilde{e} - \tilde{\vartheta}\widetilde{s} + \frac{\widetilde{p}}{\tilde{\varrho}},\ \partial_S E_{\rm int}(\tilde{\varrho}, \widetilde{S}) = \tilde{\vartheta},$$

and, consequently,

$$\begin{aligned}\int_0^T \int_{\partial\Omega} \psi \varrho_B \partial_\varrho E_{\rm int}(\tilde{\varrho}, \widetilde{S})\ [\mathbf{u}_B \cdot \mathbf{n}]^- {\rm d}\sigma_x {\rm d}t &= \int_0^T \int_{\partial\Omega} \varrho_B \widetilde{e}\ [\mathbf{u}_B \cdot \mathbf{n}]^- {\rm d}\sigma_x {\rm d}t \\ - \int_0^T \int_{\partial\Omega} \psi \varrho_B \tilde{\vartheta}\widetilde{s}\ [\mathbf{u}_B \cdot \mathbf{n}]^- {\rm d}\sigma_x {\rm d}t &+ \int_0^T \int_{\partial\Omega} \psi \varrho_B \frac{\widetilde{p}}{\tilde{\varrho}}\ [\mathbf{u}_B \cdot \mathbf{n}]^- {\rm d}\sigma_x {\rm d}t,\end{aligned}$$

$$\begin{aligned}&\int_0^T \psi \int_\Omega \Big(\varrho \partial_t \partial_\varrho E_{\rm int}(\tilde{\varrho}, \widetilde{S}) + \varrho \mathbf{u} \cdot \nabla_x \partial_\varrho E_{\rm int}(\tilde{\varrho}, \widetilde{S}) \Big)\ {\rm d}x{\rm d}t \\ &= \int_0^T \psi \int_\Omega \left(\varrho \partial_t \widetilde{e} + \varrho \mathbf{u} \cdot \nabla_x \widetilde{e} \right)\ {\rm d}x{\rm d}t - \int_0^T \psi \int_\Omega \Big(\varrho \partial_t (\tilde{\vartheta}\widetilde{s}) + \varrho \mathbf{u} \cdot \nabla_x (\tilde{\vartheta}\widetilde{s}) \Big)\ {\rm d}x{\rm d}t \\ &+ \int_0^T \psi \int_\Omega \left(\varrho \partial_t \left(\frac{\widetilde{p}}{\tilde{\varrho}} \right) + \varrho \mathbf{u} \cdot \nabla_x \left(\frac{\widetilde{p}}{\tilde{\varrho}} \right) \right)\ {\rm d}x{\rm d}t,\end{aligned}$$

and

$$\begin{aligned}&\int_0^T \psi \int_\Omega \partial_t \Big(\partial_S E_{\rm int}(\tilde{\varrho}, \widetilde{S}) \widetilde{S} + \partial_\varrho E_{\rm int}(\tilde{\varrho}, \widetilde{S}) \tilde{\varrho} - E_{\rm int}(\tilde{\varrho}, \widetilde{S}) \Big)\ {\rm d}x{\rm d}t \\ &= \int_0^T \psi \int_\Omega \partial_t (\tilde{\vartheta}\widetilde{S})\ {\rm d}x{\rm d}t + \int_0^T \psi \int_\Omega \partial_t (\tilde{\varrho}\widetilde{e})\ {\rm d}x{\rm d}t - \int_0^T \psi \int_\Omega \partial_t (\tilde{\varrho}\tilde{\vartheta}\widetilde{s})\ {\rm d}x{\rm d}t \\ &+ \int_0^T \psi \int_\Omega \partial_t \widetilde{p}\ {\rm d}x{\rm d}t - \int_0^T \psi \int_\Omega \partial_t E_{\rm int}(\tilde{\varrho}, \widetilde{S})\ {\rm d}x{\rm d}t = \int_0^T \psi \int_\Omega \partial_t \widetilde{p}\ {\rm d}x{\rm d}t.\end{aligned}$$

Consequently, the relative energy inequality (3.46) can be written in the form

$$
\begin{aligned}
&-\int_0^T \partial_t\psi \int_\Omega E\left(\varrho, S, \mathbf{m} \;\middle|\; \tilde{\varrho}, \widetilde{S}, \widetilde{\mathbf{m}}\right) \, \mathrm{d}x\mathrm{d}t \\
&+\int_0^T \int_{\partial\Omega} \psi\Big(E_{\rm int}(\varrho, S) - \partial_S E_{\rm int}(\tilde{\varrho}, \widetilde{S})S - \partial_\varrho E_{\rm int}(\tilde{\varrho}, \widetilde{S})\varrho \Big) \, [\mathbf{u}_B \cdot \mathbf{n}]^+ \mathrm{d}\sigma_x \mathrm{d}t \\
&+\int_0^T \psi \int_\Omega \frac{\tilde{\vartheta}}{\vartheta} \left(\mathbb{S} : \nabla_x \mathbf{u} - \frac{\mathbf{q} \cdot \nabla_x \vartheta}{\vartheta} \right) \, \mathrm{d}x\mathrm{d}t \\
\leq &\psi(0) \int_\Omega E\left(\varrho_0, S_0, \mathbf{m}_0 \;\middle|\; \tilde{\varrho}(0,\cdot), \widetilde{S}(0,\cdot), \widetilde{\mathbf{m}}(0,\cdot)\right) \, \mathrm{d}x \\
&+\int_0^T \psi \int_\Omega \frac{\varrho}{\tilde{\varrho}} (\mathbf{u} - \tilde{\mathbf{u}}) \cdot \nabla_x \widetilde{p} \, \mathrm{d}x\mathrm{d}t \\
&-\int_0^T \int_\Omega \psi \left(\varrho(s - \widetilde{s})\partial_t \tilde{\vartheta} + \varrho(s - \widetilde{s})\mathbf{u} \cdot \nabla_x \tilde{\vartheta} + \left(\frac{\mathbf{q}}{\vartheta}\right) \cdot \nabla_x \tilde{\vartheta} \right) \mathrm{d}x\mathrm{d}t \\
&+\int_0^T \int_{\partial\Omega} \psi \left(\frac{\tilde{\vartheta}}{\vartheta} - 1 \right) f_{i,B} \, [\mathbf{u}_B \cdot \mathbf{n}]^- \mathrm{d}\sigma_x \mathrm{d}t \\
&+\int_0^T \int_{\partial\Omega} \tilde{\vartheta} \varrho_B \left[s(\varrho_B, \vartheta) - \widetilde{s} + \frac{\widetilde{e}}{\tilde{\vartheta}} - \frac{e(\varrho_B, \vartheta)}{\vartheta} \right] \, [\mathbf{u}_B \cdot \mathbf{n}]^- \mathrm{d}\sigma_x \mathrm{d}t \\
&+\int_0^T \int_{\partial\Omega} \psi \frac{\varrho_B}{\tilde{\varrho}} \widetilde{p} \, [\mathbf{u}_B \cdot \mathbf{n}]^- \mathrm{d}\sigma_x \mathrm{d}t \\
&-\int_0^T \psi \int_\Omega \Big[\varrho(\mathbf{u} - \tilde{\mathbf{u}}) \otimes (\mathbf{u} - \tilde{\mathbf{u}}) + p\mathbb{I} - \mathbb{S} \Big] : \mathbb{D}_x \tilde{\mathbf{u}} \, \mathrm{d}x\mathrm{d}t \\
&+\int_0^T \psi \int_\Omega \varrho \Big[\mathbf{g} - \partial_t \tilde{\mathbf{u}} - (\tilde{\mathbf{u}} \cdot \nabla_x)\tilde{\mathbf{u}} - \frac{1}{\tilde{\varrho}} \nabla_x \widetilde{p} \Big] \cdot (\mathbf{u} - \tilde{\mathbf{u}}) \, \mathrm{d}x\mathrm{d}t \\
&+\int_0^T \psi \int_\Omega \left[\left(1 - \frac{\varrho}{\tilde{\varrho}} \right) \partial_t \widetilde{p} - \frac{\varrho}{\tilde{\varrho}} \mathbf{u} \cdot \nabla_x \widetilde{p} \right] \, \mathrm{d}x\mathrm{d}t
\end{aligned}
\tag{3.47}
$$

for any

$$\psi \in C^1_c[0,T), \ \psi \geq 0, \ \tilde{\mathbf{u}}|_{\partial\Omega} = \mathbf{u}_B, \ \tilde{\vartheta} > 0. \tag{3.48}$$

Summarizing we record the following result.

Theorem 4 (Relative Energy Inequality).
Let $\Omega \subset R^d$, $d = 2, 3$ be a bounded Lipschitz domain. Let $(\varrho, \vartheta, \mathbf{u})$ be a weak solution of the Navier–Stokes–Fourier system in $(0, T) \times \Omega$, $T \leq \infty$ in the sense specified in Definition 4. Let $(\tilde{\varrho}, \tilde{\vartheta}, \tilde{\mathbf{u}})$ be a trio of functions satisfying

$$
\begin{aligned}
\tilde{\varrho},\ \tilde{\vartheta} \in C^1_c([0,T) \times \overline{\Omega}),\ \tilde{\mathbf{u}} \in C^1_c([0,T) \times \overline{\Omega}; R^d),\\
\tilde{\varrho} > 0,\ \tilde{\vartheta} > 0,\ \tilde{\mathbf{u}}|_{\partial\Omega} = \mathbf{u}_B.
\end{aligned}
\tag{3.49}
$$

Then the relative energy inequality (3.47) holds, with

$$
\widetilde{p} = p(\tilde{\varrho}, \tilde{\vartheta}),\ \widetilde{e} = e(\tilde{\varrho}, \tilde{\vartheta}),\ \widetilde{s} = s(\tilde{\varrho}, \tilde{\vartheta}),\ \widetilde{S} = \tilde{\varrho} s(\tilde{\varrho}, \tilde{\vartheta}),\ \widetilde{\mathbf{m}} = \tilde{\varrho}\tilde{\mathbf{u}}.
$$

We point out that the "test functions" $(\tilde{\varrho}, \tilde{\vartheta}, \tilde{\mathbf{u}})$ in (3.47) are arbitrary, satisfying only (3.49).

3.5 Weak–Strong Uniqueness

Our goal is to establish the weak–strong uniqueness property in the class of weak solutions introduced in Definition 4. The idea is to use the strong solution $(\tilde{\varrho}, \tilde{\vartheta}, \tilde{\mathbf{u}})$ sharing the same initial and boundary data with a weak solution $(\varrho, \vartheta, \mathbf{u})$ as the trio of test functions in the relative energy inequality (3.47), use the fact that the relative energy is a Bregman distance, and apply a Gronwall type argument. With this ansatz, the relative energy inequality (3.47) yields

$$
\begin{aligned}
&\int_\Omega E\left(\varrho, S, \mathbf{m} \;\middle|\; \tilde{\varrho}, \widetilde{S}, \widetilde{\mathbf{m}}\right)(\tau, \cdot)\mathrm{d}x \\
&\quad + \int_0^\tau \int_{\partial\Omega} \left(E_{\rm int}(\varrho, S) - \partial_S E_{\rm int}(\tilde{\varrho}, \widetilde{S}) S - \partial_\varrho E_{\rm int}(\tilde{\varrho}, \widetilde{S})\varrho\right) [\mathbf{u}_B \cdot \mathbf{n}]^+ \mathrm{d}\sigma_x \mathrm{d}t \\
&\quad + \int_0^\tau \int_\Omega \frac{\tilde{\vartheta}}{\vartheta}\left(\mathbb{S} : \nabla_x \mathbf{u} - \frac{\mathbf{q} \cdot \nabla_x \vartheta}{\vartheta}\right) \mathrm{d}x\mathrm{d}t \\
&\quad \leq \int_0^\tau \int_\Omega \frac{\varrho}{\tilde{\varrho}} (\mathbf{u} - \tilde{\mathbf{u}}) \cdot \nabla_x \widetilde{p} \ \mathrm{d}x\mathrm{d}t \\
&\quad - \int_0^\tau \int_\Omega \left(\varrho(s - \widetilde{s})\partial_t \tilde{\vartheta} + \varrho(s - \widetilde{s})\mathbf{u} \cdot \nabla_x \tilde{\vartheta} + \left(\frac{\mathbf{q}}{\vartheta}\right) \cdot \nabla_x \tilde{\vartheta}\right) \mathrm{d}x\mathrm{d}t \\
&\quad + \int_0^\tau \int_{\partial\Omega} \left(\frac{\tilde{\vartheta}}{\vartheta} - 1\right) f_{i,B}\ [\mathbf{u}_B \cdot \mathbf{n}]^- \mathrm{d}\sigma_x \mathrm{d}t
\end{aligned}
$$

$$
\begin{aligned}
&+\int_0^\tau \int_{\partial\Omega} \tilde{\vartheta}\varrho_B\Big[s(\varrho_B,\vartheta) - \widetilde{s} + \frac{\widetilde{e}}{\widetilde{\vartheta}} - \frac{e(\varrho_B,\vartheta)}{\vartheta}\Big] [\mathbf{u}_B\cdot\mathbf{n}]^- \mathrm{d}\sigma_x \mathrm{d}t \\
&+\int_0^\tau \int_{\partial\Omega} \frac{\varrho_B}{\widetilde{\varrho}}\widetilde{p}\ [\mathbf{u}_B\cdot\mathbf{n}]^- \mathrm{d}\sigma_x \mathrm{d}t \\
&-\int_0^\tau \int_{\Omega} \Big[\varrho(\mathbf{u}-\tilde{\mathbf{u}})\otimes(\mathbf{u}-\tilde{\mathbf{u}}) + p\mathbb{I} - \mathbb{S}\Big] : \mathbb{D}_x\tilde{\mathbf{u}}\ \mathrm{d}x\mathrm{d}t \\
&+\int_0^\tau \int_{\Omega} \varrho\Big[\mathbf{g} - \partial_t\tilde{\mathbf{u}} - (\tilde{\mathbf{u}}\cdot\nabla_x)\tilde{\mathbf{u}} - \frac{1}{\widetilde{\varrho}}\nabla_x\widetilde{p}\Big]\cdot(\mathbf{u}-\tilde{\mathbf{u}})\ \mathrm{d}x\mathrm{d}t \\
&+\int_0^\tau \int_{\Omega} \Big[\Big(1-\frac{\varrho}{\widetilde{\varrho}}\Big)\partial_t\widetilde{p} - \frac{\varrho}{\widetilde{\varrho}}\mathbf{u}\cdot\nabla_x\widetilde{p}\Big]\ \mathrm{d}x\mathrm{d}t \qquad (3.50)
\end{aligned}
$$

for a.a. $0 \le \tau < T$. Indeed (3.50) follows from (3.47), where $\mathbb{1}_{[0,\tau]}$ is approximated by a suitable sequence of $\psi_\delta \in C^1_c[0,T)$, $\psi_\delta \nearrow \mathbb{1}_{[0,\tau]}$. The desired result

$$\int_\Omega E\left(\varrho, S, \mathbf{m}\ \Big|\ \tilde{\varrho}, \widetilde{S}, \widetilde{\mathbf{m}}\right)(\tau,\cdot)\mathrm{d}x = 0$$

will follow as soon as we can control the integrals on the right-hand side of (3.50) via a Gronwall type argument. This will be done through several steps delineated below.

3.5.1 Estimates Based on the Momentum Balance

As the strong solution $(\tilde{\varrho}, \tilde{\vartheta}, \tilde{\mathbf{u}})$ satisfies the momentum equation (1.12), we deduce

$$
\begin{aligned}
&\int_\Omega \varrho(\tilde{\mathbf{u}}-\mathbf{u})\cdot\Big(\partial_t\tilde{\mathbf{u}} + \tilde{\mathbf{u}}\cdot\nabla_x\tilde{\mathbf{u}} + \frac{1}{\widetilde{\varrho}}\nabla_x\widetilde{p} - \mathbf{g}\Big)\ \mathrm{d}x + \int_\Omega \mathbb{S} : \nabla_x\tilde{\mathbf{u}}\ \mathrm{d}x \\
&= \int_\Omega \frac{\varrho}{\widetilde{\varrho}}(\tilde{\mathbf{u}}-\mathbf{u})\cdot\mathrm{div}_x\widetilde{\mathbb{S}} + \mathbb{S} : \nabla_x\tilde{\mathbf{u}}\ \mathrm{d}x \\
&= \int_\Omega \Big(\frac{\varrho}{\widetilde{\varrho}} - 1\Big)(\tilde{\mathbf{u}}-\mathbf{u})\cdot\mathrm{div}_x\widetilde{\mathbb{S}}\ \mathrm{d}x + \int_\Omega \nabla_x(\mathbf{u}-\tilde{\mathbf{u}}) : \widetilde{\mathbb{S}} + \mathbb{S} : \nabla_x\tilde{\mathbf{u}}\ \mathrm{d}x,
\end{aligned}
$$

where we have denoted $\widetilde{\mathbb{S}}$ the viscous stress associated to the smooth solution. Note that we also need the density $\tilde{\varrho}$ associated to the strong solution to be bounded below away from zero. Substituting this in (3.50) and rearranging the remaining terms we easily obtain

$$
\begin{aligned}
&\int_\Omega E\left(\varrho, S, \mathbf{m}\ \Big|\ \tilde{\varrho}, \widetilde{S}, \widetilde{\mathbf{m}}\right)(\tau,\cdot)\mathrm{d}x \\
&\quad + \int_0^\tau \int_{\partial\Omega} \Big(E_{\mathrm{int}}(\varrho, S) - \partial_S E_{\mathrm{int}}(\tilde{\varrho}, \widetilde{S})S - \partial_\varrho E_{\mathrm{int}}(\tilde{\varrho}, \widetilde{S})\varrho\Big)\ [\mathbf{u}_B\cdot\mathbf{n}]^+ \mathrm{d}\sigma_x \mathrm{d}t
\end{aligned}
$$

$$
\begin{aligned}
&+\int_0^\tau\int_\Omega\left(\left(\frac{\tilde\vartheta}{\vartheta}-1\right)\mathbb{S}:\nabla_x\mathbf{u}+\left(1-\frac{\tilde\vartheta}{\vartheta}\right)\frac{\mathbf{q}\cdot\nabla_x\vartheta}{\vartheta}\right)\,\mathrm{d}x\mathrm{d}t\\
&+\int_0^\tau\int_\Omega\frac{\mathbf{q}}{\vartheta}\cdot\left(\nabla_x\tilde\vartheta-\nabla_x\vartheta\right)\,\mathrm{d}x\mathrm{d}t+\int_0^\tau\int_\Omega(\mathbb{S}-\widetilde{\mathbb{S}}):(\nabla_x\mathbf{u}-\nabla_x\tilde{\mathbf{u}})\,\mathrm{d}x\mathrm{d}t\\
&\leq-\int_0^\tau\int_\Omega p\mathrm{div}_x\tilde{\mathbf{u}}\,\mathrm{d}x\mathrm{d}t+\int_0^\tau\int_\Omega(\mathbf{u}-\tilde{\mathbf{u}})\cdot\nabla_x\widetilde{p}\,\mathrm{d}x\mathrm{d}t\\
&+\int_0^\tau\int_\Omega\tilde\varrho(\widetilde{s}-s)(\partial_t\tilde\vartheta+\tilde{\mathbf{u}}\cdot\nabla_x\tilde\vartheta)\mathrm{d}x\mathrm{d}t\\
&+\int_0^\tau\int_{\partial\Omega}\left(\frac{\tilde\vartheta}{\vartheta}-1\right)f_{i,B}\,[\mathbf{u}_B\cdot\mathbf{n}]^-\mathrm{d}\sigma_x\mathrm{d}t\\
&+\int_0^\tau\int_{\partial\Omega}\tilde\vartheta\varrho_B\left[s(\varrho_B,\vartheta)-\widetilde{s}+\frac{\widetilde{e}}{\tilde\vartheta}-\frac{e(\varrho_B,\vartheta)}{\vartheta}\right][\mathbf{u}_B\cdot\mathbf{n}]^-\mathrm{d}\sigma_x\mathrm{d}t\\
&+\int_0^\tau\int_{\partial\Omega}\frac{\varrho_B}{\tilde\varrho}\widetilde{p}\,[\mathbf{u}_B\cdot\mathbf{n}]^-\mathrm{d}\sigma_x\mathrm{d}t\\
&+\int_0^\tau\int_\Omega\left[\left(1-\frac{\varrho}{\tilde\varrho}\right)\partial_t\widetilde{p}-\frac{\varrho}{\tilde\varrho}\mathbf{u}\cdot\nabla_x\widetilde{p}\right]\,\mathrm{d}x\mathrm{d}t+\int_0^\tau F_1(t)\mathrm{d}t,
\end{aligned}
\tag{3.51}
$$

where we have denoted

$$
\begin{aligned}
F_1=&-\int_\Omega\varrho(\tilde{\mathbf{u}}-\mathbf{u})\otimes(\tilde{\mathbf{u}}-\mathbf{u}):\nabla_x\tilde{\mathbf{u}}\,\mathrm{d}x\\
&+\int_\Omega\left(\frac{\varrho-\tilde\varrho}{\tilde\varrho}\right)(\mathbf{u}-\tilde{\mathbf{u}})\cdot\left(\nabla_x\widetilde{p}-\mathrm{div}_x\widetilde{\mathbb{S}}\right)\,\mathrm{d}x\\
&+\int_\Omega\varrho(\widetilde{s}-s)(\mathbf{u}-\tilde{\mathbf{u}})\cdot\nabla_x\tilde\vartheta\,\mathrm{d}x+\int_\Omega(\varrho-\tilde\varrho)(\widetilde{s}-s)\left(\partial_t\tilde\vartheta+\tilde{\mathbf{u}}\cdot\nabla_x\tilde\vartheta\right)\,\mathrm{d}x.
\end{aligned}
\tag{3.52}
$$

The integral F_1 is quadratic in the differences of the weak and strong solution and therefore, at least formally, controllable by means of the relative energy.

3.5.2 Estimating Pressure

We start by a simple observation

$$
-\int_\Omega\tilde{\mathbf{u}}\cdot\nabla_x\widetilde{p}\,\mathrm{d}x=-\int_{\partial\Omega}\widetilde{p}\mathbf{u}_B\cdot\mathbf{n}\mathrm{d}\sigma_x+\int_\Omega\widetilde{p}\mathrm{div}_x\tilde{\mathbf{u}}\,\mathrm{d}x.
$$

Next, we use Gibbs' relation to deduce

$$
\widetilde{p}=\frac{\partial(\varrho e)|_S(\tilde\varrho,\widetilde{S})}{\partial S}\widetilde{S}+\frac{\partial(\varrho e)|_\varrho(\tilde\varrho,\widetilde{S})}{\partial\varrho}\tilde\varrho-\tilde\varrho\widetilde{e}.
$$

Now, as $\tilde{\varrho}$, $\tilde{\mathbf{u}}$ satisfy the equation of continuity (1.11), and, by virtue of Gibbs' relation,

$$\frac{\partial s(\tilde{\varrho},\tilde{\vartheta})|_{\vartheta}}{\partial \varrho} = -\frac{1}{\tilde{\varrho}^2}\frac{\partial p(\tilde{\varrho},\tilde{\vartheta})|_{\varrho}}{\partial \vartheta}.$$

Using the entropy equation (1.15) satisfied by the strong solution we compute

$$\begin{aligned}
&\left(1-\frac{\varrho}{\tilde{\varrho}}\right)\left(\partial_t p(\tilde{\varrho},\tilde{\vartheta}) + \tilde{\mathbf{u}}\cdot\nabla_x p(\tilde{\varrho},\tilde{\vartheta})\right) + \mathrm{div}_x\tilde{\mathbf{u}}\Big(p(\tilde{\varrho},\tilde{\vartheta}) - p(\varrho,\vartheta)\Big)\\
&=\mathrm{div}_x\tilde{\mathbf{u}}\Big(p(\tilde{\varrho},\tilde{\vartheta}) - \frac{\partial p(\tilde{\varrho},\tilde{\vartheta})|_{\vartheta}}{\partial\varrho}(\tilde{\varrho}-\varrho) - \frac{\partial p(\tilde{\varrho},\tilde{\vartheta})|_{\varrho}}{\partial\vartheta}(\tilde{\vartheta}-\vartheta) - p(\varrho,\vartheta)\Big)\\
&+\left(1-\frac{\varrho}{\tilde{\varrho}}\right)\frac{\partial p(\tilde{\varrho},\tilde{\vartheta})|_{\vartheta}}{\partial\vartheta}\left(\partial_t\tilde{\vartheta} + \tilde{\mathbf{u}}\cdot\nabla_x\tilde{\vartheta}\right) - \frac{\tilde{\vartheta}-\vartheta}{\tilde{\varrho}}\frac{\partial p(\tilde{\varrho},\tilde{\vartheta})|_{\varrho}}{\partial\vartheta}\left(\partial_t\tilde{\varrho} + \tilde{\mathbf{u}}\cdot\nabla_x\tilde{\varrho}\right)\\
&=\mathrm{div}_x\tilde{\mathbf{u}}\Big(p(\tilde{\varrho},\tilde{\vartheta}) - \frac{\partial p(\tilde{\varrho},\tilde{\vartheta})|_{\vartheta}}{\partial\varrho}(\tilde{\varrho}-\varrho) - \frac{\partial p(\tilde{\varrho},\tilde{\vartheta})|_{\varrho}}{\partial\vartheta}(\tilde{\vartheta}-\vartheta) - p(\varrho,\vartheta)\Big)\\
&-\tilde{\varrho}(\tilde{\varrho}-\varrho)\frac{\partial s(\tilde{\varrho},\tilde{\vartheta})|_{\vartheta}}{\partial\varrho}\left(\partial_t\tilde{\vartheta} + \tilde{\mathbf{u}}\cdot\nabla_x\tilde{\vartheta}\right) + \tilde{\varrho}(\tilde{\vartheta}-\vartheta)\frac{\partial s(\tilde{\varrho},\tilde{\vartheta})|_{\varrho}}{\partial\varrho}\left(\partial_t\tilde{\varrho} + \tilde{\mathbf{u}}\cdot\nabla_x\tilde{\varrho}\right)\\
&=\mathrm{div}_x\tilde{\mathbf{u}}\Big(p(\tilde{\varrho},\tilde{\vartheta}) - \frac{\partial p(\tilde{\varrho},\tilde{\vartheta})|_{\vartheta}}{\partial\varrho}(\tilde{\varrho}-\varrho) - \frac{\partial p(\tilde{\varrho},\tilde{\vartheta})|_{\varrho}}{\partial\vartheta}(\tilde{\vartheta}-\vartheta) - p(\varrho,\vartheta)\Big)\\
&-\tilde{\varrho}(\tilde{\varrho}-\varrho)\frac{\partial s(\tilde{\varrho},\tilde{\vartheta})|_{\vartheta}}{\partial\varrho}\left(\partial_t\tilde{\vartheta} + \tilde{\mathbf{u}}\cdot\nabla_x\tilde{\vartheta}\right) - \tilde{\varrho}(\tilde{\vartheta}-\vartheta)\frac{\partial s(\tilde{\varrho},\tilde{\vartheta})|_{\varrho}}{\partial\vartheta}\left(\partial_t\tilde{\vartheta} + \tilde{\mathbf{u}}\cdot\nabla_x\tilde{\vartheta}\right)\\
&-(\tilde{\vartheta}-\vartheta)\mathrm{div}_x\left(\frac{\tilde{\mathbf{q}}}{\tilde{\vartheta}}\right) + \left(1-\frac{\vartheta}{\tilde{\vartheta}}\right)\left(\tilde{\mathbb{S}}:\nabla_x\tilde{\mathbf{u}} - \frac{\tilde{\mathbf{q}}\cdot\nabla_x\tilde{\vartheta}}{\tilde{\vartheta}}\right).
\end{aligned}$$

Thus plugging the above relations in (3.51) and using the fact that $\tilde{\varrho}|_{\Gamma_{\rm in}} = \varrho_B$ we obtain

$$\begin{aligned}
&\int_\Omega E\left(\varrho, S, \mathbf{m}\;\middle|\;\tilde{\varrho}, \widetilde{S}, \widetilde{\mathbf{m}}\right)(\tau,\cdot)\mathrm{d}x\\
&+\int_0^\tau\int_{\partial\Omega}\Big(E_{\rm int}(\varrho,S) - \nabla_{S,\varrho}E_{\rm int}(\tilde{\varrho},\widetilde{S})\cdot(S-\widetilde{S},\varrho-\tilde{\varrho})\Big)\,[\mathbf{u}_B\cdot\mathbf{n}]^+\mathrm{d}\sigma_x\mathrm{d}t\\
&-\int_0^\tau\int_{\partial\Omega}E_{\rm int}(\tilde{\varrho},\widetilde{S})\,[\mathbf{u}_B\cdot\mathbf{n}]^+\mathrm{d}\sigma_x\mathrm{d}t\\
&+\int_0^\tau\int_\Omega\left(\frac{\tilde{\vartheta}}{\vartheta}-1\right)\mathbb{S}:\mathbb{D}_x\mathbf{u}\;\mathrm{d}x\mathrm{d}t + \int_0^\tau\int_\Omega\left(\frac{\vartheta}{\tilde{\vartheta}}-1\right)\left(\tilde{\mathbb{S}}:\nabla_x\tilde{\mathbf{u}}\right)\;\mathrm{d}x\mathrm{d}t\\
&+\int_0^\tau\int_\Omega\left(1-\frac{\vartheta}{\tilde{\vartheta}}\right)\frac{\tilde{\mathbf{q}}\cdot\nabla_x\tilde{\vartheta}}{\tilde{\vartheta}}\;\mathrm{d}x\mathrm{d}t + \int_0^\tau\int_\Omega\left(1-\frac{\tilde{\vartheta}}{\vartheta}\right)\frac{\mathbf{q}\cdot\nabla_x\vartheta}{\vartheta}\;\mathrm{d}x\mathrm{d}t\\
&+\int_0^\tau\int_\Omega(\mathbb{S}-\tilde{\mathbb{S}}):(\nabla_x\mathbf{u}-\nabla_x\tilde{\mathbf{u}})\;\mathrm{d}x\mathrm{d}t
\end{aligned}$$

$$+\int_0^\tau \int_\Omega \left(\frac{\mathbf{q}}{\vartheta} - \frac{\widetilde{\mathbf{q}}}{\widetilde{\vartheta}}\right) \cdot \left(\nabla_x \tilde{\vartheta} - \nabla_x \vartheta\right) \, \mathrm{d}x\mathrm{d}t$$

$$\leq \int_0^\tau \int_{\partial\Omega} \left(\frac{\tilde{\vartheta}}{\vartheta} - 1\right) f_{i,B} \, [\mathbf{u}_B \cdot \mathbf{n}]^- \mathrm{d}\sigma_x \mathrm{d}t$$

$$+\int_0^\tau \int_{\partial\Omega} \tilde{\vartheta} \varrho_B \left[s(\varrho_B, \vartheta) - \widetilde{s} + \frac{\widetilde{e}}{\widetilde{\vartheta}} - \frac{e(\varrho_B, \vartheta)}{\vartheta}\right] [\mathbf{u}_B \cdot \mathbf{n}]^- \mathrm{d}\sigma_x \mathrm{d}t$$

$$+\int_{\Gamma_{\mathrm{in}, t \leq \tau}} \widetilde{\mathbf{q}} \cdot \mathbf{n} \left(\frac{\vartheta}{\widetilde{\vartheta}} - 1\right) \, \mathrm{d}\sigma_x \mathrm{d}t + \int_0^\tau F_2(t) \mathrm{d}t, \tag{3.53}$$

where

$$F_2 = -\int_\Omega \varrho(\tilde{\mathbf{u}} - \mathbf{u}) \otimes (\tilde{\mathbf{u}} - \mathbf{u}) : \nabla_x \tilde{\mathbf{u}} \, \mathrm{d}x$$
$$+\int_\Omega \left(1 - \frac{\varrho}{\tilde{\varrho}}\right) (\mathbf{u} - \tilde{\mathbf{u}}) \cdot \left(\nabla_x \widetilde{p} - \mathrm{div}_x \widetilde{\mathbb{S}}\right) \, \mathrm{d}x$$
$$+\int_\Omega \varrho(\widetilde{s} - s)(\mathbf{u} - \tilde{\mathbf{u}}) \cdot \nabla_x \tilde{\vartheta} \, \mathrm{d}x + \int_\Omega (\varrho - \tilde{\varrho})(\widetilde{s} - s) \left(\partial_t \tilde{\vartheta} + \tilde{\mathbf{u}} \cdot \nabla_x \tilde{\vartheta}\right) \, \mathrm{d}x$$
$$-\int_\Omega \left(1 - \frac{\varrho}{\tilde{\varrho}}\right) (\mathbf{u} - \tilde{\mathbf{u}}) \cdot \nabla_x \widetilde{p} \, \mathrm{d}x$$
$$+\int_\Omega \tilde{\varrho} \left(\widetilde{s} - \frac{\partial \widetilde{s}|_\vartheta}{\partial \varrho}(\tilde{\varrho} - \varrho) - \frac{\partial \widetilde{s}|_\varrho}{\partial \vartheta}(\tilde{\vartheta} - \vartheta) - s\right) \left(\partial_t \tilde{\vartheta} + \tilde{\mathbf{u}} \cdot \nabla_x \tilde{\vartheta}\right) \, \mathrm{d}x$$
$$+\int_\Omega \mathrm{div}_x \tilde{\mathbf{u}} \Big(p(\tilde{\varrho}, \tilde{\vartheta}) - \frac{\partial p(\tilde{\varrho}, \tilde{\vartheta})|_\vartheta}{\partial \varrho}(\tilde{\varrho} - \varrho) - \frac{\partial p(\tilde{\varrho}, \tilde{\vartheta})|_\varrho}{\partial \vartheta}(\tilde{\vartheta} - \vartheta) - p(\varrho, \vartheta)\Big) \, \mathrm{d}x. \tag{3.54}$$

With the convention $\mathbf{u}|_{\partial\Omega} = \tilde{\mathbf{u}}|_{\partial\Omega}$, the first three integrals in (3.53) can be interpreted as

$$\int_\Omega E\left(\varrho, S, \mathbf{m} \;\middle|\; \tilde{\varrho}, \widetilde{S}, \widetilde{\mathbf{m}}\right)(\tau, \cdot) \, \mathrm{d}x$$
$$+\int_0^\tau \int_{\partial\Omega} E\left(\varrho, S, \mathbf{m} \;\middle|\; \tilde{\varrho}, \widetilde{S}, \widetilde{\mathbf{m}}\right) \, [\mathbf{u}_B \cdot \mathbf{n}]^+ \mathrm{d}\sigma_x \mathrm{d}t. \tag{3.55}$$

3.5.3 Boundary Integrals

Thanks to the hypothesis of thermodynamic stability, the internal energy is a convex function of the density ϱ and the total entropy S. In particular,

$$\int_{\partial\Omega} \left(E_{\mathrm{int}}(\varrho, S) - \nabla_{S,\varrho} E_{\mathrm{int}}(\tilde{\varrho}, \widetilde{S}) \cdot (S - \widetilde{S}, \varrho - \tilde{\varrho}) - E_{\mathrm{int}}(\tilde{\varrho}, \widetilde{S})\right) [\mathbf{u}_B \cdot \mathbf{n}]^+ \mathrm{d}\sigma_x$$
$$\geq 0.$$

On the other hand, as $(\tilde{\varrho}, \tilde{\mathbf{u}}, \tilde{\vartheta})$ is a strong solution,

$$\varrho_B \widetilde{e} \mathbf{u}_B \cdot \mathbf{n} + \widetilde{\mathbf{q}} \cdot \mathbf{n} = f_{i,B} [\mathbf{u}_B \cdot \mathbf{n}]^- \text{ on } \Gamma_{\rm in}.$$

Consequently, the boundary integral over $\Gamma_{\rm in}$ reads

$$\begin{aligned}
&\int_0^\tau \int_{\partial\Omega} \left(\frac{\tilde{\vartheta}}{\vartheta} - 1 \right) f_{i,B} \, [\mathbf{u}_B \cdot \mathbf{n}]^- \mathrm{d}\sigma_x \mathrm{d}t \\
&+ \int_0^\tau \int_{\partial\Omega} \tilde{\vartheta} \varrho_B \Big[s(\varrho_B, \vartheta) - \widetilde{s} + \frac{\widetilde{e}}{\tilde{\vartheta}} - \frac{e(\varrho_B, \vartheta)}{\vartheta} \Big] \, [\mathbf{u}_B \cdot \mathbf{n}]^- \mathrm{d}\sigma_x \mathrm{d}t \\
&+ \int_{\Gamma_{\rm in}, t \leq \tau} \widetilde{\mathbf{q}} \cdot \mathbf{n} \left(\frac{\vartheta}{\tilde{\vartheta}} - 1 \right) \, \mathrm{d}\sigma_x \mathrm{d}t \\
&= \int_0^\tau \int_{\partial\Omega} f_{i,B} \left(\frac{\vartheta}{\tilde{\vartheta}} + \frac{\tilde{\vartheta}}{\vartheta} - 2 \right) \, [\mathbf{u}_B \cdot \mathbf{n}]^- \mathrm{d}\sigma_x \mathrm{d}t \\
&+ \int_0^\tau \int_{\partial\Omega} \tilde{\vartheta} \varrho_B \Big[s(\varrho_B, \vartheta) - \widetilde{s} + \frac{\widetilde{e}}{\tilde{\vartheta}} - \frac{e(\varrho_B, \vartheta)}{\vartheta} \Big] \, [\mathbf{u}_B \cdot \mathbf{n}]^- \mathrm{d}\sigma_x \mathrm{d}t \\
&- \int_0^\tau \int_{\partial\Omega} \widetilde{e} \left(\frac{\vartheta}{\tilde{\vartheta}} - 1 \right) \varrho_B \, [\mathbf{u}_B \cdot \mathbf{n}]^- \mathrm{d}\sigma_x \mathrm{d}t.
\end{aligned}$$

As a consequence of Gibbs' relation (1.6), the function

$$\tilde{\vartheta} \mapsto e(\varrho_B, \tilde{\vartheta}) - \vartheta s(\varrho_B, \tilde{\vartheta})$$

is decreasing for $\tilde{\vartheta} < \vartheta$ and increasing for $\tilde{\vartheta} > \vartheta$, in particular,

$$\widetilde{e} - \vartheta \widetilde{s} = e(\varrho_B, \tilde{\vartheta}) - \vartheta s(\varrho_B, \tilde{\vartheta}) \geq e(\varrho_B, \vartheta) - \vartheta s(\varrho_B, \vartheta).$$

Consequently, since

$$\begin{aligned}
&\tilde{\vartheta} \Big[s(\varrho_B, \vartheta) - \widetilde{s} + \frac{\widetilde{e}}{\tilde{\vartheta}} - \frac{e(\varrho_B, \vartheta)}{\vartheta} \Big] = \frac{\tilde{\vartheta}}{\vartheta} \Big[\vartheta s(\varrho_B, \vartheta) - \vartheta \widetilde{s} + \frac{\vartheta \widetilde{e}}{\tilde{\vartheta}} - e(\varrho_B, \vartheta) \Big] \\
&= \frac{\tilde{\vartheta}}{\vartheta} \Big[\vartheta s(\varrho_B, \vartheta) - \vartheta \widetilde{s} + \widetilde{e} - e(\varrho_B, \vartheta) \Big] + \widetilde{e} - \frac{\tilde{\vartheta}}{\vartheta} \widetilde{e},
\end{aligned}$$

the boundary integral can be estimated as

$$\begin{aligned}
&\int_0^\tau \int_{\partial\Omega} \left(\frac{\tilde{\vartheta}}{\vartheta} - 1 \right) f_{i,B} \, [\mathbf{u}_B \cdot \mathbf{n}]^- \mathrm{d}\sigma_x \mathrm{d}t \\
&+ \int_0^\tau \int_{\partial\Omega} \tilde{\vartheta} \varrho_B \Big[s(\varrho_B, \vartheta) - \widetilde{s} + \frac{\widetilde{e}}{\tilde{\vartheta}} - \frac{e(\varrho_B, \vartheta)}{\vartheta} \Big] \, [\mathbf{u}_B \cdot \mathbf{n}]^- \mathrm{d}\sigma_x \mathrm{d}t \\
&+ \int_{\Gamma_{\rm in}, t \leq \tau} \widetilde{\mathbf{q}} \cdot \mathbf{n} \left(\frac{\vartheta}{\tilde{\vartheta}} - 1 \right) \, \mathrm{d}\sigma_x \mathrm{d}t
\end{aligned}$$

$$
= \int_0^\tau \int_{\partial\Omega} \left(f_{i,B} \left(\frac{\vartheta}{\tilde\vartheta} + \frac{\tilde\vartheta}{\vartheta} - 2 \right) - \tilde e \varrho_B \left(\frac{\vartheta}{\tilde\vartheta} + \frac{\tilde\vartheta}{\vartheta} - 2 \right) \right) \ [\mathbf{u}_B \cdot \mathbf{n}]^- \mathrm{d}\sigma_x \mathrm{d}t
$$

$$
+ \int_0^\tau \int_{\partial\Omega} \frac{\tilde\vartheta}{\vartheta} \varrho_B \Big[\vartheta s(\varrho_B, \vartheta) - \vartheta \tilde s + \tilde e - e(\varrho_B, \vartheta) \Big] \ [\mathbf{u}_B \cdot \mathbf{n}]^- \mathrm{d}\sigma_x \mathrm{d}t
$$

$$
\leq \int_{\Gamma_{\mathrm{in}}, t \leq \tau} \left(\frac{\vartheta}{\tilde\vartheta} + \frac{\tilde\vartheta}{\vartheta} - 2 \right) \tilde{\mathbf{q}} \cdot \mathbf{n} \ \mathrm{d}\sigma_x \mathrm{d}t. \tag{3.56}
$$

Going back to (3.53) we may infer that

$$
\int_\Omega E \left(\varrho, S, \mathbf{m} \ \middle| \ \tilde\varrho, \widetilde S, \widetilde{\mathbf{m}} \right) (\tau, \cdot) \mathrm{d}x
$$

$$
+ \int_0^\tau \int_\Omega \left(\frac{\tilde\vartheta}{\vartheta} - 1 \right) \mathbb{S} : \mathbb{D}_x \mathbf{u} \ \mathrm{d}x\mathrm{d}t + \int_0^\tau \int_\Omega \left(\frac{\vartheta}{\tilde\vartheta} - 1 \right) \left(\widetilde{\mathbb{S}} : \mathbb{D}_x \tilde{\mathbf{u}} \right) \ \mathrm{d}x\mathrm{d}t
$$

$$
+ \int_0^\tau \int_\Omega \left(1 - \frac{\vartheta}{\tilde\vartheta} \right) \frac{\tilde{\mathbf{q}} \cdot \nabla_x \tilde\vartheta}{\tilde\vartheta} \ \mathrm{d}x\mathrm{d}t + \int_0^\tau \int_\Omega \left(1 - \frac{\tilde\vartheta}{\vartheta} \right) \frac{\mathbf{q} \cdot \nabla_x \vartheta}{\vartheta} \ \mathrm{d}x\mathrm{d}t
$$

$$
+ \int_0^\tau \int_\Omega (\mathbb{S} - \widetilde{\mathbb{S}}) : (\mathbb{D}_x \mathbf{u} - \mathbb{D}_x \tilde{\mathbf{u}}) \ \mathrm{d}x\mathrm{d}t
$$

$$
+ \int_0^\tau \int_\Omega \left(\frac{\mathbf{q}}{\vartheta} - \frac{\tilde{\mathbf{q}}}{\tilde\vartheta} \right) \cdot \left(\nabla_x \tilde\vartheta - \nabla_x \vartheta \right) \ \mathrm{d}x\mathrm{d}t
$$

$$
\leq \int_{\Gamma_{\mathrm{in}}, t \leq \tau} \left(\frac{\vartheta}{\tilde\vartheta} + \frac{\tilde\vartheta}{\vartheta} - 2 \right) \tilde{\mathbf{q}} \cdot \mathbf{n} \ \mathrm{d}\sigma_x \mathrm{d}t + \int_0^\tau F_2(t) \mathrm{d}t, \tag{3.57}
$$

where F_2 is given by (3.54).

3.5.4 Weak–Strong Uniqueness, Conditional Result

In the derivation of the inequality (3.57), we have only used Gibbs' relation (1.16) and the hypothesis of thermodynamic stability (1.19) equivalent to the convexity of the internal energy with respect to the conservative entropy variables (ϱ, S). Unfortunately, these are not strong enough to reach any conclusion on the basis of (3.57) only. Note, in particular, that the "dissipative" terms on the right-hand side of (3.57) need not be non-negative; whence a priori of no use in the Gronwall argument. At this level of generality, we are able to show only a *conditional* result, imposing extra hypotheses on the weak solution $(\varrho, \vartheta, \mathbf{u})$.

In accordance with (1.9) and (1.10), we suppose

$$\begin{aligned} \mathbb{S} &= \mu(\varrho,\vartheta)\left(\nabla_x \mathbf{u} + \nabla_x \mathbf{u}^t - \frac{2}{d}\mathrm{div}_x \mathbf{u}\mathbb{I}\right) + \eta(\varrho,\vartheta)\mathrm{div}_x \mathbf{u}\mathbb{I}, \\ &\mu(\varrho,\vartheta) > 0,\ \eta(\varrho,\vartheta) \geq 0, \end{aligned} \tag{3.58}$$

$$\mathbf{q} = -\kappa(\varrho,\vartheta)\nabla_x \vartheta,\ \kappa(\varrho,\vartheta) > 0, \tag{3.59}$$

where the transport coefficients μ, η, and κ are continuously differentiable functions of ϱ and ϑ.

Now, a direct manipulation yields

$$\begin{aligned} &\left(\frac{\tilde{\vartheta}}{\vartheta} - 1\right)\mathbb{S} : \mathbb{D}_x \mathbf{u} + \left(\frac{\vartheta}{\tilde{\vartheta}} - 1\right)\left(\widetilde{\mathbb{S}} : \mathbb{D}_x \tilde{\mathbf{u}}\right) + (\mathbb{S} - \widetilde{\mathbb{S}}) : (\mathbb{D}_x \mathbf{u} - \mathbb{D}_x \tilde{\mathbf{u}}) \\ &\qquad = \frac{\tilde{\vartheta}}{\vartheta}\mathbb{S} : \nabla_x \mathbf{u} + \frac{\vartheta}{\tilde{\vartheta}}\widetilde{\mathbb{S}} : \nabla_x \tilde{\mathbf{u}} - \mathbb{S} : \nabla_x \tilde{\mathbf{u}} - \widetilde{\mathbb{S}} : \nabla_x \mathbf{u}, \end{aligned}$$

and, similarly,

$$\begin{aligned} &\left(1 - \frac{\vartheta}{\tilde{\vartheta}}\right)\frac{\tilde{\mathbf{q}} \cdot \nabla_x \tilde{\vartheta}}{\tilde{\vartheta}} + \left(1 - \frac{\tilde{\vartheta}}{\vartheta}\right)\frac{\mathbf{q} \cdot \nabla_x \vartheta}{\vartheta} + \left(\frac{\mathbf{q}}{\vartheta} - \frac{\tilde{\mathbf{q}}}{\tilde{\vartheta}}\right) \cdot \left(\nabla_x \tilde{\vartheta} - \nabla_x \vartheta\right) \\ &\qquad = -\frac{\vartheta}{\tilde{\vartheta}}\frac{\tilde{\mathbf{q}} \cdot \nabla_x \tilde{\vartheta}}{\tilde{\vartheta}} - \frac{\tilde{\vartheta}}{\vartheta}\frac{\mathbf{q} \cdot \nabla_x \vartheta}{\vartheta} + \frac{\mathbf{q}}{\vartheta} \cdot \nabla_x \tilde{\vartheta} + \frac{\tilde{\mathbf{q}}}{\tilde{\vartheta}} \cdot \nabla_x \vartheta. \end{aligned}$$

In view of (3.58) and (3.59), both expressions can be written in the form

$$\frac{\tilde{\vartheta}}{\vartheta}\nu|\mathbb{A}|^2 + \frac{\vartheta}{\tilde{\vartheta}}\widetilde{\nu}|\widetilde{\mathbb{A}}|^2 - (\nu + \widetilde{\nu})\mathbb{A} : \widetilde{\mathbb{A}},$$

where ν stands for a transport coefficient and $\mathbb{A}$ for the velocity or temperature gradient. More specifically, we have either

$$\begin{aligned} \nu &= \mu(\vartheta),\ \eta(\vartheta),\ \mathbb{A} = \left(\nabla_x \mathbf{u} + \nabla_x^t \mathbf{u} - \frac{2}{d}\mathrm{div}_x \mathbf{u}\mathbb{I}\right),\ \mathbb{A} = \mathrm{div}_x \mathbf{u}\mathbb{I}, \\ \widetilde{\nu} &= \mu(\tilde{\vartheta}),\ \eta(\tilde{\vartheta}),\ \widetilde{\mathbb{A}} = \left(\nabla_x \tilde{\mathbf{u}} + \nabla_x^t \tilde{\mathbf{u}} - \frac{2}{d}\mathrm{div}_x \mathbf{u}\mathbb{I}\right),\ \widetilde{\mathbb{A}} = \mathrm{div}_x \tilde{\mathbf{u}}\mathbb{I}, \end{aligned} \tag{3.60}$$

or

$$\nu = \frac{\kappa(\vartheta)}{\vartheta},\ \mathbb{A} = \nabla_x \vartheta,\ \widetilde{\nu} = \frac{\kappa(\tilde{\vartheta})}{\tilde{\vartheta}},\ \widetilde{\mathbb{A}} = \nabla_x \tilde{\vartheta}. \tag{3.61}$$

Next, compute

$$
\begin{aligned}
&\frac{\tilde{\vartheta}}{\vartheta}\nu|\mathbb{A}|^2 + \frac{\vartheta}{\tilde{\vartheta}}\widetilde{\nu}|\widetilde{\mathbb{A}}|^2 - (\nu+\widetilde{\nu})\mathbb{A} : \widetilde{\mathbb{A}} \\
&= \nu\left|\sqrt{\frac{\tilde{\vartheta}}{\vartheta}}\mathbb{A}\right|^2 + \widetilde{\nu}\left|\sqrt{\frac{\vartheta}{\tilde{\vartheta}}}\widetilde{\mathbb{A}}\right|^2 - (\nu+\widetilde{\nu})\left(\sqrt{\frac{\tilde{\vartheta}}{\vartheta}}\mathbb{A}\right) : \left(\sqrt{\frac{\vartheta}{\tilde{\vartheta}}}\widetilde{\mathbb{A}}\right) \\
&= \tilde{\vartheta}\frac{\nu}{\vartheta}\left|\mathbb{A} - \frac{\vartheta}{\tilde{\vartheta}}\widetilde{\mathbb{A}}\right|^2 + (\nu-\widetilde{\nu})\left(\mathbb{A} - \frac{\vartheta}{\tilde{\vartheta}}\widetilde{\mathbb{A}}\right) : \widetilde{\mathbb{A}} \\
&\geq \tilde{\vartheta}\frac{\nu}{\vartheta}\left|\mathbb{A} - \widetilde{\mathbb{A}}\right|^2 + 2\tilde{\vartheta}\frac{\nu}{\vartheta}\left(1 - \frac{\vartheta}{\tilde{\vartheta}}\right)\left(\mathbb{A} - \widetilde{\mathbb{A}}\right) : \widetilde{\mathbb{A}} + (\nu-\widetilde{\nu})\left(\mathbb{A} - \widetilde{\mathbb{A}}\right) : \widetilde{\mathbb{A}} \\
&+ (\nu-\widetilde{\nu})\left(1 - \frac{\vartheta}{\tilde{\vartheta}}\right)|\widetilde{\mathbb{A}}|^2. \qquad (3.62)
\end{aligned}
$$

Consequently, returning to (3.57) we obtain

$$
\begin{aligned}
&\int_\Omega E\left(\varrho, S, \mathbf{m} \;\middle|\; \tilde{\varrho}, \widetilde{S}, \widetilde{\mathbf{m}}\right)(\tau, \cdot)\mathrm{d}x \\
&+ \int_0^\tau \int_\Omega \frac{\mu\tilde{\vartheta}}{2\vartheta}\left|(\nabla_x \mathbf{u} - \nabla_x \tilde{\mathbf{u}}) + (\nabla_x \mathbf{u} - \nabla_x \tilde{\mathbf{u}})^t - \frac{2}{d}(\mathrm{div}_x \mathbf{u} - \mathrm{div}_x \tilde{\mathbf{u}})\mathbb{I}\right|^2 \mathrm{d}x\mathrm{d}t \\
&+ \int_0^\tau \int_\Omega \frac{\eta\tilde{\vartheta}}{\vartheta}\left|\mathrm{div}_x \mathbf{u} - \mathrm{div}_x \tilde{\mathbf{u}}\right|^2 \mathrm{d}x\mathrm{d}t + \int_0^\tau \int_\Omega \frac{\kappa\tilde{\vartheta}}{\vartheta^2}|\nabla_x \vartheta - \nabla_x \tilde{\vartheta}|^2 \mathrm{d}x\mathrm{d}t \\
&\leq \sum_{i=1}^3 \int_0^\tau \int_\Omega 2\tilde{\vartheta}\frac{\nu_i}{\vartheta}\left(1 - \frac{\vartheta}{\tilde{\vartheta}}\right)\left(\widetilde{\mathbb{A}}_i - \mathbb{A}_i\right) : \widetilde{\mathbb{A}}_i \,\mathrm{d}x\mathrm{d}t \\
&+ \sum_{i=1}^3 \int_0^\tau \int_\Omega (\widetilde{\nu}_i - \nu_i)\left(\mathbb{A}_i - \widetilde{\mathbb{A}}_i\right) : \widetilde{\mathbb{A}}_i \,\mathrm{d}x\mathrm{d}t \\
&+ \sum_{i=1}^3 \int_0^\tau \int_\Omega (\widetilde{\nu}_i - \nu_i)\left(1 - \frac{\vartheta}{\tilde{\vartheta}}\right)|\widetilde{\mathbb{A}}_i|^2 \,\mathrm{d}x\mathrm{d}t \\
&+ \int_{\Gamma_{\mathrm{in}}, t \leq \tau}\left(\frac{\vartheta}{\tilde{\vartheta}} + \frac{\tilde{\vartheta}}{\vartheta} - 2\right)\widetilde{\mathbf{q}} \cdot \mathbf{n} \,\mathrm{d}\sigma_x\mathrm{d}t + \int_0^\tau F_2(t)\mathrm{d}t, \qquad (3.63)
\end{aligned}
$$

where

$$
\begin{aligned}
&\nu_1 = \mu(\varrho, \vartheta),\ \widetilde{\nu}_1 = \mu(\tilde{\varrho}, \tilde{\vartheta}),\ \nu_2 = \eta(\varrho, \vartheta),\ \widetilde{\nu}_2 = \eta(\tilde{\varrho}, \tilde{\vartheta}), \\
&\nu_3 = \frac{\kappa(\varrho, \vartheta)}{\vartheta},\ \widetilde{\nu}_3 = \frac{\kappa(\tilde{\varrho}, \tilde{\vartheta})}{\tilde{\vartheta}}, \\
&\mathbb{A}_1 = \left(\nabla_x \mathbf{u} + \nabla_x \mathbf{u}^t - \frac{2}{d}\mathrm{div}_x \mathbf{u}\mathbb{I}\right),\ \widetilde{\mathbb{A}}_1 = \left(\nabla_x \tilde{\mathbf{u}} + \nabla_x \tilde{\mathbf{u}}^t - \frac{2}{d}\mathrm{div}_x \tilde{\mathbf{u}}\mathbb{I}\right), \\
&\mathbb{A}_2 = \mathrm{div}_x \mathbf{u},\ \widetilde{\mathbb{A}}_2 = \mathrm{div}_x \tilde{\mathbf{u}},\ \mathbb{A}_3 = \nabla_x \vartheta,\ \widetilde{\mathbb{A}}_3 = \nabla_x \tilde{\vartheta},
\end{aligned}
$$

and F_2 is given by (3.54).

Now *suppose* that the weak solution satisfies

$$0 < \underline{\varrho} \leq \varrho(t,x) \leq \overline{\varrho},\ 0 < \underline{\vartheta} \leq \vartheta(t,x) \leq \overline{\vartheta} \text{ a.a. in } (0,T) \times \Omega, \tag{3.64}$$

for certain constants $\underline{\varrho}$, $\overline{\varrho}$, $\underline{\vartheta}$, $\overline{\vartheta}$. Under these circumstances, we have

$$E\left(\varrho, S, \mathbf{m} \,\middle|\, \tilde{\varrho}, \widetilde{S}, \widetilde{\mathbf{m}}\right)$$
$$\approx |\varrho - \tilde{\varrho}|^2 + |S - \widetilde{S}|^2 + |\mathbf{m} - \widetilde{\mathbf{m}}|^2 \approx |\varrho - \tilde{\varrho}|^2 + |\vartheta - \tilde{\vartheta}|^2 + |\mathbf{u} - \tilde{\mathbf{u}}|^2.$$

Consequently, it is easy to check that

$$F_2(t) \lesssim \int_\Omega E\left(\varrho, S, \mathbf{m} \,\middle|\, \tilde{\varrho}, \widetilde{S}, \widetilde{\mathbf{m}}\right)(t, \cdot)\ \mathrm{d}x.$$

Similarly, the volume integrals on the right-hand side of (3.63) containing $\mathbb{A}$ can be controlled by their counterparts on the left-hand side augmented by

$$\int_0^\tau \int_\Omega E\left(\varrho, S, \mathbf{m} \,\middle|\, \tilde{\varrho}, \widetilde{S}, \widetilde{\mathbf{m}}\right)(t, \cdot)\ \mathrm{d}x\mathrm{d}t$$

as the case may be.

Finally, thanks to our hypothesis (3.64), the boundary integral can be handled by means of the trace theorem and interpolation as

$$\int_{\Gamma_{\rm in}, t \leq \tau} \left(\frac{\vartheta}{\tilde{\vartheta}} + \frac{\tilde{\vartheta}}{\vartheta} - 2 \right) \widetilde{\mathbf{q}} \cdot \mathbf{n}\ \mathrm{d}\sigma_x \mathrm{d}t \lesssim \int_0^\tau \int_{\partial\Omega} |\vartheta - \tilde{\vartheta}|^2\ \mathrm{d}\sigma_x \mathrm{d}t$$
$$\lesssim \delta \int_0^\tau \|\vartheta - \tilde{\vartheta}\|^2_{W^{1,2}(\Omega)} \mathrm{d}t + c(\delta) \int_0^\tau \|\vartheta - \tilde{\vartheta}\|^2_{L^2(\Omega)} \mathrm{d}t, \tag{3.65}$$

where $\delta > 0$ can be chosen arbitrarily small. Thus the standard Gronwall argument applied to the inequality (3.63) yields

$$E\left(\varrho, S, \mathbf{m} \,\middle|\, \tilde{\varrho}, \widetilde{S}, \widetilde{\mathbf{m}}\right) = 0 \text{ a.a. in } (0,T) \times \Omega.$$

We have shown the following *conditional* weak–strong uniqueness property.

Theorem 5 (Conditional Weak–Strong Uniqueness).
Let the thermodynamic functions p, e, and s be continuously differentiable functions of $(\varrho, \vartheta) \in (0, \infty)^2$ satisfying Gibbs' equation (1.16), *together with the hypothesis of thermodynamic stability* (1.19). *Let the transport coefficients μ, η, and κ be continuously differentiable functions of ϱ and ϑ,*

$$\mu > 0, \ \ \kappa > 0, \eta \geq 0.$$

Let $(\varrho, \vartheta, \mathbf{u})$ be a weak solution of the Navier–Stokes–Fourier system in $(0, T) \times \Omega$, $0 < T < \infty$, in the sense of Definition 4 satisfying, in addition,

$$0 < \underline{\varrho} \leq \varrho(t, x) \leq \overline{\varrho}, \ 0 < \underline{\vartheta} \leq \vartheta(t, x) \leq \overline{\vartheta} \text{ for a.a. } (t, x) \in (0, T) \times \Omega$$

for certain constants $\underline{\varrho}$, $\overline{\varrho}$, $\underline{\vartheta}$, $\overline{\vartheta}$.

Suppose that the same problem (with the same initial and boundary data) admits a strong solution $(\tilde{\varrho}, \tilde{\vartheta}, \tilde{\mathbf{u}})$ in the class

$$\tilde{\varrho}, \ \tilde{\vartheta} \in C^1([0, T] \times \overline{\Omega}), \ \tilde{\mathbf{u}} \in C^1([0, T] \times \overline{\Omega}; R^d), \ D_x^2 \tilde{\mathbf{u}}, \ D_x^2 \tilde{\vartheta} \in C([0, T] \times \overline{\Omega}).$$

Then

$$\varrho = \tilde{\varrho}, \ \vartheta = \tilde{\vartheta}, \ \mathbf{u} = \tilde{\mathbf{u}} \text{ in } [0, T] \times \overline{\Omega}.$$

To obtain an unconditional result, certain structural properties must be imposed on the constitutive relations. We therefore postpone this issue to the following chapter, where the *existence* of weak solutions is discussed.

3.6 Ballistic Energy

In the preceding part, we have used the total energy balance (3.42), together with the entropy inequality (3.41) to derive a differential inequality for the time derivative of the relative energy

$$\int_\Omega E\left(\varrho, S, \mathbf{u} \Big| \tilde{\varrho}, \widetilde{S}, \tilde{\mathbf{u}}\right) \mathrm{d}x \equiv \int_\Omega \left[\frac{1}{2}\varrho|\mathbf{u} - \tilde{\mathbf{u}}|^2 \right.$$
$$\left. + E_{\mathrm{int}}(\varrho, S) - \frac{\partial E_{\mathrm{int}}(\tilde{\varrho}, \widetilde{S})}{\partial \varrho}(\varrho - \tilde{\varrho}) - \frac{\partial E_{\mathrm{int}}(\tilde{\varrho}, \widetilde{S})}{\partial S}(S - \widetilde{S}) - E_{\mathrm{int}}(\tilde{\varrho}, \widetilde{S})\right] \mathrm{d}x.$$

Seeing that the time evolution of the integral

$$\int_\Omega \left(\frac{\partial E_{\mathrm{int}}(\tilde{\varrho}, \widetilde{S})}{\partial \varrho} \tilde{\varrho} + \frac{\partial E_{\mathrm{int}}(\tilde{\varrho}, \widetilde{S})}{\partial S} \widetilde{S} - E_{\mathrm{int}}(\tilde{\varrho}, \widetilde{S}) \right) \mathrm{d}x$$

is determined by the given functions $(\tilde{\varrho}, \widetilde{S})$, and

$$\int_{\Omega} \frac{\partial E_{\rm int}(\tilde{\varrho}, \widetilde{S})}{\partial \varrho} \varrho \ {\rm d}x$$

can be evaluated by means of the equation of continuity (3.38), the behaviour of the relative energy depends in an essential way on the time evolution of the quantity

$$\int_{\Omega} \left(\frac{1}{2} |\mathbf{u} - \tilde{\mathbf{u}}|^2 + E_{\rm int}(\varrho, S) - \frac{\partial E_{\rm int}(\tilde{\varrho}, \widetilde{S})}{\partial S} S \right) \ {\rm d}x,$$

where we easily identify the ballistic free energy

$$E_{\rm int}(\varrho, S) - \frac{\partial E_{\rm int}(\tilde{\varrho}, \widetilde{S})}{\partial S} S = \varrho e(\varrho, \vartheta) - \tilde{\vartheta} \varrho s(\varrho, \vartheta)$$

introduced in Sect. 1.3. This motivates the definition of

ballistic energy

$$E_{\tilde{\vartheta}}(\varrho, \vartheta, \mathbf{u}) \equiv \frac{1}{2} \varrho |\mathbf{u} - \mathbf{u}_B|^2 + \varrho e(\varrho, \vartheta) - \tilde{\vartheta} \varrho s(\varrho, \vartheta)$$

along with its integrated version

$$\mathcal{E}_{\tilde{\vartheta}}(\varrho, \vartheta, \mathbf{u}) \equiv \int_{\Omega} \left(\frac{1}{2} \varrho |\mathbf{u} - \mathbf{u}_B|^2 + \varrho e(\varrho, \vartheta) - \tilde{\vartheta} \varrho s(\varrho, \vartheta) \right) \ {\rm d}x.$$

Revisiting the weak formulation in Definition 4 we realize that we could have replaced the total energy balance (3.42) by the corresponding inequality for the ballistic energy. Such a modification does not seem necessary as the time evolution of the ballistic energy can be evaluated by means of (3.41) and (3.42), see (3.44). We use this alternative formulation when dealing with the Dirichlet boundary conditions for the temperature in Chap. 12.

3.7 Concluding Remarks

Similar approach to weak solutions to the Navier–Stokes–Fourier system was used in the monograph [43] in the case of conservative boundary conditions, meaning for closed fluid systems. The idea of relative energy/entropy goes back to the pioneering paper by Dafermos [25]. The concept has been developed in different contexts by many authors, see Germain [57], Mellet and Vasseur [73], Saint-Raymond [82] to name only a few. The application to the weak–strong uniqueness problem for the Navier–Stokes–Fourier system with conservative boundary conditions can be found in [42]. A nice survey on the role of convexity in the analysis of partial differential equations is presented in the recent study by Ball and Chen [2].

Chapter 4
Constitutive Theory and Weak–Strong Uniqueness Revisited

Up to now, we have imposed very general restrictions concerning the constitutive relations characterizing the material properties of the fluid. In particular, the piece of information that can be deduced from the energy and entropy balance was not sufficient to obtain an unconditional result of the weak–strong uniqueness property without the extra assumptions concerning boundedness of the density and the temperature. In this chapter, we fix the hypotheses imposed on the equation of state as well as the transport coefficients and establish a general weak–strong uniqueness principle valid in the class of finite energy weak solutions. The same set of hypotheses will guarantee the existence of global in time weak solutions for any finite energy/entropy initial data.

4.1 Constitutive Relations

The constitutive relations used in the present monograph are motivated by the properties of real *gases* rather than liquids. In particular, all transport coefficients μ, η, and κ will depend on the temperature ϑ rather than the density ϱ, cf. e.g. Becker [7]. Moreover, the equation of state will be augmented by a component corresponding to thermal radiation that has a regularizing effect necessary to overcome the so-called vacuum problem. Indeed most of the technical difficulties of the underlying mathematical theory are related to the (hypothetical) possibility that the fluid density may vanish in a certain part of the physical domain. Note that this is impossible for the strong solutions satisfying the transport equation

$$\partial_t \varrho + \mathbf{u} \cdot \nabla_x \varrho = -\varrho \mathrm{div}_x \mathbf{u}$$

E. Feireisl, A. Novotný, *Mathematics of Open Fluid Systems*, Nečas Center Series, https://doi.org/10.1007/978-3-030-94793-4_4

as long as the initial density profile ϱ_0 as well as the boundary data ϱ_B are strictly positive. As a matter of fact, it would be enough to secure the estimate

$$\mathrm{div}_x \mathbf{u} \in L^1(0,T; L^\infty(\Omega)) \tag{4.1}$$

that would guarantee the uniform upper as well as lower bound on ϱ in terms of the data. Unfortunately, in view of the results by Merle et al. [74] for the isentropic system, estimates like (4.1) are not very realistic or at least extremely difficult to obtain at least for the system far from equilibrium.

4.1.1 Equation of State

A general form of the equation of state (EOS) relating the pressure p to the standard variables (ϱ, ϑ) considered in this monograph reads

$$p(\varrho, \vartheta) = p_{\mathrm{el}}(\varrho) + p_{\mathrm{m}}(\varrho, \vartheta) + p_{\mathrm{rad}}(\vartheta). \tag{4.2}$$

A typical example is the gaseous star material in astrophysics, where $p_{\mathrm{el}} \approx \varrho^{\frac{5}{3}}$ is the pressure of the electrons relevant in the low temperature/high density regime, $p_m(\varrho, \vartheta) \approx \varrho\vartheta$ is the molecular pressure of the gas material, and $p_{\mathrm{rad}} \approx \vartheta^4$ is the radiative pressure relevant in the high temperature regime, see Battaner [5]. We will refer to (4.2) as *real gas EOS*.

Another motivation for considering (4.2) is the general form of EOS for *monoatomic gases*,

$$p_M = \frac{2}{3} \varrho e_M, \tag{4.3}$$

cf. Eliezer et al. [30]. It is a routine matter to check that (4.3) is compatible with Gibbs' relation (1.6) only if

$$p_M(\varrho, \vartheta) = \vartheta^{\frac{5}{2}} \mathcal{P}_M \left(\frac{\varrho}{\vartheta^{\frac{3}{2}}} \right), \tag{4.4}$$

where $\mathcal{P}_M \in C^1[0, \infty)$ is arbitrary. Note that the choice $\mathcal{P}_M(Z) = Z$ gives rise to the standard *Boyle–Mariotte law*

$$p_{BM}(\varrho, \vartheta) = \varrho\vartheta. \tag{4.5}$$

In accordance with (4.3), we get

$$e_M(\varrho, \vartheta) = \frac{3}{2} \frac{\vartheta^{\frac{5}{2}}}{\varrho} \mathcal{P}_M \left(\frac{\varrho}{\vartheta^{\frac{3}{2}}} \right). \tag{4.6}$$

The hypothesis of thermodynamic stability (1.19) expressed in terms of (4.4) and (4.6) translates to

$$\mathcal{P}_M'(Z) > 0 \text{ for all } Z \geq 0, \quad \frac{\frac{5}{3}\mathcal{P}_M(Z) - \mathcal{P}_M'(Z)Z}{Z} > 0 \text{ for any } Z > 0. \quad (4.7)$$

It follows from (4.7) that the function

$$Z \mapsto \frac{\mathcal{P}_M(Z)}{Z^{\frac{5}{3}}} \text{ is decreasing,}$$

and we may suppose

$$\lim_{Z\to\infty} \frac{\mathcal{P}_M(Z)}{Z^{\frac{5}{3}}} = p_\infty > 0.$$

This motivates the ansatz

$$p_{\rm el}(\varrho) = p_\infty \varrho^{\frac{5}{3}}$$

in (4.2).

As for the radiation component, it takes the standard form

$$p_{\rm rad}(\vartheta) = \frac{a}{3}\vartheta^4, \ a > 0,$$

where a is the Stefan–Boltzmann constant, see Oxenius [78].

Similarly, we write

$$e(\varrho, \vartheta) = e_{\rm el}(\varrho) + e_{\rm m}(\varrho, \vartheta) + e_{\rm rad}(\varrho, \vartheta).$$

Motivated by the previous discussion, we consider

$$p_{\rm el}(\varrho) = \overline{p}\varrho^{\frac{5}{3}}, \ \overline{p} > 0, \ e_{\rm el}(\varrho) = \frac{3}{2}\overline{p}\varrho^{\frac{2}{3}}, \quad (4.8)$$

and

$$p_{\rm rad}(\vartheta) = \frac{a}{3}\vartheta^4, \ a > 0, \ e_{\rm rad}(\varrho, \vartheta) = \frac{a}{\varrho}\vartheta^4. \quad (4.9)$$

Note that such a choice is compatible with both Gibbs' relation (1.6) and the hypothesis of thermodynamic stability (1.19). The effect of $p_{\rm el}$, $e_{\rm el}$ and $p_{\rm rad}$, $e_{\rm rad}$ is essential in the degenerate area of low temperature and low density, respectively. In particular, the presence of the radiative pressure eliminates possible oscillations of the temperature in the hypothetical vacuum zone.

As for the molecular component, the model example is the standard Boyle–Mariotte EOS (4.5). Specifically, we suppose $p_{\rm m} \in C[0, \infty)^2 \cap C^3(0, \infty)^2$ and impose the following restrictions:

$$\frac{1}{\vartheta}\left(De_{\rm m} + p_{\rm m}D\left(\frac{1}{\varrho}\right)\right) = Ds_{\rm m}, \quad (4.10)$$

meaning there exists $e_{\rm m}$ related to p_m through Gibb's equation:

$$\varrho e_{\rm m} \approx p_{\rm m}(\varrho,\vartheta); \tag{4.11}$$

$$\frac{\partial p_{\rm m}(\varrho,\vartheta)}{\partial \varrho} > 0,\ 0 < \frac{\partial e_{\rm m}(\varrho,\vartheta)}{\partial \vartheta} \leq \overline{c}_v \text{ for any } \varrho,\vartheta > 0. \tag{4.12}$$

In particular, $p_{\rm m}$ $e_{\rm m}$ satisfy the hypothesis of thermodynamic stability (1.19). In addition, we require

$$\lim_{\vartheta\to 0} e_{\rm m}(\varrho,\vartheta) = 0 \text{ for any } \varrho > 0,\ \lim_{\varrho\to 0} p_{\rm m}(\varrho,\vartheta) = 0 \text{ for any } \vartheta > 0; \tag{4.13}$$

$$\left| \frac{\partial p_{\rm m}(\varrho,\vartheta)}{\partial \vartheta} \right| \lesssim \varrho. \tag{4.14}$$

It follows from Gibbs' relation

$$\frac{\partial s_m}{\partial \varrho} = -\frac{1}{\varrho^2}\frac{\partial p_m}{\partial \vartheta};$$

whence

$$|s_m(\varrho,1)| \leq |s_m(1,1)| + \left| \int_1^{\varrho} \frac{\partial s_{\rm m}}{\partial \varrho}(z,1)\ {\rm d}z \right| \lesssim 1 + |\log(\varrho)|$$
$$|s_m(\varrho,\vartheta)| \leq |s_m(\varrho,1)| + \left| \int_1^{\vartheta} \frac{\partial s_{\rm m}}{\partial \vartheta}(\varrho,z)\ {\rm d}z \right| \lesssim (1 + |\log(\varrho)| + |\log(\vartheta)|)\,. \tag{4.15}$$

Note that the "electron" component does not interfere in the entropy, specifically,

$$s(\varrho,\vartheta) = s_{\rm m}(\varrho,\vartheta) + s_{\rm rad}(\varrho,\vartheta),\ s_{\rm rad}(\varrho,\vartheta) = \frac{4a}{3}\frac{1}{\varrho}\vartheta^3.$$

The quantity $p_{\rm el}$ is sometimes called "cold" pressure as it persists in the low temperature regime. Accordingly, we introduce the "thermal" pressure

$$p_{\rm th}(\varrho,\vartheta) = p_{\rm m}(\varrho,\vartheta) + p_{\rm rad}(\vartheta),\ e_{\rm th}(\varrho,\vartheta) = e_{\rm m}(\varrho,\vartheta) + e_{\rm rad}(\varrho,\vartheta), \tag{4.16}$$

noticing that

$$p_{\rm th} \geq 0,\ e_{\rm th} \geq 0,$$
$$p_{\rm th}(\varrho,\vartheta) \to 0,\ e_{\rm th}(\varrho,\vartheta) \to 0 \text{ as } \vartheta \to 0 \text{ for any fixed } \varrho > 0. \tag{4.17}$$

4.1.2 Transport Coefficients

For the purpose of the existence theory, we focus on the transport coefficients depending solely on the temperature. Specifically, we suppose that $\mu = \mu(\vartheta)$, $\eta = \eta(\vartheta)$, $\kappa = \kappa(\vartheta)$ are continuously differentiable functions of ϑ satisfying

$$0 < \underline{\mu}(1 + \vartheta^{\Lambda}) \leq \mu(\vartheta) \leq \overline{\mu}(1 + \vartheta^{\Lambda}), \ |\mu'(\vartheta)| \leq c \text{ for all } \vartheta \geq 0,$$
$$\text{where } \frac{1}{2} \leq \Lambda \leq 1, \tag{4.18}$$

$$0 \leq \eta(\vartheta) \leq \overline{\eta}(1 + \vartheta^{\Lambda}) \text{ for all } \vartheta \geq 0, \tag{4.19}$$

and

$$0 < \underline{\kappa}(1 + \vartheta^3) \leq \kappa(\vartheta) \leq \overline{\kappa}(1 + \vartheta^3) \text{ for all } \vartheta \geq 0 \tag{4.20}$$

for certain positive constants $\underline{\mu}$, $\overline{\mu}$, $\overline{\eta}$, $\underline{\kappa}$, $\overline{\kappa}$.

Note that $\Lambda = \frac{1}{2}$ is particularly relevant for gases, while the cubic growth of κ reflects the effect of radiation, cf. Becker [7], Oxenius [78].

4.2 Weak–Strong Uniqueness Revisited

In Sect. 3.5, we have shown a conditional weak–strong uniqueness result. Namely any bounded weak solution coincides with the strong solution sharing the same initial and boundary data. Now, we are ready to establish an unconditional weak–strong uniqueness property under the constitutive restrictions specified in the preceding section. Let $(\varrho, \vartheta, \mathbf{u})$ be a weak solution and $(\tilde{\varrho}, \tilde{\vartheta}, \tilde{\mathbf{u}})$ a strong solution of the Navier–Stokes–Fourier system emanating from the same initial and boundary data and driven by the same external force $\mathbf{g}$. First, recall the relative energy inequality derived in (3.57):

$$\int_{\Omega} E\left(\varrho, S, \mathbf{m} \,\middle|\, \tilde{\varrho}, \widetilde{S}, \widetilde{\mathbf{m}}\right)(\tau, \cdot) \mathrm{d}x$$
$$+ \int_0^{\tau} \int_{\Omega} \left(\frac{\tilde{\vartheta}}{\vartheta} - 1\right) \mathbb{S} : \mathbb{D}_x \mathbf{u} \ \mathrm{d}x\mathrm{d}t + \int_0^{\tau} \int_{\Omega} \left(\frac{\vartheta}{\tilde{\vartheta}} - 1\right) \widetilde{\mathbb{S}} : \mathbb{D}_x \tilde{\mathbf{u}} \ \mathrm{d}x\mathrm{d}t$$
$$+ \int_0^{\tau} \int_{\Omega} \left(1 - \frac{\vartheta}{\tilde{\vartheta}}\right) \frac{\tilde{\mathbf{q}} \cdot \nabla_x \tilde{\vartheta}}{\tilde{\vartheta}} \ \mathrm{d}x\mathrm{d}t + \int_0^{\tau} \int_{\Omega} \left(1 - \frac{\tilde{\vartheta}}{\vartheta}\right) \frac{\mathbf{q} \cdot \nabla_x \vartheta}{\vartheta} \ \mathrm{d}x\mathrm{d}t$$
$$+ \int_0^{\tau} \int_{\Omega} (\mathbb{S} - \widetilde{\mathbb{S}}) : (\mathbb{D}_x \mathbf{u} - \mathbb{D}_x \tilde{\mathbf{u}}) \ \mathrm{d}x\mathrm{d}t$$

$$
\begin{aligned}
&+\int_0^\tau \int_\Omega \left(\frac{\mathbf{q}}{\vartheta} - \frac{\widetilde{\mathbf{q}}}{\tilde{\vartheta}} \right) \cdot \left(\nabla_x \tilde{\vartheta} - \nabla_x \vartheta \right) \, \mathrm{d}x\mathrm{d}t \\
&\leq \int_0^\tau \int_{\partial\Omega} \left(\frac{\tilde{\vartheta}}{\vartheta} - 1 \right) f_{i,B} \, [\mathbf{u}_B \cdot \mathbf{n}]^- \mathrm{d}\sigma_x \mathrm{d}t \\
&+ \int_0^\tau \int_{\partial\Omega} \tilde{\vartheta} \varrho_B \left[s(\varrho_B, \vartheta) - \widetilde{s} + \frac{\widetilde{e}}{\tilde{\vartheta}} - \frac{e(\varrho_B, \vartheta)}{\vartheta} \right] [\mathbf{u}_B \cdot \mathbf{n}]^- \mathrm{d}\sigma_x \mathrm{d}t \\
&+ \int_{\Gamma_{\mathrm{in}}, t \leq \tau} \widetilde{\mathbf{q}} \cdot \mathbf{n} \left(\frac{\vartheta}{\tilde{\vartheta}} - 1 \right) \, \mathrm{d}\sigma_x \mathrm{d}t + \int_0^\tau F(t) \mathrm{d}t,
\end{aligned} \tag{4.21}
$$

with the remainder term

$$
\begin{aligned}
F = &- \int_\Omega \varrho (\tilde{\mathbf{u}} - \mathbf{u}) \otimes (\tilde{\mathbf{u}} - \mathbf{u}) : \nabla_x \tilde{\mathbf{u}} \, \mathrm{d}x \\
&+ \int_\Omega \left(1 - \frac{\varrho}{\tilde{\varrho}} \right) (\mathbf{u} - \tilde{\mathbf{u}}) \cdot \left(\nabla_x \widetilde{p} - \mathrm{div}_x \widetilde{\mathbb{S}} \right) \, \mathrm{d}x \\
&+ \int_\Omega \varrho (\widetilde{s} - s)(\mathbf{u} - \tilde{\mathbf{u}}) \cdot \nabla_x \tilde{\vartheta} \, \mathrm{d}x + \int_\Omega (\varrho - \tilde{\varrho})(\widetilde{s} - s) \left(\partial_t \tilde{\vartheta} + \tilde{\mathbf{u}} \cdot \nabla_x \tilde{\vartheta} \right) \, \mathrm{d}x \\
&+ \int_\Omega \left(1 - \frac{\varrho}{\tilde{\varrho}} \right) (\mathbf{u} - \tilde{\mathbf{u}}) \cdot \nabla_x \widetilde{p} \, \mathrm{d}x \\
&+ \int_\Omega \tilde{\varrho} \left(\widetilde{s} - \frac{\partial \widetilde{s}|_\vartheta}{\partial \varrho} (\tilde{\varrho} - \varrho) - \frac{\partial \widetilde{s}|_\varrho}{\partial \vartheta} (\tilde{\vartheta} - \vartheta) - s \right) \left(\partial_t \tilde{\vartheta} + \tilde{\mathbf{u}} \cdot \nabla_x \tilde{\vartheta} \right) \, \mathrm{d}x \\
&+ \int_\Omega \mathrm{div}_x \tilde{\mathbf{u}} \Big(p(\tilde{\varrho}, \tilde{\vartheta}) - \frac{\partial p(\tilde{\varrho}, \tilde{\vartheta})|_\vartheta}{\partial \varrho} (\tilde{\varrho} - \varrho) - \frac{\partial p(\tilde{\varrho}, \tilde{\vartheta})|_\varrho}{\partial \vartheta} (\tilde{\vartheta} - \vartheta) - p(\varrho, \vartheta) \Big) \, \mathrm{d}x.
\end{aligned} \tag{4.22}
$$

Similarly to Sect. 3.5, our goal is to "absorb" the integrals on the right-hand side of (4.21) by those on the left-hand side.

4.2.1 Essential vs. Residual Component

It is convenient to introduce the following sets:

$$
\begin{aligned}
\mathcal{O}_{\mathrm{ess}} = \Big\{ (t,x) \in (0,T) \times \Omega \ \Big| \\
\frac{1}{2} \inf \tilde{\varrho} \leq \varrho(t,x) \leq 2 \sup \tilde{\varrho}, \ \frac{1}{2} \inf \tilde{\vartheta} \leq \vartheta(t,x) \leq 2 \sup \tilde{\vartheta} \Big\},
\end{aligned}
$$

$$
\mathcal{O}_{\mathrm{res}} = ((0,T) \times \Omega) \setminus \mathcal{O}_{\mathrm{ess}}.
$$

Next, for any measurable function h, we set

$$
h_{\mathrm{ess}} = \mathbb{1}_{\mathcal{O}_{\mathrm{ess}}} h, \ h_{\mathrm{res}} = \mathbb{1}_{\mathcal{O}_{\mathrm{res}}} h.
$$

Now, it follows from hypotheses (4.8), (4.9), and (4.12) that the relative energy is a strictly convex function of $(\varrho, S, \mathbf{m})$ yielding immediately the estimate

$$\begin{aligned} \|[\varrho - \tilde{\varrho}]_{\rm ess}\|^2_{L^2(\Omega)} + \left\|[\vartheta - \tilde{\vartheta}]_{\rm ess}\right\|^2_{L^2(\Omega)} + \|[\mathbf{u} - \tilde{\mathbf{u}}]_{\rm ess}\|^2_{L^2(\Omega;R^d)} \\ \leq c \int_\Omega E\left(\varrho, S, \mathbf{m} \,\Big|\, \tilde{\varrho}, \widetilde{S}, \widetilde{\mathbf{m}}\right) \, {\rm d}x, \end{aligned} \tag{4.23}$$

where the constant depends on

$$\inf \tilde{\varrho}, \ \inf \tilde{\vartheta}, \ \sup \tilde{\varrho}, \ \sup \tilde{\vartheta}.$$

We also have

$$\int_\Omega \left[1 + \varrho|\mathbf{u}|^2 + \varrho|s(\varrho,\vartheta)| + \varrho e(\varrho,\vartheta)\right]_{\rm res} \, {\rm d}x \lesssim \int_\Omega E\left(\varrho, S, \mathbf{m} \,\Big|\, \tilde{\varrho}, \widetilde{S}, \widetilde{\mathbf{m}}\right) \, {\rm d}x. \tag{4.24}$$

Indeed the estimate for the total entropy follows directly from convexity whereas the bound of the internal energy follows easily from the explicit formula (1.23) relating the relative energy to the ballistic free energy.

Similarly to Sect. 3.5, the proof of weak–strong uniqueness leans on several steps specified in the forthcoming sections.

4.2.2 Temperature Gradient

Recall that $a \lesssim b$ means there exists a positive constant such that $a \leq cb$. Similarly, we define $a \gtrsim b$. We write $a \approx b$ if $a \lesssim b$ and $b \lesssim a$. Similarly to the above, the constants may depend on the strong solution $(\tilde{\varrho}, \tilde{\vartheta}, \tilde{\mathbf{u}})$.

Let us start with a crucial estimate

$$\begin{aligned} \left\|\vartheta - \tilde{\vartheta}\right\|^2_{W^{1,2}(\Omega)} \lesssim \int_\Omega \left(1 - \frac{\vartheta}{\tilde{\vartheta}}\right) \frac{\tilde{\mathbf{q}} \cdot \nabla_x \tilde{\vartheta}}{\tilde{\vartheta}} \, {\rm d}x + \int_\Omega \left(1 - \frac{\tilde{\vartheta}}{\vartheta}\right) \frac{\mathbf{q} \cdot \nabla_x \vartheta}{\vartheta} \, {\rm d}x \\ + \int_\Omega \left(\frac{\mathbf{q}}{\vartheta} - \frac{\tilde{\mathbf{q}}}{\tilde{\vartheta}}\right) \cdot \left(\nabla_x \tilde{\vartheta} - \nabla_x \vartheta\right) \, {\rm d}x + c(\tilde{\vartheta}) \int_\Omega E\left(\varrho, S, \mathbf{m} \,\Big|\, \tilde{\varrho}, \widetilde{S}, \widetilde{\mathbf{m}}\right) \, {\rm d}x. \end{aligned} \tag{4.25}$$

Indeed it follows from (3.62) with the ansatz (3.61) that

$$
\left(1-\frac{\vartheta}{\tilde{\vartheta}}\right)\frac{\tilde{\mathbf{q}}\cdot\nabla_x\tilde{\vartheta}}{\tilde{\vartheta}}+\left(1-\frac{\tilde{\vartheta}}{\vartheta}\right)\frac{\mathbf{q}\cdot\nabla_x\vartheta}{\vartheta}+\left(\frac{\mathbf{q}}{\vartheta}-\frac{\tilde{\mathbf{q}}}{\tilde{\vartheta}}\right)\cdot\left(\nabla_x\tilde{\vartheta}-\nabla_x\vartheta\right)
$$
$$
\geq \tilde{\vartheta}\frac{\kappa(\vartheta)}{\vartheta^2}\left|\nabla_x\vartheta-\nabla_x\tilde{\vartheta}\right|^2+2\frac{\kappa(\vartheta)}{\vartheta^2}(\tilde{\vartheta}-\vartheta)\left(\nabla_x\vartheta-\nabla_x\tilde{\vartheta}\right)\cdot\nabla_x\tilde{\vartheta}
$$
$$
+\left(\frac{\kappa(\vartheta)}{\vartheta}-\frac{\kappa(\tilde{\vartheta})}{\tilde{\vartheta}}\right)(\nabla_x\vartheta-\nabla_x\tilde{\vartheta})\cdot\nabla_x\tilde{\vartheta}+\left(\frac{\kappa(\vartheta)}{\vartheta}-\frac{\kappa(\tilde{\vartheta})}{\tilde{\vartheta}}\right)\left(\frac{\tilde{\vartheta}-\vartheta}{\tilde{\vartheta}}\right)|\nabla_x\tilde{\vartheta}|^2.
$$

As the heat conductivity coefficient κ satisfies (4.20), we get

$$
\left|\nabla_x\vartheta-\nabla_x\tilde{\vartheta}\right|^2\lesssim\tilde{\vartheta}\frac{\kappa(\vartheta)}{\vartheta^2}\left|\nabla_x\vartheta-\nabla_x\tilde{\vartheta}\right|^2.
$$

Consequently, by virtue of Hölder's inequality, the above lower bound yields the estimate

$$
\left[\left(1-\frac{\vartheta}{\tilde{\vartheta}}\right)\frac{\tilde{\mathbf{q}}\cdot\nabla_x\tilde{\vartheta}}{\tilde{\vartheta}}+\left(1-\frac{\tilde{\vartheta}}{\vartheta}\right)\frac{\mathbf{q}\cdot\nabla_x\vartheta}{\vartheta}+\left(\frac{\mathbf{q}}{\vartheta}-\frac{\tilde{\mathbf{q}}}{\tilde{\vartheta}}\right)\cdot\left(\nabla_x\tilde{\vartheta}-\nabla_x\vartheta\right)\right]_{\rm ess}
$$
$$
\gtrsim\left[\frac{\kappa(\vartheta)}{\vartheta^2}\right]_{\rm ess}\left|\nabla_x\vartheta-\nabla_x\tilde{\vartheta}\right|^2-c(\tilde{\vartheta})\left|[\vartheta-\tilde{\vartheta}]_{\rm ess}\right|^2. \tag{4.26}
$$

As for the residual part, we rewrite

$$
\left(1-\frac{\vartheta}{\tilde{\vartheta}}\right)\frac{\tilde{\mathbf{q}}\cdot\nabla_x\tilde{\vartheta}}{\tilde{\vartheta}}+\left(1-\frac{\tilde{\vartheta}}{\vartheta}\right)\frac{\mathbf{q}\cdot\nabla_x\vartheta}{\vartheta}+\left(\frac{\mathbf{q}}{\vartheta}-\frac{\tilde{\mathbf{q}}}{\tilde{\vartheta}}\right)\cdot\left(\nabla_x\tilde{\vartheta}-\nabla_x\vartheta\right)
$$
$$
=\tilde{\vartheta}\frac{\kappa(\vartheta)}{\vartheta^2}|\nabla_x\vartheta|^2-\frac{\kappa(\vartheta)}{\vartheta}\nabla_x\vartheta\cdot\nabla_x\tilde{\vartheta}+\vartheta\frac{\kappa(\tilde{\vartheta})|\nabla_x\tilde{\vartheta}|^2}{\tilde{\vartheta}^2}-\frac{\kappa(\tilde{\vartheta})}{\tilde{\vartheta}}\nabla_x\tilde{\vartheta}\cdot\nabla_x\vartheta.
$$

Seeing that

$$
\left|\frac{\kappa(\vartheta)}{\vartheta}\nabla_x\vartheta\cdot\nabla_x\tilde{\vartheta}\right|\leq\frac{\tilde{\vartheta}}{2}\frac{\kappa(\vartheta)}{\vartheta^2}|\nabla_x\vartheta|^2+c(\tilde{\vartheta})\kappa(\vartheta)
$$

we deduce

$$
\left(1-\frac{\vartheta}{\tilde{\vartheta}}\right)\frac{\tilde{\mathbf{q}}\cdot\nabla_x\tilde{\vartheta}}{\tilde{\vartheta}}+\left(1-\frac{\tilde{\vartheta}}{\vartheta}\right)\frac{\mathbf{q}\cdot\nabla_x\vartheta}{\vartheta}+\left(\frac{\mathbf{q}}{\vartheta}-\frac{\tilde{\mathbf{q}}}{\tilde{\vartheta}}\right)\cdot\left(\nabla_x\tilde{\vartheta}-\nabla_x\vartheta\right)
$$
$$
\geq\frac{\tilde{\vartheta}}{4}\frac{\kappa(\vartheta)}{\vartheta^2}|\nabla_x\vartheta|^2-c(\tilde{\vartheta})(\vartheta+\kappa(\vartheta)).
$$

Consequently, in view of hypothesis (4.20),

$$
\left[\left(1-\frac{\vartheta}{\tilde{\vartheta}}\right)\frac{\tilde{\mathbf{q}}\cdot\nabla_x\tilde{\vartheta}}{\tilde{\vartheta}}+\left(1-\frac{\tilde{\vartheta}}{\vartheta}\right)\frac{\mathbf{q}\cdot\nabla_x\vartheta}{\vartheta}+\left(\frac{\mathbf{q}}{\vartheta}-\frac{\tilde{\mathbf{q}}}{\tilde{\vartheta}}\right)\cdot\left(\nabla_x\tilde{\vartheta}-\nabla_x\vartheta\right)\right]_{\rm res}
$$

$$\geq \frac{\tilde{\vartheta}}{4}\left[\frac{\kappa(\vartheta)}{\vartheta^2}\right]_{\rm res} |\nabla_x \vartheta|^2 - c(\tilde{\vartheta})\left[1+\vartheta^3\right]_{\rm res}. \tag{4.27}$$

Finally, observe that

$$\left[\frac{\kappa(\vartheta)}{\vartheta^2}\right]_{\rm res} |\nabla_x \vartheta|^2 \gtrsim [1]_{\rm res}\, |\nabla_x \vartheta|^2 \gtrsim [1]_{\rm res}\, |\nabla_x \vartheta - \nabla_x \tilde{\vartheta}|^2 - c(\tilde{\vartheta})[1]_{\rm res}. \tag{4.28}$$

Summing up (4.26)–(4.28) we conclude

$$\begin{aligned}
\left\| \nabla_x \vartheta - \nabla_x \tilde{\vartheta} \right\|^2_{L^2(\Omega;R^d)} &\lesssim \int_\Omega \left(1 - \frac{\vartheta}{\tilde{\vartheta}}\right) \frac{\tilde{\mathbf{q}} \cdot \nabla_x \tilde{\vartheta}}{\tilde{\vartheta}}\, {\rm d}x \\
&+ \int_\Omega \left(1 - \frac{\tilde{\vartheta}}{\vartheta}\right) \frac{\mathbf{q} \cdot \nabla_x \vartheta}{\vartheta}\, {\rm d}x + \int_\Omega \left(\frac{\mathbf{q}}{\vartheta} - \frac{\tilde{\mathbf{q}}}{\tilde{\vartheta}}\right) \cdot \left(\nabla_x \tilde{\vartheta} - \nabla_x \vartheta\right) {\rm d}x \\
&+ c(\tilde{\vartheta}) \int_\Omega E\left(\varrho, S, \mathbf{m} \,\middle|\, \tilde{\varrho}, \widetilde{S}, \widetilde{\mathbf{m}}\right) {\rm d}x.
\end{aligned} \tag{4.29}$$

Note that, as a consequence of the hypotheses (4.2) and (4.20),

$$[1+\kappa(\vartheta)]_{\rm res} \lesssim [1 + \varrho e(\varrho,\vartheta)]_{\rm res}.$$

Now, it is easy to see that (4.29) implies (4.25) as

$$\begin{aligned}
\left\| \vartheta - \tilde{\vartheta} \right\|^2_{L^2(\Omega)} &= \left\| [\vartheta - \tilde{\vartheta}]_{\rm ess} \right\|^2_{L^2(\Omega)} + \left\| [\vartheta - \tilde{\vartheta}]_{\rm res} \right\|^2_{L^2(\Omega)} \\
&\lesssim \int_\Omega E\left(\varrho, S, \mathbf{m} \,\middle|\, \tilde{\varrho}, \widetilde{S}, \widetilde{\mathbf{m}}\right) {\rm d}x.
\end{aligned} \tag{4.30}$$

We point out that validity of the crucial estimate (4.25) depends in an essential way on the presence of the radiation component in the internal energy.

Going back to (4.21) we obtain

$$\begin{aligned}
&\int_\Omega E\left(\varrho, S, \mathbf{m} \;\middle|\; \tilde{\varrho}, \widetilde{S}, \widetilde{\mathbf{m}}\right)(\tau,\cdot){\rm d}x + \int_0^\tau \left\| \vartheta - \tilde{\vartheta} \right\|^2_{W^{1,2}(\Omega)} {\rm d}t \\
&+ \int_0^\tau \int_\Omega \left(\frac{\tilde{\vartheta}}{\vartheta} - 1\right) \mathbb{S} : \nabla_x \mathbf{u}\, {\rm d}x{\rm d}t + \int_0^\tau \int_\Omega \left(\frac{\vartheta}{\tilde{\vartheta}} - 1\right) \left(\widetilde{\mathbb{S}} : \nabla_x \tilde{\mathbf{u}}\right) {\rm d}x{\rm d}t \\
&+ \int_0^\tau \int_\Omega (\mathbb{S} - \widetilde{\mathbb{S}}) : (\mathbb{D}_x \mathbf{u} - \mathbb{D}_x \tilde{\mathbf{u}})\, {\rm d}x{\rm d}t \\
&\lesssim \int_0^\tau \int_{\partial\Omega} \left(\frac{\tilde{\vartheta}}{\vartheta} - 1\right) f_{i,B}\, [\mathbf{u}_B \cdot \mathbf{n}]^- {\rm d}\sigma_x {\rm d}t \\
&+ \int_0^\tau \int_{\partial\Omega} \tilde{\vartheta} \varrho_B \left[s(\varrho_B, \vartheta) - \widetilde{s} + \frac{\widetilde{e}}{\tilde{\vartheta}} - \frac{e(\varrho_B,\vartheta)}{\vartheta}\right] [\mathbf{u}_B \cdot \mathbf{n}]^- {\rm d}\sigma_x {\rm d}t
\end{aligned}$$

$$+\int_{\Gamma_{\rm in},t\leq\tau}\tilde{\mathbf{q}}\cdot\mathbf{n}\left(\frac{\vartheta}{\tilde{\vartheta}}-1\right)\,{\rm d}\sigma_x{\rm d}t+\int_0^\tau F(t){\rm d}t$$
$$+\int_0^\tau\int_\Omega E\left(\varrho,S,\mathbf{m}\;\middle|\;\tilde{\varrho},\widetilde{S},\widetilde{\mathbf{m}}\right)\,{\rm d}x{\rm d}t. \tag{4.31}$$

4.2.3 Velocity Gradient

We apply similar treatment to the remaining integrals on the left-hand side of (4.31) containing the velocity gradient. Suppose for a while that $\eta = 0$. Using the inequality (3.62) with the ansatz (3.60) we get

$$\left[\left(\frac{\tilde{\vartheta}}{\vartheta}-1\right)\mathbb{S}:\nabla_x\mathbf{u}+\left(\frac{\vartheta}{\tilde{\vartheta}}-1\right)\left(\widetilde{\mathbb{S}}:\nabla_x\tilde{\mathbf{u}}\right)+(\mathbb{S}-\widetilde{\mathbb{S}}):(\nabla_x\mathbf{u}-\nabla_x\tilde{\mathbf{u}})\right]_{\rm ess}$$
$$\geq\tilde{\vartheta}\left[\frac{\mu(\vartheta)}{\vartheta}\right]_{\rm ess}\left|\nabla_x(\mathbf{u}-\tilde{\mathbf{u}})+\nabla_x^t(\mathbf{u}-\tilde{\mathbf{u}})-\frac{2}{d}{\rm div}_x(\mathbf{u}-\tilde{\mathbf{u}})\mathbb{I}\right|^2$$
$$+2\left[\frac{\mu(\vartheta)}{\vartheta}\right]_{\rm ess}\left(\tilde{\vartheta}-\vartheta\right)\left(\nabla_x(\mathbf{u}-\tilde{\mathbf{u}})+\nabla_x^t(\mathbf{u}-\tilde{\mathbf{u}})-\frac{2}{d}{\rm div}_x(\mathbf{u}-\tilde{\mathbf{u}})\mathbb{I}\right):\widetilde{\mathbb{A}}$$
$$+\left[\mu(\vartheta)-\mu(\tilde{\vartheta})\right]_{\rm ess}\left(\nabla_x(\mathbf{u}-\tilde{\mathbf{u}})+\nabla_x^t(\mathbf{u}-\tilde{\mathbf{u}})-\frac{2}{d}{\rm div}_x(\mathbf{u}-\tilde{\mathbf{u}})\mathbb{I}\right):\widetilde{\mathbb{A}}$$
$$+\left[\mu(\vartheta)-\mu(\tilde{\vartheta})\right]_{\rm ess}\left(1-\frac{\vartheta}{\tilde{\vartheta}}\right)|\widetilde{\mathbb{A}}|^2,$$

where

$$\widetilde{\mathbb{A}}=\nabla_x\tilde{\mathbf{u}}+\nabla_x^t\tilde{\mathbf{u}}-\frac{2}{d}{\rm div}_x\tilde{\mathbf{u}}\mathbb{I}.$$

Similarly to the preceding part, we conclude

$$\left[\left(\frac{\tilde{\vartheta}}{\vartheta}-1\right)\mathbb{S}:\nabla_x\mathbf{u}+\left(\frac{\vartheta}{\tilde{\vartheta}}-1\right)\left(\widetilde{\mathbb{S}}:\nabla_x\tilde{\mathbf{u}}\right)+(\mathbb{S}-\widetilde{\mathbb{S}}):(\nabla_x\mathbf{u}-\nabla_x\tilde{\mathbf{u}})\right]_{\rm ess}$$
$$\gtrsim\tilde{\vartheta}\left[\frac{\mu(\vartheta)}{\vartheta}\right]_{\rm ess}\left|\nabla_x(\mathbf{u}-\tilde{\mathbf{u}})+\nabla_x^t(\mathbf{u}-\tilde{\mathbf{u}})-\frac{2}{d}{\rm div}_x(\mathbf{u}-\tilde{\mathbf{u}})\mathbb{I}\right|^2$$
$$-c(\tilde{\vartheta},\tilde{\mathbf{u}})[\vartheta-\tilde{\vartheta}]_{\rm ess}^2. \tag{4.32}$$

As for the residual component, we write

$$\left(\frac{\tilde{\vartheta}}{\vartheta}-1\right)\mathbb{S}:\nabla_x\mathbf{u}+\left(\frac{\vartheta}{\tilde{\vartheta}}-1\right)\left(\widetilde{\mathbb{S}}:\nabla_x\tilde{\mathbf{u}}\right)+(\mathbb{S}-\widetilde{\mathbb{S}}):(\nabla_x\mathbf{u}-\nabla_x\tilde{\mathbf{u}})$$
$$=\frac{\tilde{\vartheta}}{\vartheta}\mathbb{S}:\nabla_x\mathbf{u}-\mathbb{S}:\nabla_x\tilde{\mathbf{u}}+\frac{\vartheta}{\tilde{\vartheta}}\widetilde{\mathbb{S}}:\nabla_x\tilde{\mathbf{u}}-\widetilde{\mathbb{S}}:\nabla_x\mathbf{u},$$

where

$$
\begin{aligned}
|\mathbb{S}:\nabla_x\tilde{\mathbf{u}}| &= \left|\mu(\vartheta)\left(\nabla_x\mathbf{u}+\nabla_x^t\mathbf{u}-\frac{2}{d}\mathrm{div}_x\mathbf{u}\mathbb{I}\right):\nabla_x\tilde{\mathbf{u}}\right| \\
&\le \frac{\tilde{\vartheta}}{4}\frac{\mu(\vartheta)}{\vartheta}\left|\nabla_x\mathbf{u}+\nabla_x^t\mathbf{u}-\frac{2}{d}\mathrm{div}_x\mathbf{u}\mathbb{I}\right|^2 + c(\tilde{\vartheta})\vartheta\mu(\vartheta)|\nabla_x\tilde{\mathbf{u}}|^2 \\
&\le \frac{\tilde{\vartheta}}{\vartheta}\mathbb{S}:\nabla_x\mathbf{u}+c(\tilde{\vartheta},\tilde{\mathbf{u}})(1+\vartheta^2).
\end{aligned}
$$

Consequently, we obtain

$$
\begin{aligned}
&\left[\left(\frac{\tilde{\vartheta}}{\vartheta}-1\right)\mathbb{S}:\nabla_x\mathbf{u}+\left(\frac{\vartheta}{\tilde{\vartheta}}-1\right)\left(\widetilde{\mathbb{S}}:\nabla_x\tilde{\mathbf{u}}\right)+(\mathbb{S}-\widetilde{\mathbb{S}}):(\nabla_x\mathbf{u}-\nabla_x\tilde{\mathbf{u}})\right]_{\mathrm{res}} \\
&\ge \left[\frac{\tilde{\vartheta}}{4\vartheta}\right]_{\mathrm{res}}\mathbb{S}:\nabla_x\mathbf{u}-c(\tilde{\vartheta},\tilde{\mathbf{u}})[1+\vartheta^2]_{\mathrm{res}}-\widetilde{\mathbb{S}}:[\nabla_x\mathbf{u}]_{\mathrm{res}}
\end{aligned}
\tag{4.33}
$$

Now, for a given $1\le\alpha\le 2$, we get

$$
\begin{aligned}
\|\mathbb{A}\|^2_{L^\alpha(M;R^{d\times s})} &= \left\|\sqrt{\frac{\vartheta}{\mu(\vartheta)}}\sqrt{\frac{\mu(\vartheta)}{\vartheta}}\mathbb{A}\right\|^2_{L^\alpha(M;R^{d\times s})} \\
&\le \left\|\frac{\vartheta}{\mu(\vartheta)}\right\|_{L^\beta(M)}\int_M\frac{\mu(\vartheta)}{\vartheta}|\mathbb{A}|^2\,\mathrm{d}x,\ \beta=\frac{\alpha}{2-\alpha}
\end{aligned}
$$

for any measurable set $M\subset\Omega$. Choosing

$$
\mathbb{A}=\left(\nabla_x\mathbf{u}+\nabla_x^t\mathbf{u}-\frac{2}{d}\mathrm{div}_x\mathbf{u}\mathbb{I}\right),\ \alpha=\frac{8}{5-\Lambda},\ \beta=\frac{4}{1-\Lambda},
$$

where Λ is the exponent from hypothesis (4.18), we obtain

$$
\left\|\frac{\vartheta}{\mu(\vartheta)}\right\|_{L^\beta(M)}\le\left\|\frac{\vartheta}{\mu(\vartheta)}\right\|_{L^\beta(\Omega)}=\left(\int_\Omega\left(\frac{\vartheta}{\mu(\vartheta)}\right)^{\frac{4}{1-\Lambda}}\,\mathrm{d}x\right)^{\frac{1-\Lambda}{4}}
$$

In view of hypothesis (4.9),

$$
\sup_{t\in(0,T)}\int_\Omega\vartheta^4(t,\cdot)\,\mathrm{d}x\le c(\mathrm{data})
$$

for any weak solution of the Navier–Stokes–Fourier system, and, consequently,

$$
\begin{aligned}
&\left\| \nabla_x \mathbf{u} + \nabla_x^t \mathbf{u} - \frac{2}{d} \mathrm{div}_x \mathbf{u} \mathbb{I} \right\|^2_{L^\alpha(M;R^{d\times d})} \\
&\lesssim \int_M \frac{\mu(\vartheta)}{\vartheta} \left| \nabla_x \mathbf{u} + \nabla_x^t \mathbf{u} - \frac{2}{d} \mathrm{div}_x \mathbf{u} \mathbb{I} \right|^2 \, \mathrm{d}x, \ \alpha = \frac{8}{5 - \Lambda},
\end{aligned}
\tag{4.34}
$$

for any measurable set $M \subset \Omega$. Thus going back to (4.33) we conclude

$$
\begin{aligned}
&\int_\Omega \left[\left(\frac{\tilde{\vartheta}}{\vartheta} - 1 \right) \mathbb{S} : \nabla_x \mathbf{u} + \left(\frac{\vartheta}{\tilde{\vartheta}} - 1 \right) \left(\widetilde{\mathbb{S}} : \nabla_x \tilde{\mathbf{u}} \right) + (\mathbb{S} - \widetilde{\mathbb{S}}) : (\nabla_x \mathbf{u} - \nabla_x \tilde{\mathbf{u}}) \right]_{\mathrm{res}} \, \mathrm{d}x \\
&\gtrsim \left\| 1_{\mathrm{res}} \left(\nabla_x \mathbf{u} + \nabla_x^t \mathbf{u} - \frac{2}{d} \mathrm{div}_x \mathbf{u} \mathbb{I} \right) \right\|^2_{L^\alpha(\Omega;R^{d\times d})} \\
&- c(\tilde{\vartheta}, \tilde{\mathbf{u}}) [1 + \vartheta^2]_{\mathrm{res}}.
\end{aligned}
\tag{4.35}
$$

Finally, we observe

$$
\begin{aligned}
1_{\mathrm{res}} &\left(\nabla_x \mathbf{u} + \nabla_x^t \mathbf{u} - \frac{2}{d} \mathrm{div}_x \mathbf{u} \mathbb{I} \right) \\
&= 1_{\mathrm{res}} \left(\nabla_x (\mathbf{u} - \tilde{\mathbf{u}}) + \nabla_x^t (\mathbf{u} - \tilde{\mathbf{u}}) - \frac{2}{d} \mathrm{div}_x (\mathbf{u} - \tilde{\mathbf{u}}) \mathbb{I} \right) \\
&+ 1_{\mathrm{res}} \left(\nabla_x \tilde{\mathbf{u}} + \nabla_x^t \tilde{\mathbf{u}} - \frac{2}{d} \mathrm{div}_x \tilde{\mathbf{u}} \mathbb{I} \right);
\end{aligned}
$$

whence

$$
\begin{aligned}
&\int_\Omega \left[\left(\frac{\tilde{\vartheta}}{\vartheta} - 1 \right) \mathbb{S} : \nabla_x \mathbf{u} + \left(\frac{\vartheta}{\tilde{\vartheta}} - 1 \right) \left(\widetilde{\mathbb{S}} : \nabla_x \tilde{\mathbf{u}} \right) + (\mathbb{S} - \widetilde{\mathbb{S}}) : (\nabla_x \mathbf{u} - \nabla_x \tilde{\mathbf{u}}) \right]_{\mathrm{res}} \, \mathrm{d}x \\
&\gtrsim \left\| 1_{\mathrm{res}} \left(\nabla_x (\mathbf{u} - \tilde{\mathbf{u}}) + \nabla_x^t (\mathbf{u} - \tilde{\mathbf{u}}) - \frac{2}{d} \mathrm{div}_x (\mathbf{u} - \tilde{\mathbf{u}}) \mathbb{I} \right) \right\|^2_{L^\alpha(\Omega;R^{d\times d})} \\
&- c(\tilde{\vartheta}, \tilde{\mathbf{u}}) [1 + \vartheta^2]_{\mathrm{res}}
\end{aligned}
\tag{4.36}
$$

with a possibly different constant from (4.35).

Combining (4.32), (4.36) we infer that

$$
\begin{aligned}
&\int_\Omega \left(\frac{\tilde{\vartheta}}{\vartheta} - 1 \right) \mathbb{S} : \nabla_x \mathbf{u} + \left(\frac{\vartheta}{\tilde{\vartheta}} - 1 \right) \left(\widetilde{\mathbb{S}} : \nabla_x \tilde{\mathbf{u}} \right) + (\mathbb{S} - \widetilde{\mathbb{S}}) : (\nabla_x \mathbf{u} - \nabla_x \tilde{\mathbf{u}}) \mathrm{d}x \\
&\gtrsim \left\| \left(\nabla_x (\mathbf{u} - \tilde{\mathbf{u}}) + \nabla_x^t (\mathbf{u} - \tilde{\mathbf{u}}) - \frac{2}{d} \mathrm{div}_x (\mathbf{u} - \tilde{\mathbf{u}}) \mathbb{I} \right) \right\|^2_{L^\alpha(\Omega;R^{d\times d})} \\
&- c(\tilde{\vartheta}, \tilde{\mathbf{u}}) \int_\Omega E \left(\varrho, S, \mathbf{m} \middle| \tilde{\varrho}, \widetilde{S}, \widetilde{\mathbf{m}} \right) \, \mathrm{d}x.
\end{aligned}
\tag{4.37}
$$

Applying the same treatment to the component of the viscous stress corresponding to the bulk viscosity, we deduce from (4.31), (4.37), and Korn–Poincaré inequality:

$$\begin{aligned}
&\int_{\Omega} E\left(\varrho, S, \mathbf{m} \;\middle|\; \tilde{\varrho}, \widetilde{S}, \widetilde{\mathbf{m}}\right)(\tau, \cdot)\mathrm{d}x \\
&+ \int_0^{\tau} \left\|\vartheta - \tilde{\vartheta}\right\|^2_{W^{1,2}(\Omega)} \mathrm{d}t + \int_0^{\tau} \|\mathbf{u} - \tilde{\mathbf{u}}\|^2_{W^{1,\alpha}(\Omega;R^d)}\,\mathrm{d}t \\
&\lesssim \int_0^{\tau}\int_{\partial\Omega} \left(\frac{\tilde{\vartheta}}{\vartheta} - 1\right) f_{i,B}\,[\mathbf{u}_B \cdot \mathbf{n}]^- \mathrm{d}\sigma_x \mathrm{d}t \\
&+ \int_0^{\tau}\int_{\partial\Omega} \tilde{\vartheta}\varrho_B\left[s(\varrho_B, \vartheta) - \widetilde{s} + \frac{\widetilde{e}}{\tilde{\vartheta}} - \frac{e(\varrho_B, \vartheta)}{\vartheta}\right][\mathbf{u}_B \cdot \mathbf{n}]^- \mathrm{d}\sigma_x \mathrm{d}t \\
&+ \int_{\Gamma_{\mathrm{in}}, t \leq \tau} \widetilde{\mathbf{q}} \cdot \mathbf{n}\left(\frac{\vartheta}{\tilde{\vartheta}} - 1\right)\mathrm{d}\sigma_x \mathrm{d}t + \int_0^{\tau} F(t)\mathrm{d}t \\
&+ \int_0^{\tau}\int_{\Omega} E\left(\varrho, S, \mathbf{m} \;\middle|\; \tilde{\varrho}, \widetilde{S}, \widetilde{\mathbf{m}}\right)\mathrm{d}x\mathrm{d}t,\ \alpha = \frac{8}{5 - \Lambda}.
\end{aligned} \tag{4.38}$$

4.2.4 Weak–Strong Uniqueness, Unconditional Result

Our ultimate goal is to control the integrals on the right-hand side of (4.38) by those on the left-hand side.

We start with the boundary integrals. Our aim is to show

$$\begin{aligned}
&\int_0^{\tau}\int_{\partial\Omega} \left(\frac{\tilde{\vartheta}}{\vartheta} - 1\right) f_{i,B}\,[\mathbf{u}_B \cdot \mathbf{n}]^- \mathrm{d}\sigma_x \mathrm{d}t \\
&+ \int_0^{\tau}\int_{\partial\Omega} \tilde{\vartheta}\varrho_B\left[s(\varrho_B, \vartheta) - \widetilde{s} + \frac{\widetilde{e}}{\tilde{\vartheta}} - \frac{e(\varrho_B, \vartheta)}{\vartheta}\right][\mathbf{u}_B \cdot \mathbf{n}]^- \mathrm{d}\sigma_x \mathrm{d}t \\
&+ \int_{\Gamma_{\mathrm{in}}, t \leq \tau} \widetilde{\mathbf{q}} \cdot \mathbf{n}\left(\frac{\vartheta}{\tilde{\vartheta}} - 1\right)\mathrm{d}\sigma_x \mathrm{d}t \\
&= \int_0^{\tau}\int_{\partial\Omega} \left(\frac{\tilde{\vartheta}}{\vartheta} + \frac{\vartheta}{\tilde{\vartheta}} - 2\right) f_{i,B}\,[\mathbf{u}_B \cdot \mathbf{n}]^- \mathrm{d}\sigma_x \mathrm{d}t \\
&+ \int_0^{\tau}\int_{\partial\Omega} \tilde{\vartheta}\varrho_B\left[s(\varrho_B, \vartheta) - \widetilde{s} + \frac{\widetilde{e}}{\tilde{\vartheta}} - \frac{e(\varrho_B, \vartheta)}{\vartheta}\right][\mathbf{u}_B \cdot \mathbf{n}]^- \mathrm{d}\sigma_x \mathrm{d}t \\
&+ \int_0^{\tau}\int_{\partial\Omega} \varrho_B\widetilde{e}\left(1 - \frac{\vartheta}{\tilde{\vartheta}}\right)[\mathbf{u}_B \cdot \mathbf{n}]^- \mathrm{d}\sigma_x \mathrm{d}t \lesssim \int_0^{\tau}\int_{\partial\Omega} |\vartheta - \tilde{\vartheta}|^2 \mathrm{d}\sigma_x.
\end{aligned} \tag{4.39}$$

To see this, consider the function

$$\chi : \vartheta \mapsto \left(\frac{\tilde{\vartheta}}{\vartheta} + \frac{\vartheta}{\tilde{\vartheta}} - 2 \right) f_{i,B} [\mathbf{u}_B \cdot \mathbf{n}]^- \\ + \tilde{\vartheta} \varrho_B \left[s(\varrho_B, \vartheta) - \widetilde{s} + \frac{\widetilde{e}}{\tilde{\vartheta}} - \frac{e(\varrho_B, \vartheta)}{\vartheta} + \widetilde{e} \left(1 - \frac{\vartheta}{\tilde{\vartheta}} \right) \right] [\mathbf{u}_B \cdot \mathbf{n}]^- .$$

Using Gibbs' relation (1.6) we easily compute

$$\chi(\tilde{\vartheta}) = \chi'(\tilde{\vartheta}) = 0. \tag{4.40}$$

Moreover, by virtue of hypotheses (4.11)–(4.14),

$$\chi(\vartheta) \lesssim \vartheta \text{ as } \vartheta \to \infty. \tag{4.41}$$

Indeed observe that the contribution by the radiation components is

$$\left(s_{\rm rad}(\varrho_B, \vartheta) - \frac{e_{\rm rad}(\varrho_B, \vartheta)}{\vartheta} \right) \varrho_B [\mathbf{u}_B \cdot \mathbf{n}]^- = \frac{1}{3} a \vartheta^3 [\mathbf{u}_B \cdot \mathbf{n}]^- \leq 0.$$

By the same token,

$$\frac{e(\varrho_B, \vartheta)}{\vartheta} = \frac{1}{\vartheta} e_{\rm el}(\varrho_B) + \frac{1}{\vartheta} e_m(\varrho_B, \vartheta) + \frac{a}{\varrho_B} \vartheta^3,$$

and, by virtue of (4.15),

$$|s(\varrho_B, \vartheta)| \lesssim -\log(\vartheta) \text{ as } \vartheta \to 0.$$

Consequently,

$$\chi(\vartheta) \leq \frac{\tilde{\vartheta}}{\vartheta} \Big(f_{i,B} - \varrho_B e_{\rm el}(\varrho_B) - \varrho_B e_{\rm m}(\varrho_B, \vartheta) \Big) [\mathbf{u}_B \cdot \mathbf{n}]^- + c_1 \tilde{\vartheta} [\mathbf{u}_B \cdot \mathbf{n}]^- \log(\vartheta) + c_2 \tag{4.42}$$

as $\vartheta \to 0$, where c_1, c_2 are positive constants. Finally, by virtue of hypothesis (4.13),

$$e_{\rm m}(\varrho_B, \vartheta) \to 0 \text{ as } \vartheta \to 0,$$

uniformly on $\Gamma_{\rm in}$ as soon as $\inf_{\Gamma_{\rm in}} \varrho_B > 0$.

Thus if we *assume*

positivity of heat inflow:

$$\inf_{(t,x) \in \Gamma_{\rm in}} \Big\{ f_{i,B}(t,x) - \varrho_B e_{\rm el}(\varrho_B)(t,x) \Big\} > 0, \tag{4.43}$$

then

$$\chi(\vartheta) \to -\infty \text{ as } \vartheta \to 0 \tag{4.44}$$

uniformly on $\Gamma_{\rm in}$ as soon as $\inf_{\Gamma_{\rm in}} \varrho_B > 0$, $\inf_{\Gamma_{\rm in}} \tilde{\vartheta} > 0$. Combining (4.40), (4.41), and (4.44), we get the desired conclusion (4.39).

Remark 4. Hypothesis (4.43) turns out to be indispensable in the analysis of the Navier–Stokes–Fourier system. Going back to the boundary condition (3.6) we have

$$\begin{aligned} &\varrho_B e[\mathbf{u}_B \cdot \mathbf{n}]^+ + \mathbf{q} \cdot \mathbf{n} \\ &= \varrho_B e_{\rm el}(\varrho_B)[\mathbf{u}_B \cdot \mathbf{n}]^- + \varrho_B e_{\rm th}(\varrho_B, \vartheta)[\mathbf{u}_B \cdot \mathbf{n}]^- + \mathbf{q} \cdot \mathbf{n} = f_{i,B}[\mathbf{u}_B \cdot \mathbf{n}]^-. \end{aligned}$$

This can be written as a boundary condition including pure heat transport,

$$\varrho_B e_{\rm th}(\varrho_B, \vartheta)[\mathbf{u}_B \cdot \mathbf{n}]^- + \mathbf{q} \cdot \mathbf{n} = \Big(f_{i,B} - \varrho_B e_{\rm el}(\varrho_B)\Big)[\mathbf{u}_B \cdot \mathbf{n}]^-, \tag{4.45}$$

where, in accordance with hypothesis (4.43), the right-hand side is negative on $\Gamma_{\rm in}$.

Repeating the arguments used in (3.65) we may rewrite (4.38) as

$$\begin{aligned} &\int_\Omega E\left(\varrho, S, \mathbf{m} \,\middle|\, \tilde{\varrho}, \widetilde{S}, \widetilde{\mathbf{m}}\right)(\tau, \cdot) {\rm d}x \\ &+ \int_0^\tau \left\| \vartheta - \tilde{\vartheta} \right\|^2_{W^{1,2}(\Omega)} {\rm d}t + \int_0^\tau \| \mathbf{u} - \tilde{\mathbf{u}} \|^2_{W^{1,\alpha}(\Omega;R^d)} {\rm d}t \\ &\lesssim \int_0^\tau F(t) {\rm d}t + \int_0^\tau \int_\Omega E\left(\varrho, S, \mathbf{m} \,\middle|\, \tilde{\varrho}, \widetilde{S}, \widetilde{\mathbf{m}}\right) {\rm d}x{\rm d}t. \end{aligned} \tag{4.46}$$

Thus it remains to estimate the integrals in (4.22). As all integrands are quadratic in terms of the differences of components of the weak and strong solutions, their essential parts are easily controlled via (4.23).

As for the residual components, the most difficult term turns out to be

$$\int_\Omega [\varrho s(\varrho, \vartheta)]_{\rm res} (\mathbf{u} - \tilde{\mathbf{u}}) \cdot \nabla_x \tilde{\vartheta} \ {\rm d}x. \tag{4.47}$$

In view of (4.15),

$$\varrho s(\varrho, \vartheta) \lesssim \left(1 + \varrho |\log(\varrho)| + \varrho |\log(\vartheta)| + \vartheta^3\right).$$

Consequently, the most delicate step in estimating (4.47) is to control the integral

$$\int_\Omega [\varrho |\log(\vartheta)|]_{\rm res} |\mathbf{u} - \tilde{\mathbf{u}}| \ {\rm d}x.$$

Since

$$\alpha = \frac{8}{5-\Lambda} \geq \frac{16}{9},$$

the standard Sobolev embedding (for $d=3$) yields

$$W^{1,\alpha}(\Omega) \hookrightarrow L^r(\Omega),\ 1 \leq r \leq \frac{48}{11}.$$

Consequently, by Hölder's inequality,

$$\begin{aligned}
&\int_\Omega [\varrho|\log(\vartheta)|]_{\rm res}|\mathbf{u}-\tilde{\mathbf{u}}|\ {\rm d}x \leq \|[\varrho]_{\rm res}\|_{L^{\frac{5}{3}}(\Omega)}\|\log(\vartheta)\|_{L^6(\Omega)}\|\mathbf{u}-\tilde{\mathbf{u}}\|_{L^{\frac{30}{7}}(\Omega;R^d)}\\
&\leq \delta\|\mathbf{u}-\tilde{\mathbf{u}}\|^2_{W^{1,\alpha}(\Omega;R^d)} + c(\delta)\|\log(\vartheta)\|^2_{L^6(\Omega)}\|[\varrho]_{\rm res}\|^2_{L^{\frac{5}{3}}(\Omega)}\\
&\leq \delta\|\mathbf{u}-\tilde{\mathbf{u}}\|^2_{W^{1,\alpha}(\Omega;R^d)} + c(\delta)\|\log(\vartheta)\|^2_{L^6(\Omega)}\left(\int_\Omega E\left(\varrho,S,\mathbf{m}\middle|\tilde{\varrho},\widetilde{S},\widetilde{\mathbf{m}}\right)\ {\rm d}x\right)^{\frac{6}{5}}
\end{aligned}$$

for any $\delta > 0$. Now, in accordance with Definition 4, the weak solution belongs to the class

$$\begin{aligned}
\log(\vartheta) \in L^2(0,T,W^{1,2}(\Omega)) \hookrightarrow L^2(0,T,L^6(\Omega)),\\
E\left(\varrho,S,\mathbf{m}\middle|\tilde{\varrho},\widetilde{S},\widetilde{\mathbf{m}}\right) \in L^\infty(0,T;L^1(\Omega));
\end{aligned}$$

whence

$$\begin{aligned}
&c(\delta)\|\log(\vartheta)\|^2_{L^6(\Omega)}\left(\int_\Omega E\left(\varrho,S,\mathbf{m}\middle|\tilde{\varrho},\widetilde{S},\widetilde{\mathbf{m}}\right)\ {\rm d}x\right)^{\frac{6}{5}}\\
&\qquad\leq \chi\int_\Omega E\left(\varrho,S,\mathbf{m}\middle|\tilde{\varrho},\widetilde{S},\widetilde{\mathbf{m}}\right)\ {\rm d}x \text{ for some } \chi \in L^1(0,T).
\end{aligned}$$

As the remaining integrals in (4.22) can be handled in a similar way, we deduce from (4.46) the final estimate

$$\begin{aligned}
&\int_\Omega E\left(\varrho,S,\mathbf{m}\ \middle|\ \tilde{\varrho},\widetilde{S},\widetilde{\mathbf{m}}\right)(\tau,\cdot){\rm d}x\\
&+\int_0^\tau \left\|\vartheta-\tilde{\vartheta}\right\|^2_{W^{1,2}(\Omega)}\ {\rm d}t + \int_0^\tau \|\mathbf{u}-\tilde{\mathbf{u}}\|^2_{W^{1,\alpha}(\Omega;R^d)}\ {\rm d}t\\
&\lesssim \int_0^\tau \chi\int_\Omega E\left(\varrho,S,\mathbf{m}\ \middle|\ \tilde{\varrho},\widetilde{S},\widetilde{\mathbf{m}}\right)\ {\rm d}x{\rm d}t,
\end{aligned} \tag{4.48}$$

for some $\chi \in L^1(0,T)$. Thus the standard Gronwall argument yields

$$\int_\Omega E\left(\varrho,S,\mathbf{m}\ \middle|\ \tilde{\varrho},\widetilde{S},\widetilde{\mathbf{m}}\right)(\tau,\cdot)\ {\rm d}x = 0 \text{ for a.a. } \tau \in (0,T).$$

We have shown the following result.

Theorem 6 (Weak–Strong Uniqueness, Real Gas).
Let $\Omega \subset R^d$, $d = 2, 3$ be a bounded Lipschitz domain. Let the pressure $p = p(\varrho, \vartheta)$ be given by (4.2), *with the associated e, s satisfying* (4.8)–(4.14). *Let the transport coefficients μ, η, κ satisfy the structural hypotheses* (4.18)–(4.20). *In addition, suppose*

$$\inf_{(t,x)\in\Gamma_{\rm in}} \Big\{ f_{i,B}(t,x) - \varrho_B e_{\rm el}(\varrho_B)(t,x) \Big\} > 0. \tag{4.49}$$

Let $(\varrho, \vartheta, \mathbf{u})$ be a weak solution of the Navier–Stokes system in $(0,T)\times\Omega$ in the sense of Definition 4. Suppose that the same problem (with the same initial and boundary data) admits a strong solution $(\tilde\varrho, \tilde\vartheta, \tilde{\mathbf{u}})$ in the class

$$\begin{aligned} &\tilde\varrho,\ \tilde\vartheta \in C^1([0,T]\times\overline{\Omega}),\ \tilde{\mathbf{u}} \in C^1([0,T]\times\overline{\Omega}; R^d), \\ &D_x^2\tilde{\mathbf{u}},\ D_x^2\tilde\vartheta \in C([0,T]\times\overline{\Omega}) \\ &\inf_{(0,T)\times\Omega} \tilde\varrho > 0, \quad \inf_{(0,T)\times\Omega} \tilde\vartheta > 0. \end{aligned}$$

Then

$$\varrho = \tilde\varrho,\ \vartheta = \tilde\vartheta,\ \mathbf{u} = \tilde{\mathbf{u}} \text{ in } [0,T]\times\overline{\Omega}.$$

Remark 5. As the boundary of Ω is merely Lipschitz, the outer normal vector $\mathbf{n}$ exists only $\mathcal{H}^{d-1}$ a.a. on $\partial\Omega$. The set $\Gamma_{\rm in}$ is therefore defined as

$$\Gamma_{\rm in} = \Big\{ (t,x) \in [0,T]\times\partial\Omega \ \Big| \ \mathbf{n}(x) \text{ exists } \mathbf{u}_B(t,x)\cdot\mathbf{n}(x) < 0 \Big\}.$$

Alternatively, anticipating $f_{i,B}$, ϱ_B are defined on the whole boundary $\partial\Omega$ we may replace hypothesis (4.49) by

$$f_{i,B}(t,x) > \varrho_B e_{\rm el}(\varrho_B)(t,x) \text{ for all } t\in[0,T],\ x\in\partial\Omega.$$

We also point out that the *existence* of the smooth solution $(\tilde\varrho, \tilde\vartheta, \tilde{\mathbf{u}})$ in the regularity class required in Theorem 6 is known only for smooth domains with $\Gamma_{\rm in}$ separated from $\Gamma_{\rm out}$, see Valli and Zajaczkowski [89].

4.3 Hard Sphere model

Many technical problems of the mathematical theory of compressible fluids are related to rather poor a priori bounds available for the density ϱ. As predicted by the molecular theory, however, the density of *real* gases admits a natural bound

$$0 \leq \varrho \leq \overline{\varrho},$$

where the upper bound is a trivial consequence of the fact that the volume occupied by a fluid cannot be arbitrarily small. The relevant equation of state takes the form

$$p_{\rm m}(\varrho, \vartheta) = \varrho \vartheta Z(\varrho, \vartheta),$$

where Z denotes the *compressibility factor*. The exact form of Z is subject to discussion and Z is usually expressed by means of a virial series. Here, we suppose the following:

- $$Z \approx 1 \text{ for moderate values of } \varrho, \vartheta,$$

  meaning the fluid behaves as a perfect gas in the non-degenerate area;
- $$\frac{\partial Z(\varrho, \vartheta)}{\partial \varrho} > 0, \ \frac{\partial Z(\varrho, \vartheta)}{\partial \vartheta} \leq 0, \ \lim_{\vartheta \to \infty} Z(\varrho, \vartheta) = 1 \text{ for any } 0 < \varrho < \overline{\varrho},$$

  meaning, in particular, the fluid behaves as a perfect gas in the high temperature regime;
- $$\lim_{\varrho \nearrow \overline{\varrho}} Z(\varrho, \vartheta) = \infty \text{ for any } \vartheta > 0, \tag{4.50}$$

  $$\lim_{\vartheta \searrow 0} Z(\varrho, \vartheta) = \infty \text{ for any } \varrho > 0. \tag{4.51}$$

Hypothesis (4.50) is pertinent to the hard sphere model, while (4.51) represents a certain extrapolation to the low temperature regime. A simple form of Z complying with the above hypotheses reads

$$Z(\varrho, \vartheta) = 1 + \frac{A}{\vartheta (\overline{\varrho} - \varrho)^r}, \text{ with positive parameters } A > 0, r > 0.$$

Accordingly, the equation of state takes the form

$$p(\varrho, \vartheta) = p_{HS}(\varrho) + p_{\rm m}(\varrho, \vartheta), \ p_{HS}(\varrho) = \frac{A\varrho}{(\overline{\varrho} - \varrho)^r}, \ p_{\rm m}(\varrho, \vartheta) = \varrho \vartheta. \tag{4.52}$$

Note that p_{HS} plays the role of $p_{\rm el}$ as it depends solely on ϱ.

Similarly to Sect. 4.1.1, we augment the total pressure by the radiation component obtaining

hard sphere gas EOS:

$$p(\varrho, \vartheta) = p_{HS}(\varrho) + p_{\rm m}(\varrho, \vartheta) + p_{\rm rad}(\vartheta) = \frac{A\varrho}{(\overline{\varrho} - \varrho)^r} + \varrho\vartheta + \frac{a}{3}\vartheta^4. \quad (4.53)$$

Accordingly, the internal energy can be taken in the form

$$e = e_{HS}(\varrho) + e_{\rm m}(\vartheta) + e_{\rm rad}(\varrho, \vartheta),$$
$$e_{HS}(\varrho) = \int_{\overline{\varrho}/2}^{\varrho} \frac{A}{z(\overline{\varrho} - z)^r} {\rm d}z, \ e_{\rm m}(\vartheta) = \frac{3}{2}\vartheta, \ e_{\rm rad}(\varrho, \vartheta) = \frac{a}{\varrho}\vartheta^4, \quad (4.54)$$

with the associated entropy

$$s = s(\varrho, \vartheta) = \frac{3}{2}\log(\vartheta) - \log(\varrho) + \frac{4a}{3}\frac{\vartheta^3}{\varrho}. \quad (4.55)$$

Note that our choice is in agreement with both Gibbs' law (1.6) and the hypothesis of thermodynamic stability (1.19). Similarly to the preceding part, we can also write

$$p = p_{\rm th}(\varrho, \vartheta) + p_{\rm el}(\varrho), \text{ where } p_{\rm th} = \varrho\vartheta + \frac{a}{3}\vartheta^4, \ p_{\rm el} = p_{HS} = \frac{A\varrho}{(\overline{\varrho} - \varrho)^r},$$

with a similar decomposition for the internal energy.

Finally, we claim that the conditional and even unconditional weak–strong uniqueness property stated in Theorems 5, 6 still holds for the hard sphere gas model, if, in addition, the *weak* solution satisfies

$$\operatorname*{ess\,sup}_{t,x} \varrho(t, x) < \overline{\varrho}. \quad (4.56)$$

The unconditional result stated in Theorem 6, however, is more delicate for this model in the absence of (4.56). To the best of our knowledge, there is only one relevant result proved in [39] in the case of periodic boundary conditions.

4.4 Concluding Remarks

The principal hypotheses concerning the equations of state as well as transport coefficients are basically the same as in [43], where more elaborated discussion concerning the physical background is given. More about equations of state in fluid mechanics can be found in the monographs by Becker [7], Eliezer et al. [30], or Müller and Ruggeri [75].

We refer to the standard textbook by Kastler et al. [60] concerning the hard sphere pressure EOS, some specific examples can be found in Kolafa et al. [61].

Chapter 5
Existence Theory, Basic Approximation Scheme

The expected benefit of extending the class of classical solutions of the Navier–Stokes–Fourier system to more general objects—weak (distributional) solutions—is the possibility to build up a mathematical theory of global in time solutions for any physically admissible initial/boundary data far from equilibrium. As many problems related to turbulence are formulated in terms of global in time or even entire (defined for any time $t \in R$) solutions, the existence theory in the framework of weak solutions represents a suitable platform to attack them in a rigorous manner.

The weak solutions to the Navier–Stokes–Fourier system will be constructed by means of a multilevel *approximation scheme*. It is the goal of the present chapter to introduce the basic level and identify the corresponding approximate solutions. The approximate field equations are essentially the same as in [43, Chapter 3]. The main novelty here is the introduction of the non-homogeneous boundary conditions pertinent to the open fluid systems.

5.1 Approximation Scheme

Our goal is to show global in time existence of weak solutions to the Navier–Stokes–Fourier system specified in Definition 4. We focus on the real gas model introduced in Sect. 4.1.1 and the hard sphere gas model specified in Sect. 4.3. In both cases the transport coefficients μ, η, and κ comply with the constitutive relations (4.18)–(4.20). It turns out that the analysis is fairly similar for the real and hard sphere gas at lower levels of the approximation procedure. We therefore focus on the case of the real gas specifying the modifications necessary to accommodate the hard sphere model in the last level limit in Chap. 8.

E. Feireisl, A. Novotný, *Mathematics of Open Fluid Systems*, Nečas Center Series, https://doi.org/10.1007/978-3-030-94793-4_5

5.1.1 Regularization of the Constitutive Relations

Some constitutive relations are regularized/modified to fit the approximation scheme. The goal is to improve the integrability of the approximate solutions at the basic level. There are two auxiliary positive parameters ε and δ. Specifically, we consider:

- **Regularized real gas pressure EOS**

$$p_\delta(\varrho,\vartheta) = p(\varrho,\vartheta) + \delta\left(\varrho^2 + \varrho^\Gamma\right),\ \delta > 0, \tag{5.1}$$

where $\Gamma > 6$ is a sufficiently large number;

- **Regularized hard sphere gas EOS**

$$p_\delta(\varrho,\vartheta) = \varrho\vartheta + \frac{a}{3}\vartheta^4 + p_{\delta,HS}(\varrho) + \delta\varrho^2 + ([\varrho - \overline{\varrho} - 1]^+)^\Gamma, \tag{5.2}$$

with

$$p_{\delta,HS}(\varrho) = \begin{cases} \frac{A\varrho}{(\overline{\varrho}-\varrho)^r} \text{ if } 0 \le \varrho \le \overline{\varrho} - \delta, \\ a_{1,\delta}\varrho + a_{2,\delta} \text{ otherwise} \end{cases}$$

where

$$a_{1,\delta} = p'_{HS}(\overline{\varrho} - \delta),\ a_{1,\delta}(\overline{\varrho} - \delta) + a_{2,\delta} = p_{HS}(\overline{\varrho} - \delta)$$

are chosen so that $p_{HS,\delta} \in C^1[0,\infty)$;

- **Regularized internal energy/entropy for real gas**

$$e_\delta(\varrho,\vartheta) = e(\varrho,\vartheta) + \delta\vartheta,\ s_\delta(\varrho,\vartheta) = s(\varrho,\vartheta) + \delta\log(\vartheta)$$

and

$$e_{\delta,\mathrm{th}} = e_{\mathrm{th}} + \delta\vartheta;$$

- **Regularized internal energy for the hard sphere gas**

$$e_\delta(\varrho,\vartheta) = \frac{3}{2}\vartheta + \frac{a}{\varrho}\vartheta^4 + e_{\delta,HS}(\varrho),\ e_{\delta,HS}(\varrho) \equiv \int_{\overline{\varrho}/2}^{\varrho} \frac{p_{\delta,HS}(z)}{z^2}\mathrm{d}z;$$

- **Regularized viscous stress**

$$\mathbb{S}_\delta(\vartheta, \nabla_x \mathbf{u}) = (\mu(\vartheta) + \delta\vartheta)\left(\nabla_x \mathbf{u} + \nabla_x^t \mathbf{u} - \frac{2}{d}\mathrm{div}_x \mathbf{u}\mathbb{I}\right) + \eta(\vartheta)\,\mathrm{div}_x \mathbf{u}\mathbb{I}; \tag{5.3}$$

- **Regularized internal energy flux**

$$\mathbf{q}_\delta = \mathbf{q} - \delta\left(\vartheta^\Gamma + \frac{1}{\vartheta}\right)\nabla_x \vartheta.$$

Remark 6. It is worth noting that although the perturbed constitutive relations p_δ, e_δ, and s_δ still comply with the hypothesis of thermodynamic stability, the Gibbs' relation is not necessarily satisfied at the $\delta-$level.

5.1.2 Domain Approximation

In view of the real world applications, our aim is to accommodate non-smooth ideally only Lipschitz domains. However, at the basic approximation level, the spatial domain must be smooth as required by the standard theory of the parabolic equations used in the approximation scheme. We also experience some technical difficulties at the level of the equation of continuity with in/out flux boundary conditions for general Lipschitz domains. This is the reason why we focus on the class of domains that can be well approximated by smooth ones like domains with *piecewise smooth* boundary.

Definition 5 (Piecewise C^k Domain).
A bounded domain $\Omega \subset R^d$, $d = 2, 3$ is piecewise C^k, $k \geq 1$, if:

- Ω is Lipschitz.
- There is a sequence $\{\Omega_n\}_{n=1}^\infty$ of bounded domains with boundaries of class C^k such that

$$\Omega \subset \Omega_n \text{ for any } n = 1, 2, \ldots, \tag{5.4}$$

$$|\partial\Omega_n \setminus \partial\Omega|_{d-1} + |\partial\Omega \setminus \partial\Omega_n|_{d-1} \to 0 \text{ as } n \to \infty, \tag{5.5}$$

where $|\ldots|_{d-1}$ denotes the $(d-1)-$dimensional Hausdorff measure.

Note that our definition is slightly different from the standard one of piecewise C^k domains that requires the boundary $\partial\Omega$ to admit a decomposition into a finite number of C^k-manifolds. By isoperimetric inequality, hypothesis (5.5) implies

$$|\Omega_n \setminus \Omega| \to 0 \text{ as } n \to \infty. \tag{5.6}$$

5.1.3 Regularity of the Boundary Data

Besides the assumption that Ω is piecewise smooth, the boundary data should enjoy some regularity. To simplify, we suppose that $\mathbf{u}_B$, ϱ_B, $f_{i,B}$ are restrictions of smooth functions defined for all $t \in R$, $x \in R^d$. The degree of required smoothness will be specified below. In addition, we suppose that the boundary of the space time cylinder $[0,T] \times \Omega$ admits the following:

Boundary decomposition:

$$\begin{aligned}
[0,T] \times \partial\Omega &= \Gamma^0_{\text{in}} \cup \Gamma^0_{\text{out}} \cup \Gamma^0_{\text{wall}} \cup \Gamma_{\text{sing}}, \\
\Gamma^0_{\text{in}},\ \Gamma^0_{\text{out}},\ \Gamma^0_{\text{wall}} &\text{ open in } [0,T] \times \partial\Omega, \\
\Gamma^0_{\text{in}} &\subset \left\{ (t,x) \ \middle|\ \mathbf{n}(x) \text{ exists, } \mathbf{u}_B(t,x) \cdot \mathbf{n}(x) < 0 \right\}, \\
\Gamma^0_{\text{out}} &\subset \left\{ (t,x) \ \middle|\ \mathbf{n}(x) \text{ exists, } \mathbf{u}_B(t,x) \cdot \mathbf{n}(x) > 0 \right\}, \\
\Gamma^0_{\text{wall}} &\subset \left\{ (t,x) \ \middle|\ \mathbf{n}(x) \text{ exists, } \mathbf{u}_B(t,x) \cdot \mathbf{n}(x) = 0 \right\}, \\
\Gamma_{\text{sing}} &\subset [0,T] \times \partial\Omega \text{ compact, } |\Gamma_{\text{sing}}|_{\mathrm{d}t \otimes \mathrm{d}\sigma_x} = 0.
\end{aligned} \tag{5.7}$$

In particular, the component Γ_{sing} contains all singular points, where the outer normal $\mathbf{n}$ does not exist. In accordance with (5.7), this set must be compact of zero $(d-1)$-Hausdorff measure in R^d. The fact that Γ_{sing} is compact plays a crucial role in the process of renormalization of the equation of continuity discussed in detail in Appendix.

Finally, we recall the hypothesis of positivity of the heat inflow (4.43),

$$f_{i,B} > \varrho_B e_{\text{el}}(\varrho_B) \text{ in } [0,T] \times \partial\Omega \tag{5.8}$$

in the context of real gases. The same condition for the hard sphere gas reads

$$f_{i,B} > \varrho_B e_{HS}(\varrho_B) \text{ in } [0,T] \times \partial\Omega. \tag{5.9}$$

5.1.4 Galerkin Approximation

Suppose that $\Omega \subset R^d$ is piecewise C^k, with $k \geq 4$, generated by a sequence $\{\Omega_n\}_{n=1}^{\infty}$ of C^k domains as in Definition 5. Consider a finite dimensional space

$$X_n = \mathrm{span}\left\{ \mathbf{w}_i \;\middle|\; \mathbf{w}_i \in C_c^{\infty}(\Omega; R^d),\ i = 1, \ldots, n \right\}$$

where $\mathbf{w}_i$ are orthonormal with respect to the standard scalar product in L^2. Let $\Pi_n : L^2(\Omega; R^d) \to X_n$ denote the associated orthogonal projection.

5.1.5 Approximate Equation of Continuity

Consider a sequence of functions

$$z \in R \mapsto [z]_n^- \in C^{\infty}(R),\ [z]_n^- \leq [z]^-,\ [z]_n^- = \begin{cases} [z]^- \text{ if } z \leq -\frac{1}{n} \\ \text{increasing if } z \in \left[-\frac{1}{n}, \frac{1}{n}\right] \\ 0 \text{ if } z \geq \frac{1}{n}. \end{cases} \tag{5.10}$$

In other words, $[z]_n^-$ is a smooth uniform approximation of the Lipschitz function $[z]^-$.

At the basic approximation level, the equation of continuity (3.1) is replaced by its parabolic regularization

Approximate equation of continuity:

$$\partial_t \varrho + \mathrm{div}_x(\varrho \mathbf{u}) = \varepsilon \Delta_x \varrho \text{ in } (0,T) \times \Omega_n,\ \varepsilon > 0, \tag{5.11}$$

supplemented with the Robin boundary conditions

$$\varepsilon \nabla_x \varrho \cdot \mathbf{n} = (\varrho - \varrho_B)[\mathbf{u}_B \cdot \mathbf{n}]_n^- \text{ on } (0,T) \times \partial\Omega_n, \tag{5.12}$$

and the initial condition

$$\varrho(0, \cdot) = \varrho_{0,n,\delta},\ \delta > 0. \tag{5.13}$$

Note that the boundary condition (5.12) is well defined as $\partial\Omega_n$ is smooth and ϱ_B, $\mathbf{u}_B$ defined on the whole space R^{d+1}. Obviously, the boundary condition (5.12) is a smooth approximation of

$$\varepsilon\nabla_x\varrho\cdot\mathbf{n} = (\varrho-\varrho_B)[\mathbf{u}_B\cdot\mathbf{n}]^-.$$

Smoothness of $[\mathbf{u}_B\cdot\mathbf{n}]_n^-$ is needed for the standard parabolic theory to be applicable, cf. Theorem 7 below. By the same token, a smooth approximation $\varrho_{0,n,\delta}$ of the initial density ϱ_0 must comply with the standard compatibility condition

$$\varepsilon\nabla_x\varrho_{0,n,\delta}\cdot\mathbf{n} = (\varrho_{0,n,\delta}-\varrho_B(0,\cdot))[\mathbf{u}_B(0,\cdot)\cdot\mathbf{n}]_n^- \text{ on } \partial\Omega_n. \tag{5.14}$$

To construct a suitable approximation of the initial data, we consider another approximation of the initial density $\varrho_{0,\delta}\in L^\infty(\Omega)$ satisfying

$$0<\underline{\varrho}_\delta\le\varrho_{0,\delta}(x)\le\overline{\varrho}_\delta \text{ for a.a } x\in\Omega. \tag{5.15}$$

As Ω_n is of class C^k, $k\ge 4$, there is an open neighbourhood $\mathcal{U}(\partial\Omega_n)$ of radius $\omega(n)>0$ such that for each $x\in\mathcal{U}(\partial\Omega_n)$ there exists a unique nearest point $\Pi(x)\in\partial\Omega_n$. We set

$$\varrho^1_{0,n,\delta}(x) = \varrho_B(0,\Pi(x)) \text{ for } x\in\mathcal{U}(\partial\Omega_n).$$

It is easy to check that

- $\varrho^1_{0,n,\delta}$ satisfies the compatibility condition (5.14).
- $\varrho^1_{0,n,\delta}\in C^{2+\beta}$ as long as $\varrho_B(0,\cdot)\in C^{2+\beta}$, and $\partial\Omega_n$ is of class C^4.

Next, using the standard regularization procedure, we may construct a sequence $\{\varrho^2_{0,n,\delta}\}_{n=1}^\infty$, $\varrho^2_{0,n,\delta}\in C^\infty(R^d)$, such that

$$0<\underline{\varrho}_\delta\le\varrho^2_{0,n,\delta}(x)\le\overline{\varrho}_\delta,\ \varrho^2_{0,n,\delta}(x)\to\varrho_{0,\delta}(x) \text{ for a.a. } x\in K \text{ as } n\to\infty,$$

where $K\subset\Omega$ is an arbitrary compact sct. Finally, we set

$$\varrho_{0,n,\delta}(x) = (1-\chi_n(x))\varrho^1_{0,n,\delta}(x)+\chi_n(x)\varrho^1_{0,n,\delta}(x),$$

where

$$\chi_n\in C^\infty(R^d),\ 0\le\chi_n\le 1,\ \chi_n(x) = \begin{cases} 1 \text{ if } \operatorname{dist}[x;\partial\Omega_n]<\frac{1}{2}\omega(n), \\ 0 \text{ if } \operatorname{dist}[x;\partial\Omega_n]\ge\omega(n). \end{cases}$$

In view of the hypotheses (5.4), (5.5) on convergence of the approximate domains, the sequence $\{\varrho_{0,n,\delta}\}_{n=1}^\infty$ represents a suitable approximation of any initial density $\varrho_{0,\delta}$ satisfying (5.15), meaning

$$
\begin{aligned}
&\varrho_{0,n,\delta} \in C^{2+\beta}(\overline{\Omega_n}),\\
&\varrho_{0,n,\delta} \text{ satisfies the compatibility condition (5.14)},\\
&\varrho_{0,n,\delta},\ \varrho_{0,n,\delta}^{-1} \text{ bounded in } \Omega_n \text{ uniformly for } n \to \infty,\\
&\varrho_{0,n,\delta} \to \varrho_{0,\delta} \text{ in } L^r(\Omega) \text{ for any } 1 \le r < \infty.
\end{aligned} \tag{5.16}
$$

5.1.6 Approximate Momentum Balance

The approximate velocity field $\mathbf{u}$ is determined via a Faedo–Galerkin scheme based on X_n. We look for the approximate velocity field $\mathbf{u}$ in the form

$$
\mathbf{u} = \mathbf{v} + \mathbf{u}_B,\ \mathbf{v} \in C([0,T]; X_n)
$$

satisfying

Approximate momentum equation:

$$
\begin{aligned}
\int_\Omega \varrho\mathbf{u} \cdot \varphi(\tau, \cdot)\ \mathrm{d}x &= \int_\Omega \mathbf{m}_0 \cdot \varphi\ \mathrm{d}x\\
&+ \int_0^\tau \int_\Omega \Big[\varrho\mathbf{u} \cdot \partial_t \varphi + \varrho\mathbf{u} \otimes \mathbf{u} : \nabla_x \varphi + p_\delta \mathrm{div}_x \varphi - \mathbb{S}_\delta : \nabla_x \varphi\Big]\ \mathrm{d}x\\
&- \varepsilon \int_0^\tau \int_\Omega \nabla_x \varrho \cdot \nabla_x \mathbf{u} \cdot \varphi\ \mathrm{d}x\mathrm{d}t + \int_0^\tau \int_\Omega \varrho\mathbf{g} \cdot \varphi\ \mathrm{d}x\mathrm{d}t
\end{aligned} \tag{5.17}
$$

for any $\varphi \in C^1([0,T]; X_n)$.

Note carefully that the integrals are evaluated over the original domain as all test functions are compactly supported in $\Omega \subset \Omega_n$. Here, the regularized pressure p_δ and the viscous stress $\mathbb{S}_\delta$ have been introduced in (5.1), (5.2), and (5.3).

5.1.7 Internal Energy Balance

The approximation scheme should be consistent with the basic energy estimates. In the case of real gas EOS, the temperature ϑ will solve the **approximate internal energy equation:**

$$\begin{aligned}\partial_t(\varrho e_{\delta,\mathrm{th}}(\varrho,\vartheta)) &+ \mathrm{div}_x(\varrho e_{\delta,\mathrm{th}}(\varrho,\vartheta)\mathbf{u}) + \mathrm{div}_x \mathbf{q}_\delta(\vartheta,\nabla_x\vartheta) \\ &= \mathbb{S}_\delta(\vartheta,\mathbb{D}_x\mathbf{u}) : \mathbb{D}_x\mathbf{u} - p_{\mathrm{th}}(\varrho,\vartheta)\mathrm{div}_x\mathbf{u} \\ &+ \varepsilon\left[\delta\left(\Gamma\varrho^{\Gamma-2}+2\right) + \partial^2_{\varrho,\varrho}(\varrho e_{\mathrm{el}}(\varrho))\right]|\nabla_x\varrho|^2 + \delta\frac{1}{\vartheta^2} - \varepsilon\vartheta^5\end{aligned} \tag{5.18}$$

in $(0,T)\times\Omega_n$, supplemented with the Robin boundary conditions

$$\mathbf{q}_\delta\cdot\mathbf{n} + \varrho e_{\delta,\mathrm{th}}[\mathbf{u}_B\cdot\mathbf{n}]^-_n = \Big(f_{i,B} - \varrho_B e_{\mathrm{el}}(\varrho_B)\Big)[\mathbf{u}_B\cdot\mathbf{n}]^-_n \text{ in } (0,T)\times\Omega_n, \tag{5.19}$$

and the initial condition

$$\vartheta(0,\cdot) = \vartheta_{0,\delta}. \tag{5.20}$$

Here, we have

$$\frac{\partial^2(\varrho e_{\mathrm{el}}(\varrho))}{\partial\varrho^2} = \frac{5}{3}\overline{p}\varrho^{-\frac{1}{3}}.$$

In the case of hard sphere EOS, the term

$$\varepsilon\left[\delta\left(\Gamma\varrho^{\Gamma-2}+2\right) + \partial^2_{\varrho,\varrho}(\varrho e_{\mathrm{el}}(\varrho))\right]|\nabla_x\varrho|^2$$

must be replaced by

$$\varepsilon\partial^2_{\varrho,\varrho}\left[\delta\varrho^2 + \varrho\int_0^\varrho \frac{\left([z-\overline{\varrho}-1]^+\right)^\Gamma}{z^2}\mathrm{d}z + \varrho\int_{\overline{\varrho}/2}^\varrho \frac{p_{\delta,HS}(z)}{z^2}\mathrm{d}z\right]|\nabla_x\varrho|^2$$

The reader will have noticed that, unlike in (5.13), the approximation of the initial temperature need not depend on n. This is possible as we shall consider only strong, not necessarily classical solutions to (5.18), (5.19).

5.2 Solvability of the Basic Level Approximate Problem

Our first goal is to establish the existence of approximate solutions solving the problem (5.11)–(5.20) in $(0,T) \times \Omega_n$, where $0 < T < \infty$. This will be done by means of a fixed point argument passing through several steps:

- Fix a velocity field $\mathbf{v} \in C([0,T]; X_n)$ such that $\Pi_n[\varrho_0(\mathbf{v} + \mathbf{u}_B)(0,\cdot)] = \Pi_n \mathbf{m}_0$, and plug $\mathbf{u} = \mathbf{v} + \mathbf{u}_B$ in the approximate equation of continuity (5.11).
- Show that for a given $\mathbf{u} = \mathbf{v} + \mathbf{u}_B$, the approximate equation of continuity (5.11)–(5.13) admits a unique solution $\varrho = \varrho[\mathbf{v}]$ for any suitable approximation $\varrho_{0,n,\delta}$ of the initial data ϱ_0.
- Given $\varrho = \varrho[\mathbf{v}]$, identify $\vartheta = \vartheta[\varrho[\mathbf{v}], \mathbf{v}]$ as the unique solution of the approximate internal energy equation (5.18)–(5.20), with a suitable approximation $\vartheta_{0,\delta}$ of the initial data ϑ_0.
- Given $\varrho[\mathbf{v}], \vartheta[\varrho[\mathbf{v}], \vartheta[\mathbf{v}]]$ define a mapping

$$\mathcal{T} : C([0,T]; X_n) \to C([0,T]; X_n),$$

  where $\mathcal{T}[\mathbf{v}] = \mathbf{u} - \mathbf{u}_B$, where $\mathbf{u}$ is the unique solution of the Faedo–Galerkin approximation (5.17) with the initial data

$$\varrho\mathbf{u}(0,\cdot) = \Pi_n[\mathbf{m}_0].$$

- Show that $\mathcal{T}$ admits a fixed point $\mathbf{v}$.

5.2.1 Linear Parabolic Problem

The first auxiliary result that will allow us to construct solutions of the regularized equation of continuity (5.11)–(5.13) concerns general linear parabolic equations with Robin boundary conditions:

$$\partial_t V - \mathbb{A}(t,x) : \nabla_x^2 V + \mathbf{B}(t,x) \cdot \nabla_x V + C(t,x) V = f \text{ in } (0,T) \times \Omega, \quad (5.21)$$

$$\mathbf{E} \cdot \nabla_x V + D(t,x) V = g \text{ in } (0,T) \times \partial\Omega, \quad (5.22)$$

$$V(0,\cdot) = V_0 \text{ in } \Omega. \quad (5.23)$$

The following statement concerning the existence of classical solutions in the spaces of Hölder continuous functions is nowadays standard. We follow the formulation due to Lunardi [69, Theorem 5.1.21].

Theorem 7 (Existence for Linear Parabolic Problem).
Let $\Omega \subset R^d$ be a bounded domain of class $C^{2+\beta}$, $\beta > 0$. Suppose that

$$\mathbb{A} \in C^{0,\beta}([0,T] \times \overline{\Omega}; R^{d\times d}),\ \mathbf{B} \in C^{0,\beta}([0,T] \times \overline{\Omega}; R^d)$$
$$f, C \in C^{0,\beta}([0,T] \times \overline{\Omega}),$$

$$\mathbf{E} \in C^{\frac{1+\beta}{2},1+\beta}([0,T] \times \partial\Omega; R^d),\ g, D \in C^{\frac{1+\beta}{2},1+\beta}([0,T] \times \partial\Omega),$$
$$V_0 \in C^{2+\alpha}(\overline{\Omega}),$$

and

$$\mathbf{E}(0,x) \cdot \nabla_x V_0 + D(0,x) = g(0,x),\ x \in \partial\Omega. \tag{5.24}$$

Assume the uniform ellipticity condition

$$\underline{a}|\xi|^2 \leq \mathbb{A} : (\xi \otimes \xi),\ \xi \in R^d,\ \textit{for some } \underline{a} > 0$$

is satisfied.

Then the problem (5.21)–(5.23) *admits a unique classical solution* $V \in C^{1,2+\beta}([0,T] \times \overline{\Omega})$.

5.2.2 Quasilinear Parabolic Problem

Possibly the most technical step in the above delineated programme is solvability of the *nonlinear* approximate internal energy equation (5.18), with the Robin boundary conditions (5.19). Fortunately, the theory of quasilinear parabolic equations is nowadays well understood. Consider a general parabolic boundary-value problem:

$$\partial_t V - \mathbb{A}(t,x,V) : \nabla_x^2 V + B(t,x,V,\nabla_x V) = 0 \text{ in } (0,T) \times \Omega, \tag{5.25}$$

$$\mathbb{A} \cdot \nabla_x V \cdot \mathbf{n} + D(t,x,V) = 0 \text{ in } (0,T) \times \partial\Omega, \tag{5.26}$$

$$V(0,\cdot) = V_0 \text{ in } \Omega. \tag{5.27}$$

Adopting the framework of Hölder continuous functions we follow the monograph Ladyzhenskaya et al. [65, Chapter V, par. 7] and look for solutions in the Hölder class $V \in C^{1+\frac{\beta}{2},2+\beta}$, $\beta > 0$. Suppose that for any $\underline{V} < \overline{V}$ we have

$$\begin{aligned} &\mathbb{A},\ \partial_V \mathbb{A} \in C([0,T] \times \overline{\Omega} \times [\underline{V},\overline{V}]; R^{d\times d}_{\rm sym}),\\ &\nabla_x \mathbb{A} \in C([0,T] \times \overline{\Omega} \times [\underline{V},\overline{V}]; R^{d^3}),\\ &\underline{a}|\xi|^2 \le \mathbb{A} : (\xi \otimes \xi) \le \overline{a}|\xi|^2,\ \xi \in R^d, \text{ for some } \underline{a}(\underline{V},\overline{V}) > 0\\ &|\partial_V \mathbb{A}| + |\nabla_x \mathbb{A}| \le c(\underline{V},\overline{V}); \end{aligned} \tag{5.28}$$

$$\begin{aligned} &B \in C([0,T] \times \overline{\Omega} \times [\underline{V},\overline{V}] \times R^d)\\ &|B(t,x,V,\mathbf{d})| \le c(\underline{V},\overline{V}) \left(1 + |\mathbf{d}|^2\right); \end{aligned} \tag{5.29}$$

and

$$\begin{aligned} &D,\ \partial_V D \in C([0,T] \times \overline{\Omega} \times [\underline{V},\overline{V}]),\ \nabla_x D \in C([0,T] \times \overline{\Omega} \times [-\underline{V},\overline{V}]; R^d),\\ &|D| + |\nabla_x D| + |\partial_V D| \le c(\underline{V},\overline{V}). \end{aligned} \tag{5.30}$$

As we are interested in classical solutions, certain compatibility conditions on the initial datum must be imposed. Specifically, we require

$$\partial_t^j \left(\mathbb{A} \cdot \nabla_x \mathbf{V}_0 \cdot \mathbf{n} + D(t,x,\mathbf{V}_0)\right) = 0 \text{ for } t = 0,\ x \in \partial\Omega,\ j = 0,1, \tag{5.31}$$

where "$\partial_t \mathbf{V}_0$" is computed from (5.25),

$$\partial_t \mathbf{V}_0 = \mathbb{A}(0,x,V_0) : \nabla_x^2 V_0 - B(0,x,V_0,\nabla_x V_0).$$

The relevant existence theory for the problem (5.25) was developed in [65, Chapter V, Theorems 7.3, 7.4]. We report the following result.

Theorem 8 (Existence for Quasilinear Parabolic Problem).
Let $\Omega \subset R^d$ be a bounded domain of class $C^{2+\beta}$, $\beta > 0$. Under the hypotheses (5.28)–(5.30), *suppose that the problem* (5.25)–(5.27) *admits a priori bound in L^∞, specifically,*

$$\underline{V} \leq V(t,x) \leq \overline{V}$$

for any classical solution, where $\underline{V}$, $\overline{V}$ depend only on $\|V_0\|_{L^\infty(\Omega)}$. In addition to (5.28)–(5.30), *let*

$$\begin{aligned}|\partial^2_{V,V}D| + |\nabla_x\partial_V V| + |\partial^2_{V,t}D| + |\partial_t\mathbb{A}| + |\partial_t D| \leq c(\underline{V},\overline{V}),\\ |\nabla_{\mathbf{d}}B|\,(1+|\mathbf{d}|) + |\partial_V B| + |\partial_t B| \leq c(\underline{V},\overline{V})\left(1+|\mathbf{d}|^2\right),\end{aligned} \tag{5.32}$$

and

$$|\partial^2_{V,V}\mathbb{A}| + |\partial^2_{V,t}\partial_t\mathbb{A}| + |\partial_V\nabla_x\mathbb{A}| + |\nabla_x\partial_t\mathbb{A}| \leq c(\underline{V},\overline{V}) \tag{5.33}$$

Finally, suppose that for any $\overline{d} > 0$

$$\begin{aligned}\nabla_x\mathbb{A} &\in C^\beta\left([0,T]\times\overline{\Omega}\times[\underline{V},\overline{V}];R^{d^3}\right)\\ \nabla_x D &\in C^{\frac{\beta}{2},\beta}\left([0,T]\times\overline{\Omega}\times[\underline{V},\overline{V}]\right)\\ B &\in C^\beta\left([0,T]\times\overline{\Omega}\times[\underline{V},\overline{V}\times\{|\mathbf{d}|\leq\overline{d}\}]\right).\end{aligned} \tag{5.34}$$

Let the initial data $V_0 \in C^{2+\beta}(\overline{\Omega})$ satisfy the compatibility conditions (5.31).

Then the problem (5.25)–(5.27) *admits a classical solution V unique in the class $C^{1+\frac{\beta}{2},2+\beta}([0,T]\times\overline{\Omega})$.*

Very roughly indeed, one may say that the quasilinear parabolic system admits a unique classical solution if

- The spatial domain is smooth enough.
- The initial and boundary data are sufficiently smooth satisfying the relevant compatibility conditions.
- Uniform L^∞ *a priori* bounds are available.
- Coefficients in the equations are locally smooth.

Our next goal is to apply Theorems 7, 8 to solve the approximate problems (5.11), (5.12), and (5.18), (5.19), respectively.

5.2.3 Approximate Equation of Continuity

Given the assumed regularity of the initial/boundary data, we may apply Theorem 7 to solve the approximate equation of continuity (5.11)–(5.13) for a given velocity field $\mathbf{u}$.

Lemma 6 (Existence of Approximate Densities). *Let the velocity field*

$$\mathbf{u} = \mathbf{u}_B + \mathbf{v},\ \mathbf{v} \in C([0,T]; X_n)$$

be given. Let the initial datum $\varrho_{0,n,\delta} \in C^{2+\beta}$, $\beta > 0$, *satisfy the compatibility condition* (5.14).

Then there is a unique classical solution ϱ *to the problem* (5.11)–(5.13) *in the class* $C^{1,2+\beta}([0,T] \times \overline{\Omega}_n)$.

Now, multiplying (5.11) on ϱ and integrating by parts, we get the "energy" balance equation

$$\begin{aligned}
&\frac{\mathrm{d}}{\mathrm{d}t}\frac{1}{2}\int_{\Omega_n} \varrho^2 \,\mathrm{d}x + \varepsilon \int_{\Omega_n} |\nabla_x \varrho|^2 \,\mathrm{d}x - \frac{1}{2}\int_{\partial\Omega_n} (\varrho - \varrho_B)^2 [\mathbf{u}_B \cdot \mathbf{n}]_n^- \,\mathrm{d}\sigma_x \\
&\quad + \frac{1}{2}\int_{\partial\Omega_n} \varrho^2 \left(\mathbf{u}_B \cdot \mathbf{n} - [\mathbf{u}_B \cdot \mathbf{n}]_n^- \right) \mathrm{d}\sigma_x \\
&\quad = -\frac{1}{2}\int_{\partial\Omega_n} \varrho_B^2 [\mathbf{u}_B \cdot \mathbf{n}]_n^- \,\mathrm{d}\sigma_x - \frac{1}{2}\int_{\Omega_n} \varrho^2 \mathrm{div}_x \mathbf{u} \,\mathrm{d}x
\end{aligned} \tag{5.35}$$

Next, we establish a comparison principle. We say that $\overline{\varrho}$ is a *supersolution* of (5.11), (5.12), if

$$\begin{aligned}
\partial_t \overline{\varrho} + \mathrm{div}_x(\overline{\varrho}\mathbf{u}) &\geq \varepsilon \Delta_x \overline{\varrho} \text{ in } (0,T) \times \Omega_n, \\
\varepsilon \nabla_x \overline{\varrho} \cdot \mathbf{n} &\geq (\overline{\varrho} - \varrho_B)[\mathbf{u}_B \cdot \mathbf{n}]_n^- \text{ in } (0,T) \times \partial\Omega_n.
\end{aligned} \tag{5.36}$$

Similarly, $\underline{\varrho}$ is a *subsolution* if

$$\begin{aligned}
\partial_t \underline{\varrho} + \mathrm{div}_x(\underline{\varrho}\mathbf{u}) &\leq \varepsilon \Delta_x \underline{\varrho} \text{ in } (0,T) \times \Omega_n, \\
\varepsilon \nabla_x \underline{\varrho} \cdot \mathbf{n} &\leq (\underline{\varrho} - \varrho_B)[\mathbf{u}_B \cdot \mathbf{n}]_n^- \text{ in } (0,T) \times \partial\Omega_n.
\end{aligned} \tag{5.37}$$

The following result is standard.

Lemma 7 (Comparison Principle for Approximate Densities). *Let $\overline{\varrho}$ be a supersolution and $\underline{\varrho}$ a subsolution such that*

$$\underline{\varrho}(0,\cdot) \leq \overline{\varrho}(0,\cdot) \text{ in } \Omega_n.$$

Then

$$\underline{\varrho} \leq \overline{\varrho} \text{ in } [0,T] \times \Omega_n.$$

Proof. Consider $r = \underline{\varrho} - \overline{\varrho}$ and multiply the difference of the corresponding inequalities by $\mathrm{sgn}^+(r)$:

$$\begin{aligned} \partial_t [r]^+ + [r]^+ \mathrm{div}_x \mathbf{u} + \mathbf{u} \cdot \nabla_x [r]^+ &\leq \varepsilon \Delta_x r \ \mathrm{sgn}^+(r) \text{ in } (0,T) \times \Omega_n, \\ \varepsilon \nabla_x r \cdot \mathbf{n} &\leq r [\mathbf{u}_B \cdot \mathbf{n}]^-_n \text{ in } (0,T) \times \partial \Omega_n. \end{aligned} \tag{5.38}$$

Approximating sgn^+ by smooth functions, we easily justify

$$\varepsilon \int_{\Omega_n} \Delta_x r \ \mathrm{sgn}^+(r) \ \mathrm{d}x \leq \varepsilon \int_{\partial \Omega_n} \mathrm{sgn}^+(r) \nabla_x r \cdot \mathbf{n} \ \mathrm{d}\sigma_x \leq \int_{\Omega_n} [r]^+ [\mathbf{u}_B \cdot \mathbf{n}]^-_n \mathrm{d}\sigma_x.$$

On the other hand,

$$\int_{\Omega_n} \mathbf{u} \cdot \nabla_x [r]^+ \ \mathrm{d}x = \int_{\partial \Omega_n} [r]^+ \mathbf{u}_B \cdot \mathbf{n} \mathrm{d}\sigma_x - \int_{\Omega_n} [r]^+ \mathrm{div}_x \mathbf{u} \ \mathrm{d}x.$$

Thus integrating (5.38) we get the desired conclusion

$$\frac{\mathrm{d}}{\mathrm{d}t} \int_{\Omega_n} [r]^+ \ \mathrm{d}x \leq \int_{\partial \Omega_n} [r]^+ \left([\mathbf{u}_B \cdot \mathbf{n}]^-_n - \mathbf{u}_B \cdot \mathbf{n} \right) \mathrm{d}\sigma_x \leq 0,$$

cf. hypothesis (5.10).

□

Now, suppose

$$0 < \underline{\varrho} \leq \varrho_{0,n,\delta}(x), \varrho_B(t,x) \leq \overline{\varrho}, \ x \in \Omega_n, \ t \in [0,T],$$

for some positive constants $\underline{\varrho}$, $\overline{\varrho}$. Then it is easy to see that

$$\underline{\varrho} \exp(-Kt), \ \overline{\varrho} \exp(Kt), \ K \geq \|\mathrm{div}_x \mathbf{u}\|_{L^\infty((0,T) \times \Omega_n)}$$

is a subsolution, a supersolution, respectively. Consequently,

$$\underline{\varrho} \exp(-Kt) \leq \varrho(t,x) \leq \overline{\varrho} \exp(Kt). \tag{5.39}$$

Finally, our goal is to compare two solutions ϱ_1, ϱ_2 with two different velocity fields $\mathbf{u}_1$, $\mathbf{u}_2$ but the same data $\varrho_{0,n,\delta}$, $\mathbf{u}_B$, ϱ_B. It turns out that the difference $r = \varrho_1 - \varrho_2$ satisfies the equation

$$\partial_t r + \mathrm{div}_x(r\mathbf{u}_1) = \varepsilon\Delta_x r + \mathrm{div}_x\left(\varrho_2(\mathbf{u}_2 - \mathbf{u}_1)\right) \text{ in } (0,T)\times\Omega_n,$$

with the boundary conditions

$$\varepsilon\nabla_x r\cdot\mathbf{n} = r[\mathbf{u}_B\cdot\mathbf{n}]_n^- \text{ in } (0,T)\times\partial\Omega_n.$$

Consequently, evoking the energy estimates (5.35) (with $\varrho_B = 0$), we may infer that

$$\begin{aligned}
&\sup_{0\le\tau\le T}\|(\varrho_1-\varrho_2)(\tau\cdot)\|^2_{L^2(\Omega_n)} + \int_0^T\int_{\Omega_n}|\nabla_x(\varrho_1-\varrho_2)|^2\,\mathrm{d}x\mathrm{d}t\\
&\le c\Big[\|\mathbf{u}_1-\mathbf{u}_2\|^2_{L^\infty((0,T)\times\Omega;R^d)} + \|\nabla_x(\mathbf{u}_1-\mathbf{u}_2)\|^2_{L^\infty((0,T)\times\Omega;R^{d\times d})}\Big]. \qquad (5.40)
\end{aligned}$$

Here, the constant c depends on the data as well as on M,

$$\|\mathbf{u}_i\|_{L^\infty((0,T)\times\Omega;R^d)} + \|\nabla_x\mathbf{u}_i\|_{L^\infty((0,T)\times\Omega;R^{d\times d})} \le M,\ i=1,2.$$

5.2.4 Approximate Internal Energy Balance

For given ϱ, $\mathbf{u}$, our goal is to find a solution ϑ of the approximate internal energy balance (5.18), (5.19) with a given initial state $\vartheta_{0,\delta}$. We start with a comparison principle. We say that $\overline{\vartheta}$ is a *supersolution* of (5.18), (5.19) if

$$\begin{aligned}
&\partial_t(\varrho e_{\delta,\mathrm{th}}(\varrho,\overline{\vartheta})) + \mathrm{div}_x(\varrho e_{\delta,\mathrm{th}}(\varrho,\overline{\vartheta})\mathbf{u}) + \mathrm{div}_x\mathbf{q}_\delta(\overline{\vartheta},\nabla_x\overline{\vartheta})\\
&\quad\ge \mathbb{S}_\delta(\overline{\vartheta},\nabla_x\mathbf{u}):\nabla_x\mathbf{u} - p_{\mathrm{th}}(\varrho,\overline{\vartheta})\mathrm{div}_x\mathbf{u}\\
&\quad+\varepsilon\Big[\delta\left(\Gamma\varrho^{\Gamma-2}+2\right) + \partial^2_{\varrho,\varrho}(\varrho e_{\mathrm{el}}(\varrho))\Big]|\nabla_x\varrho|^2 + \delta\frac{1}{\overline{\vartheta}^2} - \varepsilon\overline{\vartheta}^5 \text{ in } (0,T)\times\Omega_n,\\
&\mathbf{q}_\delta(\overline{\vartheta},\nabla_x\overline{\vartheta})\cdot\mathbf{n} + \varrho e_{\delta,\mathrm{th}}(\varrho,\overline{\vartheta})[\mathbf{u}_B\cdot\mathbf{n}]_n^- \le \Big(f_{i,B} - \varrho_B e_{\mathrm{el}}(\varrho_B)\Big)[\mathbf{u}_B\cdot\mathbf{n}]_n^-\\
&\quad\text{in } (0,T)\times\partial\Omega_n. \qquad (5.41)
\end{aligned}$$

Similarly, we say that $\underline{\vartheta}$ is a *subsolution* if

$$\begin{aligned}
&\partial_t(\varrho e_{\delta,\mathrm{th}}(\varrho,\overline{\vartheta})) + \mathrm{div}_x(\varrho e_{\delta,\mathrm{th}}(\varrho,\overline{\vartheta})\mathbf{u}) + \mathrm{div}_x\mathbf{q}_\delta(\overline{\vartheta},\nabla_x\overline{\vartheta})\\
&\quad\le \mathbb{S}_\delta(\overline{\vartheta},\nabla_x\mathbf{u}):\nabla_x\mathbf{u} - p_{\mathrm{th}}(\varrho,\overline{\vartheta})\mathrm{div}_x\mathbf{u}\\
&\quad+\varepsilon\Big[\delta\left(\Gamma\varrho^{\Gamma-2}+2\right) + \partial^2_{\varrho,\varrho}(\varrho e_{\mathrm{el}}(\varrho))\Big]|\nabla_x\varrho|^2 + \delta\frac{1}{\overline{\vartheta}^2} - \varepsilon\overline{\vartheta}^5 \text{ in } (0,T)\times\Omega_n,\\
&\mathbf{q}_\delta(\overline{\vartheta},\nabla_x\overline{\vartheta})\cdot\mathbf{n} + \varrho e_{\delta,\mathrm{th}}(\varrho,\overline{\vartheta})[\mathbf{u}_B\cdot\mathbf{n}]_n^- \ge \Big(f_{i,B} - \varrho_B e_{\mathrm{el}}(\varrho_B)\Big)[\mathbf{u}_B\cdot\mathbf{n}]_n^-\\
&\quad\text{in } (0,T)\times\partial\Omega_n. \qquad (5.42)
\end{aligned}$$

Next, we write

$$\mathbf{q}_\delta(\vartheta, \nabla_x \vartheta) = -\nabla_x \mathcal{K}_\delta(\vartheta), \text{ where } \mathcal{K}_\delta(\vartheta) \equiv \int_1^\vartheta \left(\kappa(s) + \delta \left(s^\Gamma + \frac{1}{s} \right) \right) \mathrm{d}s.$$

Proposition 2 (Comparison Principle for Approximate Temperatures).
Let

$$\begin{aligned} &\varrho \in C([0,T]; C^2(\overline{\Omega}_n)),\ \partial_t \varrho \in C([0,T] \times \overline{\Omega}_n) \quad \inf_{(0,T)\times\Omega_n} \varrho > 0, \\ &\mathbf{u} \in C([0,T] \times \overline{\Omega}_n; R^d),\ \nabla_x \mathbf{u} \in C([0,T] \times \overline{\Omega}_n; R^{d\times d}) \end{aligned} \tag{5.43}$$

be given. Let $\overline{\vartheta}$ be a supersolution and $\underline{\vartheta}$ a subsolution belonging to the class

$$\left\{ \begin{aligned} \underline{\vartheta},\ \overline{\vartheta} \in L^2(0,T; W^{1,2}(\Omega_n)),\ \partial_t \underline{\vartheta},\ \partial_t \overline{\vartheta} \in L^2((0,T)\times\Omega_n), \\ \Delta \mathcal{K}_\delta(\underline{\vartheta}),\ \Delta \mathcal{K}_\delta(\overline{\vartheta}) \in L^2((0,T)\times\Omega_n), \end{aligned} \right\}, \tag{5.44}$$

$$\left\{ \begin{aligned} 0 < \operatorname{ess\,inf}_{(0,T)\times\Omega_n} \underline{\vartheta} \leq \operatorname{ess\,sup}_{(0,T)\times\Omega_n} \underline{\vartheta} < \infty, \\ 0 < \operatorname{ess\,inf}_{(0,T)\times\Omega_n} \overline{\vartheta} \leq \operatorname{ess\,sup}_{(0,T)\times\Omega_n} \overline{\vartheta} < \infty. \end{aligned} \right\} \tag{5.45}$$

Suppose that

$$\underline{\vartheta}(0,\cdot) \leq \overline{\vartheta}(0,\cdot) \text{ a.a. in } \Omega_n.$$

Then

$$\underline{\vartheta}(t,x) \leq \overline{\vartheta}(t,x) \text{ a.a. in } (0,T)\times\Omega_n.$$

Proof.
Multiply the difference of the corresponding differential inequalities by

$$\mathrm{sgn}^+\Big(\varrho e_{\delta,\mathrm{th}}(\varrho, \underline{\vartheta}) - e_{\delta,\mathrm{th}}(\varrho, \overline{\vartheta})\Big).$$

As e_δ is strictly increasing in ϑ, we get

$$\mathrm{sgn}^+\Big(\varrho e_{\delta,\mathrm{th}}(\varrho, \underline{\vartheta}) - \varrho e_{\delta,\mathrm{th}}(\varrho, \overline{\vartheta})\Big) = \mathrm{sgn}^+\Big(\mathcal{K}_\delta(\underline{\vartheta}) - \mathcal{K}_\delta(\overline{\vartheta})\Big).$$

Consequently, after a direct manipulation, we get

$$\Big[\partial_t\Big(\varrho e_{\delta,\mathrm{th}}(\varrho, \underline{\vartheta}) - \varrho e_{\delta,\mathrm{th}}(\varrho, \overline{\vartheta})\Big) + \mathbf{u}\cdot\nabla_x\Big(\varrho e_{\delta,\mathrm{th}}(\varrho, \underline{\vartheta}) - \varrho e_{\delta,\mathrm{th}}(\varrho, \overline{\vartheta})\Big)\Big] \times$$
$$\times\, \mathrm{sgn}^+\Big(\varrho e_{\delta,\mathrm{th}}(\varrho, \underline{\vartheta}) - \varrho e_{\delta,\mathrm{th}}(\varrho, \overline{\vartheta})\Big)$$

$$
\begin{aligned}
&- \Delta_x \Big(\mathcal{K}_\delta(\underline{\vartheta}) - \mathcal{K}_\delta(\overline{\vartheta}) \Big) \mathrm{sgn}^+ \Big(\mathcal{K}_\delta(\underline{\vartheta}) - \mathcal{K}_\delta(\overline{\vartheta}) \Big) \\
&\leq \big(\mathbb{S}_\delta(\underline{\vartheta}, \nabla_x \mathbf{u}) - \mathbb{S}_\delta(\overline{\vartheta}, \nabla_x \mathbf{u}) \big) : \nabla_x \mathbf{u} \ \mathrm{sgn}^+ \Big(\varrho e_{\delta,\mathrm{th}}(\varrho, \underline{\vartheta}) - \varrho e_{\delta,\mathrm{th}}(\varrho, \overline{\vartheta}) \Big) \\
&+ \mathrm{div}_x \mathbf{u} (p_{\mathrm{th}}(\varrho, \overline{\vartheta}) - p_{\mathrm{th}}(\varrho, \underline{\vartheta})) \ \mathrm{sgn}^+ \Big(\varrho e_{\delta,\mathrm{th}}(\varrho, \underline{\vartheta}) - \varrho e_{\delta,\mathrm{th}}(\varrho, \overline{\vartheta}) \Big) \\
&+ \delta \left(\frac{1}{\underline{\vartheta}^2} - \frac{1}{\overline{\vartheta}^2} \right) \mathrm{sgn}^+ \Big(\varrho e_{\delta,\mathrm{th}}(\varrho, \underline{\vartheta}) - \varrho e_{\delta,\mathrm{th}}(\varrho, \overline{\vartheta}) \Big) \\
&+ \varepsilon \left(\overline{\vartheta}^5 - \underline{\vartheta}^5 \right) \mathrm{sgn}^+ \Big(\varrho e_{\delta,\mathrm{th}}(\varrho, \underline{\vartheta}) - \varrho e_{\delta,\mathrm{th}}(\varrho, \overline{\vartheta}) \Big).
\end{aligned}
\tag{5.46}
$$

In view of hypothesis (5.45), we may assume that all nonlinearities are globally Lipschitz in ϑ. Consequently,

$$
\begin{aligned}
&\mathrm{sgn}^+ \Big(\varrho e_{\delta,\mathrm{th}}(\varrho, \underline{\vartheta}) - e_{\delta,\mathrm{th}}(\varrho, \overline{\vartheta}) \Big) \times \\
&\times \Big[\Big(\partial_t \Big(\varrho e_{\delta,\mathrm{th}}(\varrho, \underline{\vartheta}) - \varrho e_{\delta,\mathrm{th}}(\varrho, \overline{\vartheta}) \Big) + \nabla_x \Big(\varrho e_{\delta,\mathrm{th}}(\varrho, \underline{\vartheta}) - \varrho e_{\delta,\mathrm{th}}(\varrho, \overline{\vartheta}) \Big) \cdot \mathbf{u} \Big] \\
&= \partial_t \Big[\varrho e_{\delta,\mathrm{th}}(\varrho, \underline{\vartheta}) - \varrho e_{\delta,\mathrm{th}}(\varrho, \overline{\vartheta}) \Big]^+ + \nabla_x \Big[\varrho e_{\delta,\mathrm{th}}(\varrho, \underline{\vartheta}) - \varrho e_{\delta,\mathrm{th}}(\varrho, \overline{\vartheta}) \Big]^+ \cdot \mathbf{u}.
\end{aligned}
$$

Moreover, it is easy to check that

$$
\int_{\Omega_n} \Delta_x w \ \mathrm{sgn}^+(w) \ \mathrm{d}x \leq \int_{\partial \Omega_n} \mathrm{sgn}^+(w) \ \nabla_x w \cdot \mathbf{n} \ \mathrm{d}\sigma_x \ \text{for any } w \in W^{2,2}.
$$

Consequently, we may integrate (5.46) obtaining

$$
\begin{aligned}
&\int_{\Omega_n} \Big[\varrho e_{\delta,\mathrm{th}}(\varrho, \underline{\vartheta}) - \varrho e_{\delta,\mathrm{th}}(\varrho, \overline{\vartheta}) \Big]^+ (\tau, \cdot) \ \mathrm{d}x \\
&+ \int_0^\tau \int_{\partial \Omega_n} \Big[\varrho e_{\delta,\mathrm{th}}(\varrho, \underline{\vartheta}) - e_{\delta,\mathrm{th}}(\varrho, \overline{\vartheta}) \Big]^+ \mathbf{u}_B \cdot \mathbf{n} \mathrm{d}\sigma_x \mathrm{d}t \\
&+ \int_0^\tau \int_{\partial \Omega_n} \mathrm{sgn}^+ \Big(\varrho e_{\delta,\mathrm{th}}(\varrho, \underline{\vartheta}) - \varrho e_{\delta,\mathrm{th}}(\varrho, \overline{\vartheta}) \Big) \times \\
&\times \Big(\mathbf{q}_\delta(\underline{\vartheta}, \nabla_x \underline{\vartheta}) - \mathbf{q}_\delta(\overline{\vartheta}, \nabla_x \overline{\vartheta})) \Big) \cdot \mathbf{n} \ \mathrm{d}\sigma_x \mathrm{d}t \\
&\leq c \int_0^\tau \int_{\Omega_n} (1 + |\mathrm{div}_x \mathbf{u}|) \Big[\varrho e_{\delta,\mathrm{th}}(\varrho, \underline{\vartheta}) - \varrho e_{\delta,\mathrm{th}}(\varrho, \overline{\vartheta}) \Big]^+ \ \mathrm{d}x \ \mathrm{d}t
\end{aligned}
\tag{5.47}
$$

for a.a. $\tau > 0$.

Now, by virtue of the boundary conditions (5.41), (5.42),

$$
\begin{aligned}
&\int_{\partial \Omega_n} [\varrho e_{\delta,\mathrm{th}}(\varrho, \underline{\vartheta}) - \varrho e_{\delta,\mathrm{th}}(\varrho, \overline{\vartheta})]^+ \mathbf{u}_B \cdot \mathbf{n} \mathrm{d}\sigma_x \mathrm{d}t \\
&+ \int_{\partial \Omega_n} \mathrm{sgn}^+ \Big(\varrho e_{\delta,\mathrm{th}}(\varrho, \underline{\vartheta}) - \varrho e_{\delta,\mathrm{th}}(\varrho, \overline{\vartheta}) \Big) \Big(\mathbf{q}_\delta(\underline{\vartheta}, \nabla_x \underline{\vartheta}) - \mathbf{q}_\delta(\overline{\vartheta}, \nabla_x \overline{\vartheta})) \Big) \cdot \mathbf{n} \mathrm{d}\sigma_x
\end{aligned}
$$

$$\geq \int_{\partial\Omega_n} [\varrho e_{\delta,\text{th}}(\varrho,\underline{\vartheta}) - \varrho e_{\delta,\text{th}}(\varrho,\overline{\vartheta})\Big] [\mathbf{u}_B \cdot \mathbf{n}]_n^- \mathrm{d}\sigma_x$$

$$+ \int_{\Omega_n} \text{sgn}^+ \Big(\varrho e_{\delta,\text{th}}(\varrho,\underline{\vartheta}) - \varrho e_{\delta,\text{th}}(\varrho,\overline{\vartheta})\Big) \Big(\mathbf{q}_\delta(\underline{\vartheta}, \nabla_x \underline{\vartheta}) - \mathbf{q}_\delta(\overline{\vartheta}, \nabla_x \overline{\vartheta}))\Big) \cdot \mathbf{n} \mathrm{d}\sigma_x$$

$$\geq 0.$$

Thus (5.47) reduces to

$$\int_{\Omega_n} \Big[\varrho e_{\delta,\text{th}}(\varrho,\underline{\vartheta}) - \varrho e_{\delta,\text{th}}(\varrho,\overline{\vartheta})\Big]^+ (\tau) \ \mathrm{d}x$$

$$\leq c \int_0^\tau \int_\Omega (1 + |\text{div}_x \mathbf{u}|) \Big[\varrho e_{\delta,\text{th}}(\varrho,\underline{\vartheta}) - \varrho e_{\delta,\text{th}}(\varrho,\overline{\vartheta})\Big]^+ \ \mathrm{d}x \ \mathrm{d}t,$$

and the desired conclusion follows by means of the standard Gronwall argument.

□

Remark 7. The same result holds for the Dirichlet boundary conditions considered in Chap. 12,

$$\vartheta|_{\partial\Omega_n} = \vartheta_B.$$

Indeed the boundary inequality for the supersolution must by replaced by

$$\overline{\vartheta} \geq \vartheta_B \text{ on } \partial\Omega,$$

whereas the subsolution satisfies

$$\underline{\vartheta} \leq \vartheta_B \text{ on } \partial\Omega.$$

The proof can be carried over without essential modifications.

As a direct consequence of Proposition 2, we deduce uniqueness of strong solutions.

Corollary 1 (Uniqueness).
For given ϱ, $\mathbf{u}$ *satisfying* (5.43), *and initial data* $\vartheta(0,\cdot) = \vartheta_0$, *the problem* (5.18), (5.19) *admits at most one strong solution* ϑ *in the class* (5.44), (5.45).

The comparison principle yields useful uniform bounds on strong solutions in terms of the initial data. In particular, the temperature remains strictly positive on any compact interval $[0, T]$. Specifically, we report the following result.

Proposition 3 (Minimum/Maximum Principle).
Let ϱ, $\mathbf{u}$ be given belonging to the class (5.43). Let

$$0 < \underline{f} \leq \Big(f_{i,B}(t,x) - \varrho_B e_{\rm el}(\varrho_B) \Big) \leq \overline{f} \ \text{for a.a.} \ (t,x) \in (0,T) \times \partial\Omega_n. \tag{5.48}$$

Suppose that ϑ is a strong solution of (5.18), (5.19) in the class (5.44), (5.45) such that

$$0 < \underline{\vartheta}_0 \leq \vartheta(0,\cdot) \leq \overline{\vartheta}_0.$$

Then there exist constants $\underline{\vartheta}$, $\overline{\vartheta}$,

$$0 < \underline{\vartheta} \leq \underline{\vartheta}_0 \leq \overline{\vartheta}_0 \leq \overline{\vartheta}$$

depending solely on

$$\varepsilon,\ \delta,\quad \sup_{(0,T)\times\Omega_n} (|\mathbf{u}| + |\nabla_x \mathbf{u}|) \quad \sup_{(0,T)\times\Omega_n} (|\varrho| + |\varrho^{-1}| + |\partial_t \varrho| + |\nabla_x \varrho|),\ \underline{f},\ \overline{f}$$

such that

$$\underline{\vartheta} \leq \vartheta(t,x) \leq \overline{\vartheta} \ \text{for a.a.} \ (t,x) \in (0,T) \times \Omega_n.$$

Remark 8. Hypothesis (5.48) should be compared with a similar one required for the validity of the weak–strong uniqueness principle in Theorem 6.

Proof.
It is easy to check that a constant function $\underline{\vartheta}$ is a subsolution if:

$$\begin{aligned}
\frac{\delta}{\underline{\vartheta}^2} \geq \Big[& \varepsilon \underline{\vartheta}^5 + p_{\rm m}(\varrho, \underline{\vartheta}) {\rm div}_x \mathbf{u} + p_{\rm rad}(\underline{\vartheta}) {\rm div}_x \mathbf{u} \\
& + \varrho \frac{\partial e_{\rm m}(\varrho, \underline{\vartheta})}{\partial \varrho} \Big(\partial_t \varrho + \mathbf{u} \cdot \nabla_x \varrho \Big) + \Big(e_{\rm m}(\varrho, \underline{\vartheta}) + a \underline{\vartheta}^4 + \delta \underline{\vartheta} \Big) \Big(\partial_t \varrho + {\rm div}_x (\varrho \mathbf{u}) \Big) \\
& - \mathbb{S}_\delta(\underline{\vartheta}, \nabla_x \mathbf{u}) : \nabla_x \mathbf{u} - \varepsilon \left[\delta (\Gamma \varrho^{\Gamma - 2} + 2) + \partial^2_{\varrho,\varrho} (\varrho e_{\rm el}(\varrho)) |\nabla_x \varrho|^2 \right] \\
& \text{in } (0,T) \times \Omega,
\end{aligned} \tag{5.49}$$

and

$$\varrho e_{\delta,{\rm th}}(\varrho, \underline{\vartheta}) [\mathbf{u}_B \cdot \mathbf{n}]^-_n \geq \Big(f_{i,B} - \varrho_B e_{\rm el}(\varrho_B) \Big) [\mathbf{u}_B \cdot \mathbf{n}]^-_n \ \text{on} \ \Gamma_{\rm in}. \tag{5.50}$$

In view of the structural hypotheses (4.2), it is easy to see that (5.49) holds if $\underline{\vartheta} > 0$ is chosen small enough. By the same token, we deduce (5.50) from (5.48) seeing that

$$\varrho e_{\delta,{\rm th}}(\varrho, \vartheta) \to 0 \text{ as } \vartheta \to 0$$

uniformly for ϱ bounded above and below away from zero, cf. (4.17). Thus the desired conclusion follows from the comparison principle stated in Proposition 2.

The upper bound is deduced in the same way using $\varepsilon\overline{\vartheta}^5$ as the dominating term in (5.49).

□

Finally, we derive suitable *a priori* bounds on solutions to (5.18), (5.19) in the energy norm.

Lemma 8 (A Priori Energy Bounds for Temperature).
Under the hypotheses of Proposition 3, let ϑ be a strong solution of (5.18), (5.19) *in the class* (5.44), (5.45) *emanating from the initial state*

$$\vartheta_{0,\delta} \in W^{1,2}(\Omega_n), \ 0 < \underline{\vartheta}_0 \leq \vartheta_{0,\delta} \leq \overline{\vartheta}_0.$$

Then

$$\operatorname*{ess\ sup}_{t\in(0,T)} \|\vartheta\|^2_{W^{1,2}(\Omega_n)} + \int_0^T \left(\|\partial_t\vartheta\|^2_{L^2(\Omega_n)} + \|\Delta_x\mathcal{K}_\delta(\vartheta)\|^2_{L^2(\Omega_n)} \right) \mathrm{d}t \leq c, \tag{5.51}$$

where c depends on

$$\varepsilon,\ \delta,\ \sup_{(0,T)\times\Omega_n} (|\mathbf{u}| + |\nabla_x\mathbf{u}|) \quad \sup_{(0,T)\times\Omega_n} (|\varrho| + |\varrho^{-1}| + |\partial_t\varrho| + |\nabla_x\varrho|),$$
$$\sup_{(0,T)\times\partial\Omega_n} |\partial_t f_{i,B}|,\ \sup_{(0,T)\times\partial\Omega_n} |\partial_t\mathbf{u}_B|,\ \sup_{(0,T)\times\partial\Omega_n} |\partial_t\varrho_B|$$

and on $\|\vartheta_{0,\delta}\|_{W^{1,2}(\Omega_n)}$.

Proof.
Multiplying equation (5.18) on ϑ and integrating the resulting expression by parts, we get

$$\begin{aligned}
&\frac{1}{2}\int_0^T\int_{\Omega_n} \frac{\partial(\varrho e_{\delta,\mathrm{th}})}{\partial\vartheta}(\varrho,\vartheta)\partial_t(\vartheta)^2 \ \mathrm{d}x\mathrm{d}t - \int_0^T\int_{\Omega_n} \varrho e_{\delta,\mathrm{th}}(\varrho,\vartheta)\nabla_x\vartheta\cdot\mathbf{u} \ \mathrm{d}x\mathrm{d}t \\
&+ \int_0^T\int_{\partial\Omega_n} \Big(f_{i,B} - \varrho_B e_{\mathrm{el}}(\varrho_B) \Big)\vartheta \ [\mathbf{u}_B\cdot\mathbf{n}]_n^- \mathrm{d}\sigma_x\mathrm{d}t \\
&+ \int_0^T\int_{\partial\Omega_n} \vartheta\varrho e_{\delta,\mathrm{th}} \ \left(\mathbf{u}_B\cdot\mathbf{n} - [\mathbf{u}_B\cdot\mathbf{n}]_n^-\right) \mathrm{d}\sigma_x\mathrm{d}t \\
&+ \int_0^T\int_{\Omega_n} \kappa_\delta(\vartheta)|\nabla_x\vartheta|^2 \ \mathrm{d}x\mathrm{d}t = \int_0^T\int_{\Omega_n} F_1(t,x)\vartheta \ \mathrm{d}x\mathrm{d}t,
\end{aligned} \tag{5.52}$$

where

$$F_1 = -\frac{\partial(\varrho e_{\delta,\mathrm{th}})}{\partial\vartheta}(\varrho,\vartheta)\partial_t\varrho + \mathbb{S}_\delta(\vartheta,\nabla_x\mathbf{u}) : \nabla_x\mathbf{u}$$
$$+ \varepsilon\left[\delta(\Gamma\varrho^{\Gamma-2}+2) + \partial^2_{\varrho,\varrho}(\varrho e_{\mathrm{el}}(\varrho))\right]|\nabla_x\varrho|^2 - p_{\mathrm{th}}(\varrho,\vartheta)\mathrm{div}_x\mathbf{u} + \delta\frac{1}{\vartheta^2} - \varepsilon\vartheta^5.$$

Similarly, multiplying (5.18) on $\partial_t\mathcal{K}_\delta(\vartheta)$ yields

$$\begin{aligned}
&\left[\int_{\Omega_n}\frac{1}{2}|\nabla_x\mathcal{K}_\delta(\vartheta)|^2\,\mathrm{d}x\right]_{t=0}^{t=T} + \int_0^T\int_{\Omega_n}\varrho\kappa_\delta(\vartheta)\frac{\partial e_{\delta,\mathrm{th}}}{\partial\vartheta}(\varrho,\vartheta)|\partial_t\vartheta|^2\,\mathrm{d}x\mathrm{d}t\\
&+\int_0^T\int_{\partial\Omega_n}\partial_t\mathcal{K}_\delta(\vartheta)\Big(f_{i,B} - \varrho_B e_{\mathrm{el}}(\varrho_B)\Big)\,[\mathbf{u}_B\cdot\mathbf{n}]_n^-\mathrm{d}\sigma_x\mathrm{d}t\\
&+\int_0^T\int_{\partial\Omega_n}\partial_t\mathcal{K}_\delta(\vartheta)\varrho e_{\delta,\mathrm{th}}\,\left(\mathbf{u}_B\cdot\mathbf{n} - [\mathbf{u}_B\cdot\mathbf{n}]_n^-\right)\mathrm{d}\sigma_x\mathrm{d}t\\
&+\int_0^T\int_{\Omega_n}\varrho\frac{\partial e_{\delta,\mathrm{th}}}{\partial\vartheta}(\varrho,\vartheta)\,\partial_t\vartheta\nabla_x\mathcal{K}_\delta(\vartheta)\cdot\mathbf{u}\,\mathrm{d}x\mathrm{d}t = \int_0^T\int_{\Omega_n}F_2(t,x)\partial_t\vartheta\,\mathrm{d}x\mathrm{d}t,
\end{aligned} \tag{5.53}$$

where

$$\begin{aligned}
F_2 = &-\kappa_\delta(\vartheta)\Big(\partial_\varrho[\varrho e_{\delta,\mathrm{th}}](\varrho,\vartheta)\partial_t\varrho - \partial_\varrho[\varrho e_{\delta,\mathrm{th}}](\varrho,\vartheta)\nabla_x\varrho\cdot\mathbf{u}\\
&-\varrho e_{\delta,\mathrm{th}}(\varrho,\vartheta)\mathrm{div}_x\mathbf{u}\Big) + \mathbb{S}_\delta(\vartheta,\nabla_x\mathbf{u}) : \nabla_x\mathbf{u}\\
&+\varepsilon\left[\delta(\Gamma\varrho^{\Gamma-2}+2) + \partial^2_{\varrho,\varrho}(\varrho e_{\mathrm{el}}(\varrho))\right]|\nabla_x\varrho|^2\\
&-p_{\mathrm{th}}(\varrho,\vartheta)\mathrm{div}_x\mathbf{u} + \delta\frac{1}{\vartheta^2} - \varepsilon\vartheta^5.
\end{aligned}$$

In view of the uniform bounds already established, both F_1 and F_2 are bounded in $L^\infty((0,T)\times\Omega_n)$. By the same token, the boundary integrals

$$\begin{aligned}
&\int_0^T\int_{\partial\Omega_n}(f_{i,B} - \varrho_B p_{\mathrm{el}}(\varrho_B))\,\vartheta\,[\mathbf{u}_B\cdot\mathbf{n}]_n^-\mathrm{d}\sigma_x\mathrm{d}t,\\
&\int_0^T\int_{\partial\Omega_n}\varrho e_\delta\,\left(\mathbf{u}_B\cdot\mathbf{n} - [\mathbf{u}_B\cdot\mathbf{n}]_n^-\right)\mathrm{d}\sigma_x\mathrm{d}t,\\
&\int_0^T\int_{\partial\Omega_n}\partial_t\mathcal{K}_\delta(\vartheta)\Big(f_{i,B} - \varrho_B p_{\mathrm{el}}(\varrho_B)\Big)\,[\mathbf{u}_B\cdot\mathbf{n}]_n^-\mathrm{d}\sigma_x\mathrm{d}t
\end{aligned}$$

are bounded as long as $\sup|\partial_t f_{i,B}|$, $\sup|\partial_t\varrho_B|$ are controlled.

Finally, the integral

$$\int_0^T\int_{\partial\Omega_n}\partial_t\mathcal{K}_\delta(\vartheta)\varrho e_{\delta,\mathrm{th}}\,\left(\mathbf{u}_B\cdot\mathbf{n} - [\mathbf{u}_B\cdot\mathbf{n}]_n^-\right)\mathrm{d}\sigma_x\mathrm{d}t$$

can be handled as follows. Introducing a function

$$\chi(\varrho,\vartheta) = \varrho \int_1^{\vartheta} \kappa_\delta(s) e_{\delta,\mathrm{th}}(\varrho,s)\mathrm{d}s$$

we compute

$$\partial_t \chi(\varrho,\vartheta) = \partial_t \varrho \left(\int_1^{\vartheta} \kappa_\delta(s) e_{\delta,\mathrm{th}}(\varrho,s)\mathrm{d}s + \int_1^{\vartheta} \kappa_\delta(s) \frac{\partial e_{\delta,\mathrm{th}}}{\partial \varrho}(\varrho,s)\mathrm{d}s \right) \\ + \partial_t \mathcal{K}(\vartheta) \varrho e_{\delta,\mathrm{th}}.$$

Consequently,

$$\int_0^T \int_{\partial \Omega_n} \partial_t \mathcal{K}_\delta(\vartheta) \varrho e_{\delta,\mathrm{th}} \ \left(\mathbf{u}_B \cdot \mathbf{n} - [\mathbf{u}_B \cdot \mathbf{n}]_n^- \right) \mathrm{d}\sigma_x \mathrm{d}t$$
$$= \int_0^T \int_{\partial \Omega_n} \partial_t \chi(\varrho,\vartheta) \ \left(\mathbf{u}_B \cdot \mathbf{n} - [\mathbf{u}_B \cdot \mathbf{n}]_n^- \right) \mathrm{d}\sigma_x \mathrm{d}t$$
$$- \int_0^T \int_{\partial \Omega_n} \partial_t \varrho \left(\int_1^{\vartheta} \kappa_\delta(s) e_{\delta,\mathrm{th}}(\varrho,s)\mathrm{d}s \right) \ \left(\mathbf{u}_B \cdot \mathbf{n} - [\mathbf{u}_B \cdot \mathbf{n}]_n^- \right) \mathrm{d}\sigma_x \mathrm{d}t$$
$$- \int_0^T \int_{\partial \Omega_n} \partial_t \varrho \left(\int_1^{\vartheta} \kappa_\delta(s) \frac{\partial e_{\delta,\mathrm{th}}}{\partial \varrho}(\varrho,s)\mathrm{d}s \right) \ \left(\mathbf{u}_B \cdot \mathbf{n} - [\mathbf{u}_B \cdot \mathbf{n}]_n^- \right) \mathrm{d}\sigma_x \mathrm{d}t,$$

and the integral is controlled as long as $\sup |\partial_t \mathbf{u}_B|$ is bounded.

Thus summing up (5.52), (5.53) and using Young's inequality and Gronwall's lemma, we may infer that

$$\operatorname*{ess\ sup}_{t \in (0,T)} \| \nabla_x \mathcal{K}_\delta(\vartheta) \|^2_{L^2(\Omega_n;R^3)} + \int_0^T \| \partial_t \vartheta \|^2_{L^2(\Omega_n)} \ \mathrm{d}t \le c.$$

Finally, computing $\Delta_x \mathcal{K}_\delta(\vartheta)$ from equation (5.18) we conclude the proof.

□

Having established the comparison principle and the *a priori* bounds, we can apply Theorem 8 to show the *existence* of strong solutions to the approximate internal energy problem (5.18)–(5.20) as long as the functions ϱ, $\mathbf{u}$ are given and smooth enough. Of course, this cannot be done in a straightforward manner as the coefficients do not, in general, enjoy the regularity properties required by Theorem 8. A rather lengthy but nowadays well understood regularization procedure detailed in [43, Section 3.4.2] must be applied. As a result, we claim the approximate internal energy balance admits global in time strong solutions.

Proposition 4 (Existence of Approximate Temperatures). *Let*

$$\varrho \in C([0,T]; C^2(\overline{\Omega}_n)),\ \partial_t \varrho \in C([0,T] \times \overline{\Omega}_n) \quad \inf_{(0,T)\times\Omega_n} \varrho > 0,$$

$$\mathbf{u} \in C([0,T] \times \overline{\Omega}_n; R^d),\ \nabla_x \mathbf{u} \in C([0,T] \times \overline{\Omega}_n; R^{d\times d})$$

be given. Let

$$0 < \underline{f} \leq \Big(f_{i,B} - \varrho_B e_{\rm el}(\varrho_B) \Big)(t,x) \leq \overline{f} \ \textit{for}\ (t,x) \in (0,T) \times \partial\Omega_n.$$

Then for any initial temperature

$$\vartheta_0 \in W^{1,2} \cap L^\infty(\Omega),\ 0 < \underline{\vartheta_0} \leq \vartheta_0 \leq \overline{\vartheta_0} \ \textit{a.a. in}\ \Omega,$$

the approximate internal energy equation (5.18)*, endowed with the boundary conditions* (5.19)*, admits a unique strong solution in the class*

$$\vartheta \in L^\infty(0,T; W^{1,2}(\Omega_n)),\ \vartheta \in L^2(0,T; W^{2,2}(\Omega_n)),\ \partial_t \vartheta \in L^2((0,T)\times\Omega_n),$$

$$0 < \underline{\vartheta} \leq \vartheta(t,x) \leq \overline{\vartheta} \ \textit{for a.a.}\ (t,x) \in (0,T) \times \Omega_n$$

for certain constants $\underline{\vartheta}$*,* $\overline{\vartheta}$*.*

Remark 9. Note that for the strong solutions belonging to the class specified above, the boundary conditions (5.19) can be interpreted in the strong sense, specifically a.a. with respect to $dt \times d\sigma$ on $(0,T) \times \partial\Omega$.

Finally, we observe that, similarly to (5.40), the solution ϑ considered as a function of the parameters $\varrho, \mathbf{u}$ is continuous. More specifically, consider $\mathbf{u}$, $(\mathbf{u} - \mathbf{u}_B) \in C([0,T]; X_n)$ and $\varrho = \varrho[\mathbf{u}]$—the unique solution of the approximate equation of continuity (5.11), with (5.12), (5.13). Then the unique solution $\vartheta = \vartheta[\varrho[\mathbf{u}], \mathbf{u}]$ of the approximate internal energy equation (5.18), (5.19), with $\vartheta(0,\cdot) = \vartheta_0$ can be associated with a mapping

$$\mathbf{u} \in C([0,T]; X_n + \mathbf{u}_B) \mapsto \vartheta \in Y,$$

where

$$Y = \left\{ \begin{array}{c} \partial_t \vartheta \in L^2((0,T) \times \Omega_n),\ \Delta_x \mathcal{K}_\delta \in L^2((0,T) \times \Omega_n), \\ \vartheta \in L^\infty(0,T; W^{1,2}(\Omega_n)),\ \vartheta, \vartheta^{-1} \in L^\infty((0,T) \times \Omega_n) \end{array} \right\}$$

The mapping $\mathbf{u} \mapsto \vartheta[\mathbf{u}]$ maps bounded sets in $C([0,T]; X_n + \mathbf{u}_B)$ into bounded sets in Y and is continuous with values in $L^2(0,T; W^{1,2}(\Omega_n))$, [43, Chapter 3, Section 3.4.2].

5.3 Existence of Basic Approximate Solutions

With the results established in the preceding part of this chapter, the existence of the approximate solutions at the basic level can be shown by means of a fixed point argument detailed in [43, Chapter 3, Sections 3.4.3, 3.4.4]. We have the following result.

Theorem 9 (Existence of Zero-th Level Approximate Solutions). *Suppose that $\Omega_n \subset R^d$ is a bounded domain of class C^4. Let the initial data*

$$\varrho_{0,n,\delta} \in C^{2+\beta}(\Omega_n) \text{ satisfying the compatibility conditions (5.14)},$$
$$\vartheta_0 \in W^{1,2}(\Omega_n),\ 0 < \underline{\vartheta}_0 \leq \vartheta_0 \leq \overline{\vartheta}_0,$$
$$\mathbf{m}_0 \in L^1(\Omega_n; R^d),$$

be given. Finally, suppose

$$\inf_{[0,T]\times\partial\Omega_n} \Big\{ f_{i,B} - \varrho_B e_{\rm el}(\varrho_B) \Big\} > 0$$

in the case of real gas EOS, or

$$\inf_{[0,T]\times\partial\Omega_n} \Big\{ f_{i,B} - \varrho_B e_{HS}(\varrho_B) \Big\} > 0$$

in the case of hard sphere EOS.

Then the approximate problem (5.11)–(5.13), (5.17), (5.18)–(5.20) *admits a solution $(\varrho, \vartheta, \mathbf{u})$ in $(0,T) \times \Omega_n$.*

Remark 10. In the case of the hard sphere gas, we suppose

$$0 < \inf_{[0,T]\times\partial\Omega_n} \varrho_B \leq \sup_{[0,T]\times\partial\Omega_n} \varrho_B < \overline{\varrho}.$$

Consequently, $\varrho_B e_{HS}(\varrho_B) = \varrho_B e_{\delta,HS}(\varrho_B)$ provided $\delta > 0$ is small enough.

Chapter 6
Vanishing Galerkin Limit and Domain Approximation

Having established the existence of the zeroth level approximate solutions in Theorem 9, our first task is to perform the limit $n \to \infty$ in the Faedo–Galerkin approximation. Simultaneously, we suppose that $\Omega \subset R^d$—a bounded Lipschitz domain—is approximated by smooth bounded domains $\Omega_n \subset R^d$ as stated in (5.4), (5.5), specifically,

$$\begin{aligned} &\Omega_n \text{ of class } C^k, \ k \geq 4, \ \Omega \subset \Omega_n, \\ &|\partial\Omega_n \setminus \partial\Omega|_{d-1} + |\partial\Omega \setminus \partial\Omega_n|_{d-1} \to 0. \end{aligned} \tag{6.1}$$

We consider the sequence of approximate solutions $\{\varrho_n, \vartheta_n, \mathbf{u}_n\}_{n=1}^{\infty}$ obtained in Theorem 9 in the preceding section satisfying the following system of equations:

- **Approximate equation of continuity**

$$\begin{aligned} \partial_t \varrho_n + \mathrm{div}_x(\varrho_n \mathbf{u}_n) &= \varepsilon \Delta_x \varrho_n \text{ in } (0,T) \times \Omega_n, \\ \varepsilon \nabla_x \varrho \cdot \mathbf{n} &= (\varrho - \varrho_B)[\mathbf{u}_B \cdot \mathbf{n}]_n^- \text{ in } (0,T) \times \partial\Omega_n, \\ \varrho_n(0,\cdot) &= \varrho_{0,n,\delta}, \end{aligned} \tag{6.2}$$

where

$$0 < \underline{\varrho}_\delta \leq \varrho_{0,n,\delta}(x) \leq \overline{\varrho}_\delta, \ x \in \Omega_n, \varrho_{0,n,\delta} \to \varrho_{0,\delta} \text{ in } L^r(\Omega), \ 1 \leq r < \infty.$$

E. Feireisl, A. Novotný, *Mathematics of Open Fluid Systems*, Nečas Center Series, https://doi.org/10.1007/978-3-030-94793-4_6

- **Approximate momentum balance.**

$$\mathbf{u}_n = \mathbf{v}_n + \mathbf{u}_B,\ \mathbf{v}_n \in C^1([0,T];X_n),$$

and

$$\int_\Omega \varrho_n \mathbf{u}_n \cdot \boldsymbol{\varphi}\ \mathrm{d}x(\tau,\cdot) = \int_\Omega \mathbf{m}_0 \cdot \boldsymbol{\varphi}\ \mathrm{d}x$$
$$+ \int_0^\tau \int_\Omega \Big[\varrho_n \mathbf{u}_n \cdot \partial_t \boldsymbol{\varphi} + \varrho_n \mathbf{u}_n \otimes \mathbf{u}_n : \nabla_x \boldsymbol{\varphi} + p_\delta \mathrm{div}_x \boldsymbol{\varphi} - \mathbb{S}_\delta : \nabla_x \boldsymbol{\varphi} \Big]\ \mathrm{d}x$$
$$- \varepsilon \int_0^\tau \int_\Omega \nabla_x \varrho_n \cdot \nabla_x \mathbf{u}_n \cdot \boldsymbol{\varphi}\ \mathrm{d}x\mathrm{d}t + \int_0^\tau \int_\Omega \varrho_n \mathbf{g} \cdot \boldsymbol{\varphi}\ \mathrm{d}x\mathrm{d}t \tag{6.3}$$

for any $\boldsymbol{\varphi} \in C^1([0,T];X_n)$.

Recall that $\boldsymbol{\varphi}(t,\cdot) \in C_c^\infty(\Omega; R^d)$; therefore, Ω_n can be replaced by Ω in (6.3).

- **Approximate internal energy equation.**

$$\partial_t(\varrho e_{\delta,\mathrm{th}}(\varrho_n, \vartheta_n)) + \mathrm{div}_x \Big(\varrho e_{\delta,\mathrm{th}}(\varrho_n, \vartheta_n) \mathbf{u}_n \Big) - \Delta_x \mathcal{K}_\delta(\vartheta_n)$$
$$= \mathbb{S}_\delta : \mathbb{D}_x \mathbf{u} - p_{\mathrm{th}}(\varrho, \vartheta) \mathrm{div}_x \mathbf{u}$$
$$+ \varepsilon \left[\delta \left(\Gamma \varrho^{\Gamma - 2} + 2 \right) + \partial^2_{\varrho,\varrho} (\varrho_B e_{\mathrm{el}}(\varrho_B)) \right] |\nabla_x \varrho|^2 + \delta \frac{1}{\vartheta^2} - \varepsilon \vartheta^5$$
$$\text{a.a. in } (0,T) \times \Omega_n,$$
$$- \nabla_x \mathcal{K}_\delta(\vartheta_n) \cdot \mathbf{n} + \varrho e_{\delta,\mathrm{th}} [\mathbf{u}_B \cdot \mathbf{n}]_n^- = \Big(f_{i,B} - \varrho_B e_{\mathrm{el}}(\varrho_B) \Big) [\mathbf{u}_B \cdot \mathbf{n}]_n^-$$
$$\text{a.a. in } (0,T) \times \partial \Omega_n,$$
$$\vartheta_n(0,\cdot) = \vartheta_0 \in W^{1,2} \cap L^\infty(\Omega_n),$$
$$0 < \underline{\vartheta}_0 \le \vartheta_{0,\delta}(x) \text{ for a.a. } x \in \Omega_n, \tag{6.4}$$

where

$$\mathcal{K}_\delta(\vartheta) = \int_1^\vartheta \left(\kappa(s) + \delta \left(s^\Gamma + \frac{1}{s} \right) \right) \mathrm{d}s.$$

Recall that the approximate equation of continuity (6.2) is satisfied in the classical sense, while the approximate internal energy equation (6.4) holds in the strong sense, meaning all relevant derivatives exist a.a. in $(0,T) \times \Omega_n$.

6.1 Weak Formulation, Entropy, and Total Energy Balance

As the uniform bounds on higher order derivatives cannot be retained in the limit $n \to \infty$, we have to pass to the weak formulation of the problem.

6.1.1 Equation of Continuity

Multiplying (6.2) on a test function $\varphi \in C^1_c([0,T) \times \overline{\Omega})$, we easily deduce

$$\begin{aligned}\int_0^T \int_{\Omega_n} \Big(\varrho_n \partial_t \varphi + \varrho_n \mathbf{u}_n \cdot \nabla_x \varphi - \varepsilon \nabla_x \varrho_n \cdot \nabla_x \varphi \Big) \, dxdt \\ = \int_0^T \int_{\partial\Omega_n} \varphi \varrho_B \, [\mathbf{u}_B \cdot \mathbf{n}]_n^- \mathrm{d}\sigma_x \mathrm{d}t \\ + \int_0^T \int_{\partial\Omega_n} \varphi \varrho \, \left(\mathbf{u}_B \cdot \mathbf{n} - [\mathbf{u}_B \cdot \mathbf{n}]_n^- \right) \mathrm{d}\sigma_x \mathrm{d}t \\ - \int_{\Omega_n} \varphi_{0,n,\delta} \varphi(0, \cdot) \, dx. \end{aligned} \tag{6.5}$$

6.1.2 Kinetic Energy Balance

Multiplying the approximate equation of continuity (6.2) on $P'(\varrho_n)$, we obtain its renormalized version

$$\begin{aligned}\partial_t P(\varrho_n) + \mathrm{div}_x \left(P(\varrho_n) \mathbf{u}_n \right) + \Big(P'(\varrho_n) \varrho_n - P(\varrho_n) \Big) \mathrm{div}_x \mathbf{u}_n \\ = \varepsilon \mathrm{div}_x \left(P'(\varrho_n) \nabla_x \varrho_n \right) - \varepsilon P''(\varrho_n) |\nabla_x \varrho_n|^2 \text{ in } (0,T) \times \Omega_n, \\ \varepsilon P'(\varrho_n) \nabla_x \varrho_n \cdot \mathbf{n} = P'(\varrho_n)(\varrho_n - \varrho_B)[\mathbf{u}_B \cdot \mathbf{n}]_n^- \text{ in } (0,T) \times \partial\Omega_n. \end{aligned} \tag{6.6}$$

Consequently, we compute

$$\begin{aligned}&\int_0^T \psi(t) \int_{\Omega_n} \left[\delta \left(\varrho_n^\Gamma + \varrho_n^2 \right) + p_{\rm el}(\varrho_n) \right] \mathrm{div}_x \mathbf{u}_n \, dxdt \\ &= \int_0^T \partial_t \psi \int_{\Omega_n} \left[\delta \left(\frac{1}{\Gamma - 1} \varrho_n^\Gamma + \varrho_n^2 \right) + \varrho_n e_{\rm el}(\varrho_n) \right] \, dxdt \\ &+ \psi(0) \int_{\Omega_n} \left[\delta \left(\frac{1}{\Gamma - 1} \varrho_{0,n,\delta}^\Gamma + \varrho_{0,n,\delta}^2 \right) + \varrho_{0,n,\delta} e_{\rm el}(\varrho_{0,n,\delta}) \right] \, dxdt \\ &- \int_0^T \psi \int_{\partial\Omega_n} \left[\delta \left(\frac{1}{\Gamma - 1} \varrho_n^\Gamma + \varrho_n^2 \right) + \varrho_n e_{\rm el}(\varrho_n) \right] \, [\mathbf{u}_B \cdot \mathbf{n}]_n^- \mathrm{d}\sigma_x \mathrm{d}t \\ &- \int_0^T \psi \int_{\partial\Omega_n} \left[\delta \left(\frac{1}{\Gamma - 1} \varrho_n^\Gamma + \varrho_n^2 \right) + \varrho_n e_{\rm el}(\varrho_n) \right] \, \left(\mathbf{u}_B \cdot \mathbf{n} - [\mathbf{u}_B \cdot \mathbf{n}]_n^- \right) \mathrm{d}\sigma_x \mathrm{d}t \\ &+ \int_0^T \psi \int_{\partial\Omega_n} \left[\delta \left(\frac{\Gamma}{\Gamma - 1} \varrho_n^{\Gamma - 1} + 2\varrho_n \right) + \partial_\varrho (\varrho_n e_{\rm el}(\varrho_n)) \right] \times \\ &\quad \times (\varrho_n - \varrho_B)[\mathbf{u}_B \cdot \mathbf{n}]_n^- \mathrm{d}\sigma_x \, \mathrm{d}t \end{aligned}$$

$$-\varepsilon\int_0^T \psi\int_{\Omega_n}\left[\delta\left(\Gamma\varrho_n^{\Gamma-2}+2\right)+\partial^2_{\varrho,\varrho}(\varrho_n e_{\rm el}(\varrho_n))\right]|\nabla_x\varrho_n|^2\,{\rm d}x \tag{6.7}$$

for any $\psi \in C^1_c[0,T)$.

Next, mimicking the steps of Sect. 3.2.1, we use $\psi(t)(\mathbf{u}_n - \mathbf{u}_B) \in C^1([0,T];X_n)$ as a test function in the momentum equation (6.3) to derive a kinetic energy balance. Splitting $p = p_{\rm th} + p_{\rm el}$ and using (6.7), we obtain

$$\begin{aligned}
&\left[\int_{\Omega_n}\left[\frac{1}{2}\varrho_n|\mathbf{u}_n-\mathbf{u}_B|^2+\delta\left(\frac{1}{\Gamma-1}\varrho_n^\Gamma+\varrho_n^2\right)+\varrho_n e_{\rm el}(\varrho_n)\right]\psi\,{\rm d}x\right]_{t=0}^{t=\tau}\\
&-\int_0^\tau\partial_t\psi\int_{\Omega_n}\left[\frac{1}{2}\varrho_n|\mathbf{u}_n-\mathbf{u}_B|^2+\delta\left(\frac{1}{\Gamma-1}\varrho_n^\Gamma+\varrho_n^2\right)+\varrho_n e_{\rm el}(\varrho_n)\right]\,{\rm d}x{\rm d}t\\
&+\int_0^\tau\psi\int_{\Omega_n}\mathbb{S}_\delta(\vartheta_n,\nabla_x\mathbf{u}_n):\mathbb{D}_x\mathbf{u}\,{\rm d}x{\rm d}t\\
&+\int_0^\tau\psi\int_{\partial\Omega_n}\left[\delta\left(\frac{1}{\Gamma-1}\varrho_n^\Gamma+\varrho_n^2\right)+\varrho_n e_{\rm el}(\varrho_n)\right]\left(\mathbf{u}_B\cdot\mathbf{n}-[\mathbf{u}_B\cdot\mathbf{n}]_n^-\right){\rm d}\sigma_x{\rm d}t\\
&-\delta\int_0^\tau\psi\int_{\partial\Omega_n}\frac{1}{\Gamma-1}\left[\varrho_B^\Gamma-\Gamma\varrho_n^{\Gamma-1}(\varrho_B-\varrho_n)-\varrho_n^\Gamma\right][\mathbf{u}_B\cdot\mathbf{n}]_n^-{\rm d}\sigma_x{\rm d}t\\
&-\delta\int_0^\tau\psi\int_{\partial\Omega_n}(\varrho_n-\varrho_b)^2\,[\mathbf{u}_B\cdot\mathbf{n}]_n^-{\rm d}\sigma_x{\rm d}t\\
&-\int_0^\tau\psi\int_{\partial\Omega_n}\left[\varrho_B e_{\rm el}(\varrho_B)-\partial_\varrho(\varrho e_{\rm el}(\varrho_n))(\varrho_B-\varrho_n)-\varrho_n e_{\rm el}(\varrho_n)\right][\mathbf{u}_B\cdot\mathbf{n}]_n^-{\rm d}\sigma_x{\rm d}t\\
&+\varepsilon\int_0^\tau\psi\int_{\Omega_n}\left[\delta\left(\Gamma\varrho_n^{\Gamma-2}+2\right)+\partial^2_{\varrho,\varrho}(\varrho_n e_{\rm el}(\varrho_n))\right]|\nabla_x\varrho_n|^2\,{\rm d}x{\rm d}t\\
&=-\int_0^\tau\psi\int_{\Omega_n}\left[\varrho_n\mathbf{u}_n\otimes\mathbf{u}_n+p_\delta(\varrho_n,\vartheta_n)\mathbb{I}\right]:\mathbb{D}_x\mathbf{u}_B\,{\rm d}x{\rm d}t\\
&+\frac{1}{2}\int_0^\tau\psi\int_{\Omega_n}\varrho_n\mathbf{u}_n\cdot\nabla_x|\mathbf{u}_B|^2\,{\rm d}x{\rm d}t+\int_0^\tau\psi\int_{\Omega_n}p_{\rm th}(\varrho_n,\vartheta_n){\rm div}_x\mathbf{u}_n\,{\rm d}x{\rm d}t\\
&+\int_0^\tau\psi\int_{\Omega_n}\mathbb{S}_\delta(\vartheta_n,\nabla_x\mathbf{u}_n):\mathbb{D}_x\mathbf{u}_B\,{\rm d}x{\rm d}t+\int_0^\tau\psi\int_{\Omega_n}\varrho_n\mathbf{g}\cdot(\mathbf{u}_n-\mathbf{u}_B)\,{\rm d}x{\rm d}t\\
&-\int_0^\tau\psi\int_{\partial\Omega_n}\varrho_n e_{\rm el}(\varrho_n)\,[\mathbf{u}_B\cdot\mathbf{n}]_n^-{\rm d}\sigma_x{\rm d}t\\
&-\delta\int_0^\tau\psi\int_{\partial\Omega_n}\left(\frac{1}{\Gamma-1}\varrho_B^\Gamma+\varrho_B^2\right)[\mathbf{u}_B\cdot\mathbf{n}]_n^-{\rm d}\sigma_x{\rm d}t\\
&+\varepsilon\int_0^\tau\psi\int_{\Omega_n}\nabla_x\varrho_n\cdot\nabla_x(\mathbf{u}_n-\mathbf{u}_B)\cdot\mathbf{u}_B\,{\rm d}x{\rm d}t.
\end{aligned}\tag{6.8}$$

6.1.3 Total Energy Balance

Now, we multiply the internal energy equation (6.4) by $\psi = \psi(t)$, integrate over Ω_n, and sum with the kinetic energy equation (6.8) to obtain the total energy balance:

$$
\begin{aligned}
&\left[\int_{\Omega_n}\left[\frac{1}{2}\varrho_n|\mathbf{u}_n-\mathbf{u}_B|^2+\delta\left(\frac{1}{\Gamma-1}\varrho_n^\Gamma+\varrho_n^2\right)+\varrho_n e_\delta(\varrho_n,\vartheta_n)\right]\psi\ \mathrm{d}x\right]_{t=0}^{t=\tau}\\
&-\int_0^\tau\partial_t\psi\int_{\Omega_n}\left[\frac{1}{2}\varrho_n|\mathbf{u}_n-\mathbf{u}_B|^2+\delta\left(\frac{1}{\Gamma-1}\varrho_n^\Gamma+\varrho_n^2\right)+\varrho_n e_\delta\right]\ \mathrm{d}x\mathrm{d}t\\
&+\int_0^\tau\psi\int_{\partial\Omega_n}\varrho_n e_\delta(\varrho_n,\vartheta_n)\ \left(\mathbf{u}_B\cdot\mathbf{n}-[\mathbf{u}_B\cdot\mathbf{n}]_n^-\right)\mathrm{d}\sigma_x\mathrm{d}t\\
&+\delta\int_0^\tau\int_{\partial\Omega_n}\left(\frac{1}{\Gamma-1}\varrho_n^\Gamma+\varrho_n^2\right)\ \left(\mathbf{u}_B\cdot\mathbf{n}-[\mathbf{u}_B\cdot\mathbf{n}]_n^-\right)\mathrm{d}\sigma_x\mathrm{d}t\\
&+\int_0^\tau\psi\int_{\partial\Omega_n}f_{i,B}\ [\mathbf{u}_B\cdot\mathbf{n}]_n^-\mathrm{d}\sigma_x\mathrm{d}t\\
&-\delta\int_0^\tau\psi\int_{\partial\Omega_n}\frac{1}{\Gamma-1}\left[\varrho_B^\Gamma-\Gamma\varrho_n^{\Gamma-1}(\varrho_B-\varrho_n)-\varrho_n^\Gamma\right]\ [\mathbf{u}_B\cdot\mathbf{n}]_n^-\mathrm{d}\sigma_x\mathrm{d}t\\
&-\int_0^\tau\psi\int_{\partial\Omega_n}\Big[\varrho_B e_{\mathrm{el}}(\varrho_B)-\partial_\varrho(\varrho_n e_{\mathrm{el}}(\varrho_n))(\varrho_B-\varrho_n)-\varrho_n e_{\mathrm{el}}(\varrho_n)\Big]\times\\
&\quad\times[\mathbf{u}_B\cdot\mathbf{n}]^-\mathrm{d}\sigma_x\ \mathrm{d}t\\
&-\delta\int_0^\tau\psi\int_{\partial\Omega_n}(\varrho_n-\varrho_B)^2\ [\mathbf{u}_B\cdot\mathbf{n}]_n^-\mathrm{d}\sigma_x\mathrm{d}t\\
&=-\int_0^\tau\psi\int_{\Omega_n}\left[\varrho_n\mathbf{u}_n\otimes\mathbf{u}_n+p_\delta(\varrho_n,\vartheta_n)\mathbb{I}-\mathbb{S}_\delta(\vartheta_n,\nabla_x\mathbf{u}_n)\right]:\mathbb{D}_x\mathbf{u}_B\ \mathrm{d}x\mathrm{d}t\\
&+\frac{1}{2}\int_0^\tau\psi\int_{\Omega_n}\varrho_n\mathbf{u}_n\cdot\nabla_x|\mathbf{u}_B|^2\ \mathrm{d}x\mathrm{d}t\\
&+\int_0^\tau\psi\int_{\Omega_n}\varrho_n(\mathbf{g}-\partial_t\mathbf{u}_B)\cdot(\mathbf{u}_n-\mathbf{u}_B)\ \mathrm{d}x\mathrm{d}t\\
&+\int_0^\tau\psi\int_{\Omega_n}\left(\delta\frac{1}{\vartheta_n^2}-\varepsilon\vartheta_n^5\right)\ \mathrm{d}x\mathrm{d}t\\
&-\delta\int_0^\tau\psi\int_{\partial\Omega_n}\left(\frac{1}{\Gamma-1}\varrho_B^\Gamma+\varrho_B^2\right)\ [\mathbf{u}_B\cdot\mathbf{n}]_n^-\mathrm{d}\sigma_x\mathrm{d}t\\
&+\varepsilon\int_0^\tau\psi\int_{\Omega_n}\nabla_x\varrho_n\cdot\nabla_x(\mathbf{u}_n-\mathbf{u}_B)\cdot\mathbf{u}_B\ \mathrm{d}x\mathrm{d}t
\end{aligned}
\tag{6.9}
$$

for any $\psi\in C_c^1[0,T)$.

6.1.4 Entropy Inequality

At this stage, the internal energy equation (6.4) holds a.a., and the approximate temperatures ϑ_n are bounded below away from zero. Consequently, the entropy balance can be derived in a standard way dividing (6.4) on ϑ. Using the approximate equation of continuity (6.2), we get

$$\varrho_n \frac{1}{\vartheta_n} \partial_t e_{\delta,\text{th}} + \frac{1}{\vartheta_n} \varrho_n \mathbf{u}_n \cdot \nabla_x e_{\delta,\text{th}} + \varepsilon e_\delta \frac{1}{\vartheta_n} \Delta_x \varrho_n + \text{div}_x \left(\frac{\mathbf{q}_\delta(\vartheta_n, \nabla_x \vartheta_n)}{\vartheta_n} \right)$$
$$= \frac{1}{\vartheta_n} \left(\mathbb{S}_\delta : \mathbb{D}_x \mathbf{u}_n - \frac{\mathbf{q}_\delta \cdot \nabla_x \vartheta_n}{\vartheta_n} + \frac{\delta}{\vartheta_n^2} \right) - \frac{p_{\text{th}}(\varrho_n, \vartheta_n)}{\vartheta_n} \text{div}_x \mathbf{u}_n$$
$$+ \varepsilon \frac{1}{\vartheta_n} \left[\delta \left(\Gamma \varrho_n^{\Gamma - 2} + 2 \right) + \partial^2_{\varrho,\varrho} (\varrho_n e_{\text{el}}(\varrho_n)) \right] |\nabla_x \varrho_n|^2 - \varepsilon \vartheta_n^4.$$

Denoting

$$s_\delta(\varrho, \vartheta) = s(\varrho, \vartheta) + \delta \log(\vartheta),$$

we may use Gibbs' relation (1.16) to deduce the entropy balance equation

$$\partial_t(\varrho_n s_\delta(\varrho_n, \vartheta_n)) + \text{div}_x(\varrho_n s_\delta(\varrho_n, \vartheta_n)\mathbf{u}_n) + \text{div}_x \left(\frac{\mathbf{q}_\delta(\vartheta_n, \nabla_x \vartheta_n)}{\vartheta_n} \right)$$
$$= \frac{1}{\vartheta_n} \left(\mathbb{S}_\delta : \mathbb{D}_x \mathbf{u}_n - \frac{\mathbf{q}_\delta \cdot \nabla_x \vartheta_n}{\vartheta_n} + \frac{\delta}{\vartheta_n^2} \right)$$
$$+ \varepsilon \frac{1}{\vartheta_n} \left[\delta \left(\Gamma \varrho_n^{\Gamma - 2} + 2 \right) + \partial^2_{\varrho,\varrho} (\varrho_n e_{\text{el}}(\varrho_n)) \right] |\nabla_x \varrho_n|^2 - \varepsilon \vartheta_n^4$$
$$- \varepsilon \Delta_x \varrho_n \left(\frac{e_{\delta,\text{th}}}{\vartheta_n} - s_\delta + \frac{p_{\text{th}}}{\varrho_n \vartheta_n} \right).$$

To get a weak formulation, multiply the equation by $\varphi \in C^1_c([0,T) \times \overline{\Omega}_n)$ and integrate by parts:

$$- \int_{\Omega_n} \varrho_{0,n,\delta} s_\delta(\varrho_{0,n,\delta}, \vartheta_{0,\delta}) \varphi(0, \cdot) \ \mathrm{d}x - \int_0^T \int_{\Omega_n} \varrho_n s_\delta \partial_t \varphi \ \mathrm{d}x\mathrm{d}t$$
$$- \int_0^T \int_{\Omega_n} \varrho_n s_\delta \mathbf{u}_n \cdot \nabla_x \varphi \ \mathrm{d}x\mathrm{d}t + \int_0^T \int_{\partial \Omega_n} \varphi \varrho_n s_\delta \mathbf{u}_B \cdot \mathbf{n} \ \mathrm{d}\sigma_x \mathrm{d}t$$
$$- \int_0^T \int_{\Omega_n} \frac{\mathbf{q}_\delta(\vartheta_n, \nabla_x \vartheta_n)}{\vartheta_n} \cdot \nabla_x \varphi \ \mathrm{d}x\mathrm{d}t + \int_0^T \int_{\partial \Omega_n} \varphi \left(\frac{\mathbf{q}_\delta}{\vartheta_n} \right) \cdot \mathbf{n} \mathrm{d}\sigma_x \mathrm{d}t$$
$$= \int_0^T \int_{\Omega_n} \frac{\varphi}{\vartheta_n} \left(\mathbb{S}_\delta : \nabla_x \mathbf{u}_n - \frac{\mathbf{q}_\delta \cdot \nabla_x \vartheta_n}{\vartheta_n} + \frac{\delta}{\vartheta_n^2} \right) \ \mathrm{d}x\mathrm{d}t$$
$$+ \varepsilon \int_0^T \int_{\Omega_n} \frac{\varphi}{\vartheta_n} \left[\delta \left(\Gamma \varrho_n^{\Gamma - 2} + 2 \right) + \partial^2_{\varrho,\varrho} (\varrho_n e_{\text{el}}(\varrho_n)) \right] |\nabla_x \varrho_n|^2 \ \mathrm{d}x\mathrm{d}t$$

$$
\begin{aligned}
&- \varepsilon \int_0^T \int_{\Omega_n} \varphi \vartheta_n^4 \, \mathrm{d}x\mathrm{d}t \\
&+ \varepsilon \int_0^T \int_{\Omega_n} \nabla_x \varrho_n \cdot \nabla_x \left[\varphi \left(\frac{e_{\delta,\mathrm{th}}}{\vartheta_n} - s_\delta + \frac{p_{\mathrm{th}}}{\varrho_n \vartheta_n} \right) \right] \, \mathrm{d}x\mathrm{d}t \\
&- \varepsilon \int_0^T \int_{\partial\Omega_n} \varphi \nabla_x \varrho_n \cdot \mathbf{n} \left(\frac{e_{\delta,\mathrm{th}}}{\vartheta_n} - s_\delta + \frac{p_{\mathrm{th}}}{\varrho_n \vartheta_n} \right) \mathrm{d}\sigma_x \mathrm{d}t.
\end{aligned} \tag{6.10}
$$

In addition, we express the boundary integrals using the boundary conditions,

$$
\begin{aligned}
&\int_0^T \int_{\partial\Omega_n} \varphi \frac{1}{\vartheta_n} \mathbf{q}_\delta \cdot \mathbf{n} \, \mathrm{d}\sigma_x \mathrm{d}t \\
&\quad = \int_0^T \int_{\partial\Omega_n} \varphi \frac{1}{\vartheta_n} \left(f_{i,B} - \varrho_B e_{\mathrm{el}}(\varrho_B) - \varrho_n e_{\delta,\mathrm{th}} \right) \, [\mathbf{u}_B \cdot \mathbf{n}]_n^- \mathrm{d}\sigma_x \mathrm{d}t,
\end{aligned}
$$

and

$$
\begin{aligned}
&-\varepsilon \int_0^T \int_{\partial\Omega_n} \varphi \nabla_x \varrho_n \cdot \mathbf{n} \left(\frac{e_{\delta,\mathrm{th}}}{\vartheta_n} - s_\delta + \frac{p_{\mathrm{th}}}{\varrho_n \vartheta_n} \right) \mathrm{d}\sigma_x \mathrm{d}t \\
&\quad = - \int_0^T \int_{\partial\Omega_n} \varphi (\varrho_n - \varrho_B) \left(\frac{e_{\delta,\mathrm{th}}}{\vartheta_n} - s_\delta + \frac{p_{\mathrm{th}}}{\varrho_n \vartheta_n} \right) \, [\mathbf{u}_B \cdot \mathbf{n}]_n^- \mathrm{d}\sigma_x \mathrm{d}t.
\end{aligned}
$$

Regrouping all surface integrals on the right-hand side of (6.10), we get

$$
\begin{aligned}
&- \int_0^T \int_{\partial\Omega_n} \varphi \varrho_n s_\delta \mathbf{u}_B . \cdot \mathbf{n} \mathrm{d}\sigma_x \mathrm{d}t \\
&\quad - \int_0^T \int_{\partial\Omega_n} \varphi \frac{1}{\vartheta_n} \left(f_{i,B} - \varrho_B e_{\mathrm{el}}(\varrho_B) - \varrho_n e_{\delta,\mathrm{th}} \right) \, [\mathbf{u}_B \cdot \mathbf{n}]_n^- \mathrm{d}\sigma_x \mathrm{d}t \\
&\quad - \int_0^T \int_{\partial\Omega_n} \varphi (\varrho_n - \varrho_B) \left(\frac{e_{\delta,\mathrm{th}}}{\vartheta_n} - s_\delta + \frac{p_{\mathrm{th}}}{\varrho_n \vartheta_n} \right) \, [\mathbf{u}_B \cdot \mathbf{n}]_n^- \mathrm{d}\sigma_x \mathrm{d}t \\
&\quad = - \int_0^T \int_{\partial\Omega_n} \varphi \frac{1}{\vartheta_n} f_{i,B} \, [\mathbf{u}_B \cdot \mathbf{n}]_n^- \mathrm{d}\sigma_x \mathrm{d}t \\
&\quad - \int_0^T \int_{\partial\Omega_n} \varphi \varrho_n s_\delta \, \left(\mathbf{u}_B \cdot \mathbf{n} - [\mathbf{u}_B \cdot \mathbf{n}]_n^- \right) \mathrm{d}\sigma_x \mathrm{d}t \\
&\quad + \int_0^T \int_{\partial\Omega_n} \varphi \left[\varrho_B \left(\frac{e_{\delta,\mathrm{th}}}{\vartheta_n} - s_\delta + \frac{p_{\mathrm{th}}}{\varrho_n \vartheta_n} \right) - \frac{p_{\mathrm{th}}}{\vartheta_n} \right] \, [\mathbf{u}_B \cdot \mathbf{n}]_n^- \mathrm{d}\sigma_x \mathrm{d}t \\
&\quad + \int_0^T \int_{\partial\Omega_n} \varphi \frac{1}{\vartheta_n} \varrho_B e_{\mathrm{el}}(\varrho_B) \, [\mathbf{u}_B \cdot \mathbf{n}]_n^- \mathrm{d}\sigma_x \mathrm{d}t.
\end{aligned} \tag{6.11}
$$

Now, as e_{th}, p_{th}, and s satisfy the hypothesis of thermodynamic stability, we may use convexity of the function

$$E_{\rm int,th}(\varrho, S) = \varrho e_{\rm th}(\varrho, S)$$

obtaining

$$\begin{aligned}
\varrho_B e_{\rm th}(\varrho_B, S_B) &\geq \varrho e_{\rm th}(\varrho, S) + \left(e_{\rm th} - \vartheta s + \frac{p_{\rm th}}{\varrho} \right) (\varrho_B - \varrho) + \vartheta (S_B - S) \\
&= \left(e_{\rm th} - \vartheta s + \frac{p_{\rm th}}{\varrho} \right) \varrho_B + \varrho e_{\rm th} - \left(e_{\rm th} - \vartheta s + \frac{p_{\rm th}}{\varrho} \right) \varrho + \vartheta S_B - \varrho \vartheta s \\
&= \left(e_{\rm th} - \vartheta s + \frac{p_{\rm th}}{\varrho} \right) \varrho_B - p_{\rm th} + \vartheta S_B,
\end{aligned}$$

where we have set $S_B = \varrho_B s(\varrho_B, \vartheta)$. Consequently,

$$\frac{1}{\vartheta} \left(e_{\rm th} - \vartheta s + \frac{p_{\rm th}}{\varrho} \right) \varrho_B - \frac{1}{\vartheta} p_{\rm th} \leq \frac{1}{\vartheta} \varrho_B e_{\rm th}(\varrho_B, \vartheta) - \varrho_B s(\varrho_B, \vartheta).$$

Thus if $\varphi \geq 0$, the last two integrals in (6.11) can be estimated as

$$\begin{aligned}
&\int_0^T \int_{\partial \Omega_n} \varphi \left[\varrho_B \left(\frac{e_{\delta,\rm th}}{\vartheta_n} - s_\delta + \frac{p_{\rm th}}{\varrho_n \vartheta_n} \right) - \frac{p_{\rm th}}{\vartheta_n} \right] [\mathbf{u}_B \cdot \mathbf{n}]_n^- {\rm d}\sigma_x {\rm d}t \\
&+ \int_0^T \int_{\partial \Omega_n} \varphi \frac{1}{\vartheta_n} \varrho_B e_{\rm el}(\varrho_B) \; [\mathbf{u}_B \cdot \mathbf{n}]_n^- {\rm d}\sigma_x {\rm d}t \\
&\geq \delta \int_0^T \int_{\partial \Omega_n} \varphi \varrho_B (1 - \log(\vartheta_n)) \; [\mathbf{u}_B \cdot \mathbf{n}]_n^- {\rm d}\sigma_x \\
&+ \int_0^T \int_{\partial \Omega_n} \varphi \frac{1}{\vartheta_n} \varrho_B e_{\rm th}(\varrho_B, \vartheta) - \varrho_B s(\varrho_B, \vartheta) \; [\mathbf{u}_B \cdot \mathbf{n}]_n^- {\rm d}\sigma_x {\rm d}t \\
&+ \int_0^T \int_{\partial \Omega_n} \varphi \frac{1}{\vartheta_n} \varrho_B e_{\rm el}(\varrho_B) \; [\mathbf{u}_B \cdot \mathbf{n}]_n^- {\rm d}\sigma_x {\rm d}t \\
&= \delta \int_0^T \int_{\partial \Omega_n} \varphi \varrho_B (1 - \log(\vartheta_n)) \; [\mathbf{u}_B \cdot \mathbf{n}]_n^- {\rm d}\sigma_x \\
&+ \int_0^T \int_{\partial \Omega_n} \varphi \frac{1}{\vartheta_n} \varrho_B e(\varrho_B, \vartheta_n) - \varrho_B s(\varrho_B, \vartheta_n) \; [\mathbf{u}_B \cdot \mathbf{n}]_n^- {\rm d}\sigma_x {\rm d}t
\end{aligned}$$

Summarizing the previous discussion, we may rewrite (6.10) as an *approximate entropy inequality*,

$$\begin{aligned}
&\left[\int_{\Omega_n} \varrho_n s_\delta(\varrho_n, \vartheta_n) \varphi \; {\rm d}x \right]_{t=0}^{t=\tau} \\
&- \int_0^\tau \int_{\Omega_n} \varrho_n s_\delta \partial_t \varphi \; {\rm d}x{\rm d}t - \int_0^\tau \int_{\Omega_n} \varrho_n s_\delta \mathbf{u}_n \cdot \nabla_x \varphi \; {\rm d}x{\rm d}t \\
&+ \int_0^\tau \int_{\partial \Omega_n} \varphi \varrho_n s_\delta \; \left(\mathbf{u}_B \cdot \mathbf{n} - [\mathbf{u}_B \cdot \mathbf{n}]_n^- \right) {\rm d}\sigma_x {\rm d}t
\end{aligned}$$

$$
\begin{aligned}
&- \int_0^\tau \int_{\Omega_n} \frac{\mathbf{q}_\delta(\vartheta_n, \nabla_x \vartheta_n)}{\vartheta_n} \cdot \nabla_x \varphi \, \mathrm{d}x\mathrm{d}t \\
&\geq \int_0^\tau \int_{\Omega_n} \frac{\varphi}{\vartheta_n} \left(\mathbb{S}_\delta : \nabla_x \mathbf{u}_n - \frac{\mathbf{q}_\delta \cdot \nabla_x \vartheta_n}{\vartheta_n} + \frac{\delta}{\vartheta_n^2} \right) \mathrm{d}x\mathrm{d}t \\
&+ \varepsilon \int_0^\tau \int_{\Omega_n} \frac{\varphi}{\vartheta_n} \Big[\delta \left(\Gamma \varrho_n^{\Gamma - 2} + 2 \right) + \partial^2_{\varrho,\varrho} (\varrho_n e_{\rm el}(\varrho_n)) \Big] |\nabla_x \varrho_n|^2 \, \mathrm{d}x\mathrm{d}t \\
&- \varepsilon \int_0^\tau \int_{\Omega_n} \varphi \vartheta_n^4 \, \mathrm{d}x\mathrm{d}t - \int_0^\tau \int_{\partial\Omega_n} \varphi \frac{1}{\vartheta_n} f_{i,B} \, [\mathbf{u}_B \cdot \mathbf{n}]_n^- \mathrm{d}\sigma_x \mathrm{d}t \\
&+ \varepsilon \int_0^\tau \int_{\Omega_n} \nabla_x \varrho_n \cdot \nabla_x \left[\varphi \left(\frac{e_{\delta,\rm th}}{\vartheta_n} - s_\delta + \frac{p_{\rm th}}{\varrho_n \vartheta_n} \right) \right] \mathrm{d}x\mathrm{d}t \\
&+ \delta \int_0^\tau \int_{\partial\Omega_n} \varphi \varrho_B (1 - \log(\vartheta_n)) \, [\mathbf{u}_B \cdot \mathbf{n}]_n^- \mathrm{d}\sigma_x \\
&+ \int_0^\tau \int_{\partial\Omega_n} \varphi \left(\frac{1}{\vartheta_n} \varrho_B e(\varrho_B, \vartheta_n) - \varrho_B s(\varrho_B, \vartheta_n) \right) [\mathbf{u}_B \cdot \mathbf{n}]_n^- \mathrm{d}\sigma_x \mathrm{d}t \qquad (6.12)
\end{aligned}
$$

for a.a. $0 < \tau < T$ and any $\varphi \in C^1_c([0,T) \times \overline{\Omega}_n)$, $\varphi \geq 0$.

6.2 Uniform Bounds

Our goal is to derive uniform bounds independent of the Galerkin parameter $n \to \infty$. At this stage the bounds may still depend on the remaining approximation parameters ε and δ.

First, we reorganize several terms in the approximate entropy inequality (6.4). Using Gibbs' relation, we easily compute

$$
\begin{aligned}
&\nabla_x \varrho_n \cdot \nabla_x \left[\varphi \left(\frac{e_{\delta,\rm th}}{\vartheta_n} - s_\delta + \frac{p_{\rm th}}{\varrho_n \vartheta_n} \right) \right] \\
&\quad = \nabla_x \varrho_n \cdot \nabla_x \varphi \left(\frac{e_{\delta,\rm th}}{\vartheta_n} - s_\delta + \frac{p_{\rm th}}{\varrho_n \vartheta_n} \right) \\
&\quad + \varphi |\nabla_x \varrho_n|^2 \frac{1}{\vartheta_n \varrho_n} \frac{\partial p_{\rm th}}{\partial \varrho} + \varphi \nabla_x \varrho_n \cdot \nabla_x \vartheta_n \left[\frac{1}{\varrho_n \vartheta_n} \frac{\partial p_{\rm th}}{\partial \vartheta} - \frac{1}{\vartheta_n^2} \left(e_{\delta,\rm th} + \frac{p_{\rm th}}{\varrho_n} \right) \right] \\
&\quad = \nabla_x \varrho_n \cdot \nabla_x \varphi \left(\frac{e_{\delta,\rm th}}{\vartheta_n} - s_\delta + \frac{p_{\rm th}}{\varrho_n \vartheta_n} \right) \\
&\quad + \varphi |\nabla_x \varrho_n|^2 \frac{1}{\vartheta_n \varrho_n} \frac{\partial p(\varrho_n, \vartheta_n)}{\partial \varrho} - \varphi |\nabla_x \varrho_n|^2 \frac{1}{\vartheta_n} \partial^2_{\varrho,\varrho} (\varrho_n e_{\rm el}(\varrho_n)) \\
&\quad + \varphi \nabla_x \varrho_n \cdot \nabla_x \vartheta_n \left[\frac{1}{\varrho_n \vartheta_n} \frac{\partial p_{\rm th}}{\partial \vartheta} - \frac{1}{\vartheta_n^2} \left(e_{\delta,\rm th} + \frac{p_{\rm th}}{\varrho_n} \right) \right],
\end{aligned}
$$

where we have used the relation

$$\partial^2_{\varrho,\varrho}(\varrho e_{\mathrm{el}}(\varrho)) = \frac{1}{\varrho}\partial_\varrho p_{\mathrm{el}}(\varrho).$$

Consequently, revisiting the entropy inequality (6.4) and using monotonicity of the pressure with respect to ϱ, we get

$$\begin{aligned}
&\varepsilon \int_0^\tau \int_{\Omega_n} \frac{\varphi}{\vartheta_n} \left[\delta\left(\Gamma\varrho_n^{\Gamma-2} + 2\right) + \partial^2_{\varrho,\varrho}(\varrho_n e_{\mathrm{el}}(\varrho_n))\right] |\nabla_x \varrho_n|^2\Big) \, \mathrm{d}x\mathrm{d}t \\
&+ \varepsilon \int_0^\tau \int_{\Omega_n} \nabla_x \varrho_n \cdot \nabla_x \left[\varphi\left(\frac{e_{\delta,\mathrm{th}}}{\vartheta_n} - s_\delta + \frac{p_{\mathrm{th}}}{\varrho_n\vartheta_n}\right)\right] \, \mathrm{d}x\mathrm{d}t \\
&\geq \varepsilon \int_0^\tau \int_{\Omega_n} \frac{\varphi}{\vartheta_n}\delta\left(\Gamma\varrho_n^{\Gamma-2} + 2\right)|\nabla_x\varrho_n|^2\Big) \, \mathrm{d}x\mathrm{d}t \\
&+ \varepsilon \int_0^\tau \int_{\Omega_n} \nabla_x \varrho_n \cdot \nabla_x \varphi \left(\frac{e_{\delta,\mathrm{th}}}{\vartheta_n} - s_\delta + \frac{p_{\mathrm{th}}}{\varrho_n\vartheta_n}\right) \, \mathrm{d}x\mathrm{d}t \\
&+ \varepsilon \int_0^\tau \int_{\Omega_n} \varphi \nabla_x \varrho_n \cdot \nabla_x \vartheta_n \left[\frac{1}{\varrho_n\vartheta_n}\frac{\partial p_{\mathrm{th}}}{\partial\vartheta} - \frac{1}{\vartheta_n^2}\left(e_{\delta,\mathrm{th}} + \frac{p_{\mathrm{th}}}{\varrho_n}\right)\right] \, \mathrm{d}x\mathrm{d}t
\end{aligned}$$

whenever $\varphi \geq 0$. Moreover,

$$\begin{aligned}
&\left|\frac{1}{\varrho_n\vartheta_n}\frac{\partial p_{\mathrm{th}}}{\partial\vartheta} - \frac{1}{\vartheta_n^2}\left(e_{\delta,\mathrm{th}} + \frac{p_{\mathrm{th}}}{\varrho_n}\right)\right| \\
&= \left|\frac{1}{\varrho_n\vartheta_n}\frac{\partial p_{\mathrm{m}}(\varrho_n\vartheta_n)}{\partial\vartheta} - \frac{1}{\vartheta_n^2}\left(e_{\mathrm{m}}(\varrho_n\vartheta_n) + \delta\vartheta_n + \frac{p_{\mathrm{m}}(\varrho_n,\vartheta_n)}{\varrho_n}\right)\right| \lesssim \frac{1}{\vartheta_n},
\end{aligned}$$

where we have used hypothesis (4.14) to control $\frac{\partial p_{\mathrm{m}}}{\partial\vartheta}$. Now, we can estimate

$$\varepsilon\frac{1}{\vartheta_n}|\nabla_x\varrho_n \cdot \nabla_x\vartheta_n| \leq \frac{1}{2}\frac{\varepsilon}{\delta}\frac{|\nabla_x\vartheta_n|^2}{\vartheta_n} + \frac{1}{2}\varepsilon\delta\frac{|\nabla_x\varrho_n|^2}{\vartheta_n},$$

and, similarly,

$$\varepsilon\varrho_n^{\frac{2}{3}}\frac{1}{\vartheta_n^2}|\nabla_x\varrho_n \cdot \nabla_x\vartheta_n| \leq \frac{1}{2}\frac{\varepsilon}{\delta}\frac{|\nabla_x\vartheta_n|^2}{\vartheta_n^3} + \frac{1}{2}\varepsilon\delta\varrho_n^{\frac{4}{3}}\frac{|\nabla_x\varrho_n|^2}{\vartheta_n}.$$

Thus if $0 < \varepsilon \leq \delta^2$, we get

$$\begin{aligned}
&\varepsilon\left|\int_0^\tau \int_{\Omega_n} \varphi \nabla_x \varrho_n \cdot \nabla_x \vartheta_n \left[\frac{1}{\varrho_n\vartheta_n}\frac{\partial p_{\mathrm{th}}}{\partial\vartheta} - \frac{1}{\vartheta_n^2}\left(e_{\delta,\mathrm{th}} + \frac{p_{\mathrm{th}}}{\varrho_n}\right)\right] \, \mathrm{d}x\mathrm{d}t\right| \\
&\qquad \leq \frac{1}{2}\varepsilon \int_0^\tau \int_{\Omega_n} \frac{\varphi}{\vartheta_n}\delta\left(\Gamma\varrho_n^{\Gamma-2} + 2\right)|\nabla_x\varrho_n|^2\Big) \, \mathrm{d}x\mathrm{d}t \\
&\qquad + \frac{\delta}{2}\int_0^\tau \int_{\Omega_n} \frac{\varphi}{\vartheta_n^2}\left(\vartheta_n^\Gamma + \frac{1}{\vartheta_n}\right)|\nabla_x\vartheta_n|^2 \, \mathrm{d}x\mathrm{d}t
\end{aligned}$$

provided $\varphi \geq 0$.

Finally, we observe

$$\begin{aligned}\frac{e_{\delta,\mathrm{th}}}{\vartheta_n} - s_\delta + \frac{p_{\mathrm{th}}}{\varrho_n \vartheta_n} \\ = \frac{e_{\mathrm{m}}(\varrho_n, \vartheta_n)}{\vartheta_n} + \delta - s_{\mathrm{m}}(\varrho_n, \vartheta_n) - \delta \log(\vartheta_n) + \frac{p_{\mathrm{m}}(\varrho_n, \vartheta_n)}{\varrho_n \vartheta_n}.\end{aligned}$$

Thus the approximate entropy inequality (6.4) reduces to

$$\begin{aligned}
&\left[\int_{\Omega_n} \varrho_n s_\delta \varphi \, \mathrm{d}x\right]_{t=0}^{t=\tau} \\
&- \int_0^\tau \int_{\Omega_n} \varrho_n s_\delta \partial_t \varphi \, \mathrm{d}x\mathrm{d}t - \int_0^\tau \int_{\Omega_n} \varrho_n s_\delta \mathbf{u}_n \cdot \nabla_x \varphi \, \mathrm{d}x\mathrm{d}t \\
&+ \int_0^\tau \int_{\partial\Omega_n} \varphi \varrho_n s_\delta \left(\mathbf{u}_B \cdot \mathbf{n} - [\mathbf{u}_B \cdot \mathbf{n}]_n^-\right) \mathrm{d}\sigma_x \mathrm{d}t \\
&- \int_0^\tau \int_{\Omega_n} \frac{\mathbf{q}_\delta(\vartheta_n, \nabla_x \vartheta_n)}{\vartheta_n} \cdot \nabla_x \varphi \, \mathrm{d}x\mathrm{d}t \\
&\geq \int_0^\tau \int_{\Omega_n} \frac{\varphi}{\vartheta_n} \left(\mathbb{S}_\delta : \nabla_x \mathbf{u}_n - \frac{\mathbf{q} \cdot \nabla_x \vartheta_n}{\vartheta_n} + \frac{\delta}{\vartheta_n^2}\right) \mathrm{d}x\mathrm{d}t \\
&+ \frac{\delta}{2} \int_0^\tau \int_{\Omega_n} \frac{\varphi}{\vartheta_n^2} \left(\vartheta_n^\Gamma + \frac{1}{\vartheta_n}\right) |\nabla_x \vartheta_n|^2 \, \mathrm{d}x\mathrm{d}t \\
&+ \frac{\varepsilon}{2} \int_0^\tau \int_{\Omega_n} \frac{\varphi}{\vartheta_n} \delta \left(\Gamma \varrho_n^{\Gamma-2} + 2\right) |\nabla_x \varrho_n|^2 - \vartheta_n^4 \Big) \, \mathrm{d}x\mathrm{d}t \\
&+ \varepsilon \int_0^\tau \int_{\Omega_n} \nabla_x \varrho_n \cdot \nabla_x \varphi \left(\frac{e_{\mathrm{m}}(\varrho_n, \vartheta_n)}{\vartheta_n} - s_{\mathrm{m}}(\varrho_n, \vartheta_n) + \frac{p_{\mathrm{m}}(\varrho_n, \vartheta_n)}{\varrho_n \vartheta_n}\right) \mathrm{d}x\mathrm{d}t \\
&+ \varepsilon \int_0^\tau \int_{\Omega_n} \nabla_x \varrho_n \cdot \nabla_x \varphi \left(\delta - \delta \log(\vartheta_n)\right) \mathrm{d}x\mathrm{d}t \\
&+ \int_0^\tau \int_{\partial\Omega_n} \varphi \left[\delta \varrho_B (1 - \log(\vartheta_n)) - \frac{1}{\vartheta_n} f_{i,b}\right] [\mathbf{u}_B \cdot \mathbf{n}]_n^- \mathrm{d}\sigma_x \mathrm{d}t \\
&+ \int_0^\tau \int_{\partial\Omega_n} \varphi \varrho_B \left(\frac{e(\varrho_B, \vartheta_n)}{\vartheta_n} - s(\varrho_B, \vartheta_n)\right) [\mathbf{u}_B \cdot \mathbf{n}]_n^- \mathrm{d}\sigma_x \mathrm{d}t
\end{aligned} \tag{6.13}$$

for any $\varphi \in C_c^1([0, T) \times \overline{\Omega}_n)$, $\varphi \geq 0$.

Similarly, the energy balance (6.9) can be converted into inequality neglecting several non-negative integrals on the left-hand side:

$$\begin{aligned}
&\left[\int_{\Omega_n} \left[\frac{1}{2} \varrho_n |\mathbf{u}_n - \mathbf{u}_B|^2 + \delta \left(\frac{1}{\Gamma - 1} \varrho_n^\Gamma + \varrho_n^2\right) + \varrho_n e_\delta\right] \psi \, \mathrm{d}x\right]_{t=0}^{t=\tau} \\
&- \int_0^\tau \partial_t \psi \int_{\Omega_n} \left[\frac{1}{2} \varrho_n |\mathbf{u}_n - \mathbf{u}_B|^2 + \delta \left(\frac{1}{\Gamma - 1} \varrho_n^\Gamma + \varrho_n^2\right) + \varrho_n e_\delta\right] \mathrm{d}x\mathrm{d}t
\end{aligned}$$

$$
\begin{aligned}
&+\int_0^\tau \psi \int_{\partial\Omega_n} \varrho_n e_\delta \ \left(\mathbf{u}_B\cdot\mathbf{n} - [\mathbf{u}_B\cdot\mathbf{n}]_n^-\right) \mathrm{d}\sigma_x \mathrm{d}t \\
&+\delta \int_0^\tau \int_{\partial\Omega_n} \left(\frac{1}{\Gamma-1}\varrho_n^\Gamma + \varrho_n^2\right) \ \left(\mathbf{u}_B\cdot\mathbf{n} - [\mathbf{u}_B\cdot\mathbf{n}]_n^-\right) \mathrm{d}\sigma_x \mathrm{d}t \\
&+\int_0^\tau \psi \int_{\partial\Omega_n} f_{i,B} \ [\mathbf{u}_B\cdot\mathbf{n}]_n^- \mathrm{d}\sigma_x \mathrm{d}t \\
&\leq -\int_0^\tau \psi \int_{\Omega_n} \left[\varrho_n(\mathbf{u}_n - \mathbf{u}_B)\otimes(\mathbf{u}_n - \mathbf{u}_B) + p_\delta \mathbb{I}\right] : \nabla_x \mathbf{u}_B \ \mathrm{d}x\mathrm{d}t \\
&+\frac{1}{2}\int_0^\tau \psi \int_{\Omega_n} \varrho_n \mathbf{u}_n \cdot \nabla_x |\mathbf{u}_B|^2 \ \mathrm{d}x\mathrm{d}t \\
&+\int_0^\tau \psi \int_{\Omega_n} \mathbb{S}_\delta(\vartheta_n, \nabla_x \mathbf{u}_n) : \nabla_x \mathbf{u}_B \ \mathrm{d}x\mathrm{d}t \\
&+\int_0^\tau \psi \int_{\Omega_n} \varrho_n(\mathbf{g} - \partial_t \mathbf{u}_B + \mathbf{u}_B\cdot\nabla_x \mathbf{u}_B)\cdot(\mathbf{u}_n - \mathbf{u}_B) \ \mathrm{d}x\mathrm{d}t \\
&+\int_0^\tau \psi \int_{\Omega_n} \left(\delta \frac{1}{\vartheta_n^2} - \varepsilon \vartheta_n^5\right) \ \mathrm{d}x\mathrm{d}t \\
&-\delta \int_0^\tau \psi \int_{\partial\Omega_n} \left(\frac{1}{\Gamma-1}\varrho_B^\Gamma + \varrho_B^2\right) \ [\mathbf{u}_B\cdot\mathbf{n}]_n^- \mathrm{d}\sigma_x \mathrm{d}t \\
&+\varepsilon \int_0^\tau \psi \int_{\Omega_n} \nabla_x \varrho_n \cdot \nabla_x(\mathbf{u}_n - \mathbf{u}_B)\cdot \mathbf{u}_B \ \mathrm{d}x\mathrm{d}t
\end{aligned} \tag{6.14}
$$

for any $\psi \in C^1[0,T)$, $\psi \geq 0$.

Now, subtracting the entropy balance (6.13) from (6.14), we obtain

$$
\begin{aligned}
&\int_{\Omega_n} \left[\frac{1}{2}\varrho_n|\mathbf{u}_n - \mathbf{u}_B|^2 + \varrho_n e_\delta - \varrho_n s_\delta + \delta\left(\frac{1}{\Gamma-1}\varrho_n^\Gamma + \varrho_n^2\right)\right](\tau,\cdot) \ \mathrm{d}x \\
&+\int_0^\tau \int_{\partial\Omega_n} (\varrho_n e_\delta - \varrho_n s_\delta) \ \left(\mathbf{u}_B\cdot\mathbf{n} - [\mathbf{u}_B\cdot\mathbf{n}]_n^-\right) \mathrm{d}\sigma_x \mathrm{d}t \\
&+\int_0^\tau \int_{\partial\Omega_n} \left(1 - \frac{1}{\vartheta_n}\right) f_{i,B} \ [\mathbf{u}_B\cdot\mathbf{n}]_n^- \mathrm{d}\sigma_x \mathrm{d}t \\
&+\int_0^\tau \int_{\Omega_n} \frac{1}{\vartheta_n}\left(\mathbb{S}_\delta(\vartheta_n, \nabla_x \mathbf{u}_n) : \nabla_x \mathbf{u}_n - \frac{\mathbf{q}\cdot\nabla_x\vartheta_n}{\vartheta_n}\right) \ \mathrm{d}x\mathrm{d}t \\
&+\frac{\delta}{2}\int_0^\tau \int_{\Omega_n} \frac{1}{\vartheta_n^2}\left(\vartheta_n^\Gamma + \frac{1}{\vartheta_n}\right)|\nabla_x \vartheta_n|^2 \ \mathrm{d}x\mathrm{d}t \\
&+\frac{\varepsilon\delta}{2}\int_0^\tau \int_{\Omega_n} \frac{1}{\vartheta_n}\left(\Gamma \varrho_n^{\Gamma-2} + 2\right)|\nabla_x \varrho_n|^2 - \vartheta_n^4\right) \ \mathrm{d}x\mathrm{d}t \\
&+\delta \int_0^\tau \int_{\Omega_n} \left(\frac{1}{\vartheta_n^3} - \frac{1}{\vartheta_n^2}\right) \ \mathrm{d}x\mathrm{d}t + \varepsilon \int_0^\tau \int_{\Omega_n} (\vartheta_n^5 - \vartheta_n^4) \ \mathrm{d}x\mathrm{d}t
\end{aligned}
$$

$$
\begin{aligned}
&\leq \int_{\Omega_n} \left[\frac{1}{2} \frac{|\mathbf{m}_0|^2}{\varrho_{0,n,\delta}} - \mathbf{m}_0 \cdot \mathbf{u}_B + \frac{1}{2} \varrho_{0,n,\delta} |\mathbf{u}_B|^2 \right] \mathrm{d}x \\
&+ \int_{\Omega_n} \left[\varrho_{0,n,\delta} e_\delta(\varrho_{0,\delta}, \vartheta_{0,\delta}) - \varrho_{0,n,\delta} s_\delta(\varrho_{0,n,\delta}, \vartheta_{0,\delta}) \right] \mathrm{d}x \\
&+ \delta \int_{\Omega_n} \left(\frac{1}{\Gamma - 1} \varrho_{0,n,\delta}^\Gamma + \varrho_{0,n,\delta}^2 \right) \mathrm{d}x \\
&- \int_0^\tau \int_{\Omega_n} \left[\varrho_n (\mathbf{u}_n - \mathbf{u}_B) \otimes (\mathbf{u}_n - \mathbf{u}_B) + p_\delta \mathbb{I} \right] : \nabla_x \mathbf{u}_B \, \mathrm{d}x\mathrm{d}t \\
&+ \int_0^\tau \int_{\Omega_n} \mathbb{S}_\delta(\vartheta_n, \nabla_x \mathbf{u}_n) : \nabla_x \mathbf{u}_B \, \mathrm{d}x\mathrm{d}t \\
&+ \int_0^\tau \int_{\Omega_n} \varrho_n (\mathbf{g} - \partial_t \mathbf{u}_B + \mathbf{u}_B \cdot \nabla_x \mathbf{u}_B) \cdot (\mathbf{u}_n - \mathbf{u}_B) \, \mathrm{d}x\mathrm{d}t \\
&- \delta \int_0^\tau \int_{\partial \Omega_n} \left(\frac{1}{\Gamma - 1} \varrho_B^\Gamma + \varrho_B^2 \right) [\mathbf{u}_B \cdot \mathbf{n}]_n^- \mathrm{d}\sigma_x \mathrm{d}t \\
&- \delta \int_0^\tau \int_{\partial \Omega_n} \varrho_B (1 - \log(\vartheta_n)) \, [\mathbf{u}_B \cdot \mathbf{n}]_n^- \mathrm{d}\sigma_x \mathrm{d}t \\
&- \int_0^\tau \int_{\partial \Omega_n} \left(\frac{e(\varrho_B, \vartheta_n)}{\vartheta_n} - s(\varrho_B, \vartheta_n) \right) \varrho_B \, [\mathbf{u}_B \cdot \mathbf{n}]_n^- \mathrm{d}\sigma_x \mathrm{d}t \\
&+ \varepsilon \int_0^\tau \psi \int_{\Omega_n} \nabla_x \varrho_n \cdot \nabla_x (\mathbf{u}_n - \mathbf{u}_B) \cdot \mathbf{u}_B \, \mathrm{d}x\mathrm{d}t
\end{aligned} \tag{6.15}
$$

for a.a. $0 < \tau < T$.

We show that all integrals on the right-hand side of (6.15) may be controlled by the left-hand side via a Gronwall's argument.

First, we recall the density energy balance (5.35),

$$
\begin{aligned}
&\int_{\Omega_n} \varrho_n^2(\tau, \cdot) \mathrm{d}x + \varepsilon \int_0^\tau \int_{\Omega_n} |\nabla_x \varrho_n|^2 \, \mathrm{d}x\mathrm{d}t - \int_0^\tau \int_{\partial \Omega_n} (\varrho_n - \varrho_B)^2 \, [\mathbf{u}_B \cdot \mathbf{n}]_n^- \mathrm{d}\sigma_x \mathrm{d}t \\
&\quad + \int_0^\tau \int_{\partial \Omega_n} \varrho_n^2 [\mathbf{u}_B \cdot \mathbf{n}]^+ \, \left(\mathbf{u}_B \cdot \mathbf{n} - [\mathbf{u}_B \cdot \mathbf{n}]_n^- \right) \mathrm{d}\sigma_x \mathrm{d}t \\
&\quad \leq \int_{\Omega_n} \varrho_{0,n,\delta}^2 \, \mathrm{d}x - \int_0^\tau \int_{\partial \Omega_n} \varrho_B^2 \, [\mathbf{u}_B \cdot \mathbf{n}]_n^- \mathrm{d}\sigma_x \mathrm{d}t \\
&\quad - \int_0^\tau \int_{\Omega_n} \varrho_n^2 \mathrm{div}_x \mathbf{u}_n \, \mathrm{d}x\mathrm{d}t
\end{aligned} \tag{6.16}
$$

Next, the integrals

$$
\begin{aligned}
&\int_{\Omega_n} \left[\frac{1}{2} \frac{|\mathbf{m}_0|^2}{\varrho_{0,n,\delta}} - \mathbf{m}_0 \cdot \mathbf{u}_B + \frac{1}{2} \varrho_{0,n,\delta} |\mathbf{u}_B|^2 \right] \mathrm{d}x \\
&+ \int_{\Omega_n} \left[\varrho_{0,n,\delta} e_\delta(\varrho_{0,n,\delta}, \vartheta_{0,\delta}) - \varrho_{0,n,\delta} s_\delta(\varrho_{0,n,\delta}, \vartheta_{0,\delta}) \right] \mathrm{d}x
\end{aligned}
$$

$$+ \delta \int_{\Omega_n} \left(\frac{1}{\Gamma - 1} \varrho_{0,n,\delta}^{\Gamma} + \varrho_{0,n,\delta}^2 \right) \, \mathrm{d}x$$

as well as

$$\int_{\Omega_n} \varrho_{0,n,\delta}^2 \, \mathrm{d}x$$

are bounded in terms of the initial data.

Now, we check easily

$$\left| \int_{\Omega_n} [\varrho_n (\mathbf{u}_n - \mathbf{u}_B) \otimes (\mathbf{u}_n - \mathbf{u}_B) + p_\delta \mathbb{I}] : \nabla_x \mathbf{u}_B \, \mathrm{d}x \right|$$
$$\lesssim \int_{\Omega_n} \left[\frac{1}{2} \varrho_n |\mathbf{u}_n - \mathbf{u}_B|^2 + \varrho_n e_\delta - \varrho_n s_\delta + \delta \left(\frac{1}{\Gamma - 1} \varrho_n^\Gamma + \varrho_n^2 \right) \right] \, \mathrm{d}x,$$

and, similarly,

$$\left| \int_{\Omega_n} \varrho_n (\mathbf{g} - \partial_t \mathbf{u}_B + \mathbf{u}_B \cdot \nabla_x \mathbf{u}_B) \cdot (\mathbf{u}_n - \mathbf{u}_B) \, \mathrm{d}x \right|$$
$$\lesssim \int_{\Omega_n} \left[\frac{1}{2} \varrho_n |\mathbf{u}_n - \mathbf{u}_B|^2 + \varrho_n e_\delta - \varrho_n s_\delta + \delta \left(\frac{1}{\Gamma - 1} \varrho_n^\Gamma + \varrho_n^2 \right) \right] \, \mathrm{d}x.$$

To control the viscous stress tensor, we have

$$\begin{aligned}
|\mathbb{S}_\delta(\vartheta_n, \nabla_x \mathbf{u}_n)| &\le (\mu(\vartheta_n) + \delta \vartheta_n) \left| \nabla_x \mathbf{u}_n + \nabla_x^t \mathbf{u}_n - \frac{2}{d} \mathrm{div}_x \mathbf{u}_n \mathbb{I} \right| \\
&\quad + \eta(\vartheta_n) |\mathrm{div}_x \mathbf{u}_n| \\
&\le \frac{\omega}{4\vartheta_n} \left((\mu(\vartheta_n) + \delta \vartheta_n) \left| \nabla_x \mathbf{u}_n + \nabla_x^t \mathbf{u}_n - \frac{2}{d} \mathrm{div}_x \mathbf{u}_n \mathbb{I} \right|^2 \right) \\
&\quad + \frac{\omega}{4\vartheta_n} \eta(\vartheta_n) |\mathrm{div}_x \mathbf{u}_n|^2 + c(\omega) \vartheta_n (\mu(\vartheta_n) + \delta \vartheta_n + \eta(\vartheta_n)) \\
&\le \frac{\omega}{\vartheta_n} \mathbb{S}_\delta(\vartheta_n, \nabla_x \mathbf{u}_n) : \nabla_x \mathbf{u}_n + c(\omega, \delta, \overline{\mu}, \overline{\eta})(1 + \vartheta_n^2)
\end{aligned}$$

for any $\omega > 0$. Using the above inequality, we get

$$\begin{aligned}
\left| \int_{\Omega_n} \mathbb{S}_\delta(\vartheta_n, \nabla_x \mathbf{u}_n) : \nabla_x \mathbf{u}_B \, \mathrm{d}x \right| &\le \int_{\Omega_n} \frac{1}{2\vartheta_n} \mathbb{S}_\delta(\vartheta_n, \nabla_x \mathbf{u}_n) : \nabla_x \mathbf{u}_n \, \mathrm{d}x \\
&\quad + c_1(\delta) \int_{\Omega_n} (1 + \vartheta_n^2) \, \mathrm{d}x \\
&\le \int_{\Omega_n} \frac{1}{2\vartheta_n} \mathbb{S}_\delta(\vartheta_n, \nabla_x \mathbf{u}_n) : \nabla_x \mathbf{u}_n \, \mathrm{d}x \\
&\quad + c_2(\delta) \left(1 + \int_{\Omega_n} \varrho_n e_n \, \mathrm{d}x \right).
\end{aligned}$$

As for the boundary integrals, we realize that

$$
\begin{aligned}
& -\delta \int_0^\tau \int_{\partial\Omega_n} \varrho_B(1-\log(\vartheta_n))\ [\mathbf{u}_B\cdot\mathbf{n}]_n^- \mathrm{d}\sigma_x \mathrm{d}t \\
& -\int_0^\tau \int_{\partial\Omega_n} \left(\frac{e(\varrho_B,\vartheta_n)}{\vartheta_n} - s(\varrho_B,\vartheta_n) \right) \varrho_B\ [\mathbf{u}_B\cdot\mathbf{n}]_n^- \mathrm{d}\sigma_x \mathrm{d}t \\
& = \frac{1}{3}\int_0^\tau \int_{\partial\Omega_n} \vartheta_n^3\ [\mathbf{u}_B\cdot\mathbf{n}]_n^- \mathrm{d}\sigma_x \mathrm{d}t \\
& -\int_0^\tau \int_{\partial\Omega_n} \left(\frac{e_{\mathrm{m}}(\varrho_B,\vartheta_n)+e_{\mathrm{el}}(\varrho_B)}{\vartheta_n} - s_{\mathrm{m}}(\varrho_B,\vartheta_n) \right) \varrho_B\ [\mathbf{u}_B\cdot\mathbf{n}]_n^- \mathrm{d}\sigma_x \mathrm{d}t \\
& -\delta \int_0^\tau \int_{\partial\Omega_n} \varrho_B(1-\log(\vartheta_n))\ [\mathbf{u}_B\cdot\mathbf{n}]_n^- \mathrm{d}\sigma_x \mathrm{d}t \\
& = \frac{1}{3}\int_0^\tau \int_{\partial\Omega_n} \vartheta_n^3\ [\mathbf{u}_B\cdot\mathbf{n}]_n^- \mathrm{d}\sigma_x \mathrm{d}t - \int_0^\tau \int_{\partial\Omega_n} \frac{1}{\vartheta_n}\varrho_B e_{\mathrm{el}}(\varrho_B)\ [\mathbf{u}_B\cdot\mathbf{n}]_n^- \mathrm{d}\sigma_x \mathrm{d}t \\
& -\int_0^\tau \int_{\partial\Omega_n} \left(\frac{e_{\mathrm{m}}(\varrho_B,\vartheta_n)}{\vartheta_n} - s_{\mathrm{m}}(\varrho_B,\vartheta_n) \right) \varrho_B\ [\mathbf{u}_B\cdot\mathbf{n}]_n^- \mathrm{d}\sigma_x \mathrm{d}t \\
& -\delta \int_0^\tau \int_{\partial\Omega_n} \varrho_B(1-\log(\vartheta_n))\ [\mathbf{u}_B\cdot\mathbf{n}]_n^- \mathrm{d}\sigma_x \mathrm{d}t. \qquad (6.17)
\end{aligned}
$$

Note carefully that

$$
\int_0^\tau \int_{\partial\Omega_n} \vartheta_n^3\ [\mathbf{u}_B\cdot\mathbf{n}]_n^- \mathrm{d}\sigma_x \mathrm{d}t \leq 0;
$$

whence this integral can be put on the left-hand side of (6.15) providing boundary estimates on ϑ_n.

Finally, the rightmost integral in (6.15) reads

$$
\begin{aligned}
& \varepsilon \int_0^\tau \psi \int_{\Omega_n} \nabla_x\varrho_n \cdot \nabla_x(\mathbf{u}_n-\mathbf{u}_B)\cdot\mathbf{u}_B\ \mathrm{d}x\mathrm{d}t \\
& \quad \leq \frac{1}{2}\varepsilon \int_0^\tau \int_{\Omega_n} |\nabla_x\varrho_n|^2\ \mathrm{d}x\mathrm{d}t + c(\mathbf{u}_B)\varepsilon \int_0^\tau \int_{\Omega_n} |\nabla_x(\mathbf{u}_n-\mathbf{u}_B)|^2\ \mathrm{d}x\mathrm{d}t, \qquad (6.18)
\end{aligned}
$$

while (6.16) gives rise to

$$
\begin{aligned}
& \int_{\Omega_n} \varrho_n^2(\tau,\cdot)\ \mathrm{d}x + \varepsilon \int_0^\tau \int_{\Omega_n} |\nabla_x\varrho_n|^2\ \mathrm{d}x\mathrm{d}t \\
& \qquad - \int_0^\tau \int_{\partial\Omega_n} (\varrho_n-\varrho_B)^2\ [\mathbf{u}_B\cdot\mathbf{n}]_n^- \mathrm{d}\sigma_x \mathrm{d}t \\
& \qquad + \int_0^\tau \int_{\partial\Omega_n} \varrho_n^2\ \left(\mathbf{u}_B\cdot\mathbf{n} - [\mathbf{u}_B\cdot\mathbf{n}]_n^-\right) \mathrm{d}\sigma_x \mathrm{d}t
\end{aligned}
$$

$$\lesssim \left(1 + \omega^2 \int_0^\tau \int_{\Omega_n} |\mathrm{div}_x \mathbf{u}_n|^2 \, \mathrm{d}x\mathrm{d}t + c(\omega) \int_0^\tau \int_{\Omega_n} \varrho_n^4 \, \mathrm{d}x\mathrm{d}t\right) \tag{6.19}$$

for any $\omega > 0$.

Thus adding (6.19) to (6.15) and using the previous bounds, notably (6.17), (6.18), we deduce the final estimate

$$\begin{aligned}
&\int_{\Omega_n} \left[\frac{1}{2}\varrho_n |\mathbf{u}_n - \mathbf{u}_B|^2 + \varrho_n e_\delta - \varrho_n s_\delta + \delta\left(\frac{1}{\Gamma - 1}\varrho_n^\Gamma + \varrho_n^2\right)\right](\tau, \cdot)\,\mathrm{d}x \\
&+ \int_0^\tau \int_{\partial\Omega_n} (\varrho_n e_\delta - \varrho_n s_\delta)\,\left(\mathbf{u}_B \cdot \mathbf{n} - [\mathbf{u}_B \cdot \mathbf{n}]_n^-\right) \mathrm{d}\sigma_x \mathrm{d}t \\
&- \int_0^\tau \int_{\partial\Omega_n} \left[\frac{1}{\vartheta_n}(f_{i,B} - \varrho_B e_{\mathrm{el}}(\varrho_B)) + \frac{1}{3}\vartheta_n^3\right] [\mathbf{u}_B \cdot \mathbf{n}]_n^- \mathrm{d}\sigma_x \mathrm{d}t \\
&+ \int_0^\tau \int_{\partial\Omega_n} \varrho_n^2\,\left(\mathbf{u}_B \cdot \mathbf{n} - [\mathbf{u}_B \cdot \mathbf{n}]_n^-\right) \mathrm{d}\sigma_x \mathrm{d}t - \int_0^\tau \int_{\partial\Omega_n} \varrho_n^2\,[\mathbf{u}_B \cdot \mathbf{n}]_n^- \mathrm{d}\sigma_x \mathrm{d}t \\
&+ \frac{1}{2}\int_0^\tau \int_{\Omega_n} \frac{1}{\vartheta_n}\left(\mathbb{S}_\delta(\vartheta_n, \nabla_x \mathbf{u}_n) : \nabla_x \mathbf{u}_n - \frac{\mathbf{q} \cdot \nabla_x \vartheta_n}{\vartheta_n}\right) \mathrm{d}x\mathrm{d}t \\
&+ \frac{\delta}{2}\int_0^\tau \int_{\Omega_n} \frac{1}{\vartheta_n^2}\left(\vartheta_n^\Gamma + \frac{1}{\vartheta_n}\right) |\nabla_x \vartheta_n|^2 \,\mathrm{d}x\mathrm{d}t \\
&+ \frac{\varepsilon\delta}{2}\int_0^\tau \int_{\Omega_n} \frac{1}{\vartheta_n}\left(\Gamma\varrho_n^{\Gamma-2} + 2\right) |\nabla_x \varrho_n|^2 \,\mathrm{d}x\mathrm{d}t + \frac{1}{2}\varepsilon \int_0^\tau \int_{\Omega_n} |\nabla_x \varrho_n|^2 \,\mathrm{d}x\mathrm{d}t \\
&+ \delta \int_0^\tau \int_{\Omega_n} \frac{1}{\vartheta_n^3} \,\mathrm{d}x\mathrm{d}t + \varepsilon \int_0^\tau \int_{\Omega_n} \vartheta_n^5 \,\mathrm{d}x\mathrm{d}t \\
&\leq c(\delta, \mathrm{data})\Bigg(1 + \\
&+ \int_0^\tau \int_{\Omega_n} \left[\frac{1}{2}\varrho_n |\mathbf{u}_n - \mathbf{u}_B|^2 + \varrho_n e_\delta - \varrho_n s_\delta + \delta\left(\frac{1}{\Gamma - 1}\varrho_n^\Gamma + \varrho_n^2\right)\right] \mathrm{d}x\mathrm{d}t \\
&+ \int_0^\tau \int_{\partial\Omega_n} \varrho_B\Big(\delta \log(\vartheta_n) + s_{\mathrm{m}}(\varrho_B, \vartheta_n)\Big)\,[\mathbf{u}_B \cdot \mathbf{n}]_n^- \mathrm{d}\sigma_x \mathrm{d}t \\
&- \int_0^\tau \int_{\partial\Omega_n} \frac{1}{\vartheta_n}\varrho_B e_{\mathrm{m}}(\varrho_B, \vartheta_n)\,[\mathbf{u}_B \cdot \mathbf{n}]_n^- \mathrm{d}\sigma_x \mathrm{d}t\Bigg).
\end{aligned} \tag{6.20}$$

In order to apply Gronwall's lemma, we have to control the last two boundary integrals on the right-hand side. To this end, we use hypothesis (5.48). On the one hand, hypothesis (5.48) says

$$0 < \underline{f} \leq f_{i,B}(t, x) - \varrho_B e_{\mathrm{el}}(\varrho_B) \text{ for any } (t, x) \in \Gamma_{\mathrm{in}}.$$

On the other hand,

$$\lim_{\vartheta \to 0} \varrho e_{\mathrm{m}}(\varrho, \vartheta) = 0 \text{ for any fixed } \varrho \geq 0.$$

As ϱ_B is fixed, there is a constant $\underline{\vartheta}(\varrho_B)$ such that

$$|\varrho_B e_{\rm m}(\varrho_B, \vartheta)| \leq \frac{1}{2}\underline{f} \text{ as soon as } 0 < \vartheta \leq \underline{\vartheta}(\varrho_B).$$

Since

$$\frac{1}{\vartheta_n} |\varrho_B e_{\rm m}(\varrho_B, \vartheta_n)| \stackrel{<}{\sim} 1$$

we conclude that the last in integral in (6.20) is dominated by

$$-\frac{1}{2}\int_0^T \int_{\partial\Omega_n} \left[\frac{1}{\vartheta_n}\Big(f_{i,B} - \varrho_B e_{\rm el}(\varrho_B)\Big) + \frac{1}{3}\vartheta_n^3\right] [\mathbf{u}_B \cdot \mathbf{n}]_n^- {\rm d}\sigma_x {\rm d}t.$$

Moreover, similar arguments can be used to control

$$\int_{\partial\Omega_n} \varrho_B\Big(\delta\log(\vartheta_n) + s_{\rm m}(\varrho_B, \vartheta_n)\Big) [\mathbf{u}_B\cdot\mathbf{n}]_n^- {\rm d}\sigma_x \stackrel{<}{\sim} \int_{\partial\Omega_n} [\log(\vartheta_n)]^- [\mathbf{u}_B\cdot\mathbf{n}]_n^- {\rm d}\sigma_x.$$

Going back to (6.20), we collect the following bounds, uniform for $n \to \infty$:

$$\sup_{\tau\in[0,T]} \|\varrho_n(\tau,\cdot)\|_{L^{\frac{5}{3}}(\Omega_n)} + \sup_{\tau\in[0,T]} \|\vartheta_n(\tau,\cdot)\|_{L^4(\Omega_n)} \stackrel{<}{\sim} 1, \tag{6.21}$$

$$\sup_{\tau\in[0,T]} \|\sqrt{\varrho_n}\mathbf{u}_n(\tau,\cdot)\|_{L^2(\Omega_n;R^d)} \stackrel{<}{\sim} 1, \tag{6.22}$$

$$\sup_{\tau\in[0,T]} \delta\int_{\Omega_n} \varrho_n^\Gamma(\tau,\cdot)\ {\rm d}x \stackrel{<}{\sim} 1, \tag{6.23}$$

$$\delta\int_0^T\int_{\Omega_n} \frac{1}{\vartheta_n^3}\ {\rm d}x{\rm d}t + \varepsilon\int_0^T\int_{\Omega_n} \vartheta_n^5\ {\rm d}x{\rm d}t \stackrel{<}{\sim} 1, \tag{6.24}$$

$$\delta\int_0^T\int_{\Omega_n} \left[|\nabla_x\mathbf{u}_n|^2 + \left(\frac{1}{\vartheta_n^3} + \vartheta_n^{\Gamma-2}|\nabla_x\vartheta_n|^2\right)\right]\ {\rm d}x{\rm d}t \stackrel{<}{\sim} 1, \tag{6.25}$$

$$\varepsilon\delta\int_0^T\int_{\Omega_n} \frac{1}{\vartheta_n}\left(\varrho_n^{\Gamma-2} + 1\right)|\nabla_x\varrho_n|^2\ {\rm d}x{\rm d}t \stackrel{<}{\sim} 1 \tag{6.26}$$

$$\varepsilon\int_0^T\int_{\Omega_n} |\nabla_x\varrho_n|^2\ {\rm d}x{\rm d}t \stackrel{<}{\sim} 1 \tag{6.27}$$

$$-\int_0^T\int_{\partial\Omega_n} \varrho_n^2\ [\mathbf{u}_B\cdot\mathbf{n}]_n^- {\rm d}\sigma_x{\rm d}t \stackrel{<}{\sim} 1 \tag{6.28}$$

$$\int_0^T\int_{\partial\Omega_n} \varrho_n^2\ \left(\mathbf{u}_B\cdot\mathbf{n} - [\mathbf{u}_B\cdot\mathbf{n}]_n^-\right) {\rm d}\sigma_x{\rm d}t \stackrel{<}{\sim} 1, \tag{6.29}$$

$$-\int_0^T\int_{\partial\Omega_n} \left(\frac{1}{\vartheta_n} + \vartheta_n^3\right)\ [\mathbf{u}_B\cdot\mathbf{n}]_n^- {\rm d}\sigma_x{\rm d}t \stackrel{<}{\sim} 1, \tag{6.30}$$

$$\int_0^T \int_{\partial\Omega_n} \Big(\varrho_n e_\delta(\varrho_n) - \varrho_n s_\delta(\varrho_n, \vartheta_n) \Big) \left(\mathbf{u}_B \cdot \mathbf{n} - [\mathbf{u}_B \cdot \mathbf{n}]_n^- \right) \mathrm{d}\sigma_x \mathrm{d}t \lesssim 1. \tag{6.31}$$

In addition, revisiting the original kinetic energy balance, we realize that all terms on the right-hand side of (6.8) are bounded, and we have

$$\varepsilon\delta \int_0^T \int_{\Omega_n} \varrho_n^{\Gamma-2} |\nabla_x \varrho_n|^2 \, \mathrm{d}x\mathrm{d}t \lesssim 1. \tag{6.32}$$

This bound is important to control the pressure term at the zeroth order level of approximation.

6.3 Convergence $n \to \infty$

Now we have everything at hand to perform the asymptotic limit $n \to \infty$. In view of (6.21), (6.23), (6.25), and the hypothesis $\Omega \subset \Omega_n$, we may suppose

$$\begin{aligned}
&\varrho_n \to \varrho \text{ weakly-(*) in } L^\infty(0,T; L^\Gamma(\Omega)),\\
&\vartheta_n \to \vartheta \text{ weakly-(*) in } L^\infty(0,T; L^4(\Omega)),\\
&\mathbf{u}_n \to \mathbf{u} \text{ weakly in } L^2(0,T; W^{1,2}(\Omega; R^d))
\end{aligned} \tag{6.33}$$

passing to suitable subsequences as the case may be.

6.3.1 Approximate Equation of Continuity

Using (6.22), (6.23), we deduce

$$\varrho_n \mathbf{u}_n \to \overline{\varrho \mathbf{u}} \text{ weakly-(*) in } L^\infty(0,T; L^{\frac{2\Gamma}{\Gamma+1}}(\Omega; R^d)).$$

Here and hereafter, the symbol $\overline{B(\varrho, \vartheta, \mathbf{u})}$ will denote the weak limit of the composition $\{B(\varrho_n, \vartheta_n, \mathbf{u}_n)\}_{n=1}^\infty$ of a weakly converging sequence $\{\varrho_n, \vartheta_n, \mathbf{u}_n\}_{n=1}^\infty$ with a nonlinear function B. Strictly speaking, we should consider a new subsequence each time when we perform such a limit. However, using separability of the space of continuous functions defined on a compact set, one can show that it is possible to choose a single subsequence such that the weak limit

$$B(\varrho_n, \vartheta_n, \mathbf{u}_n) \to \overline{B(\varrho, \vartheta, \mathbf{u})} \text{ weakly in } L^1$$

holds for any compactly supported continuous function B, cf. the proof of the Fundamental Theorem of the theory of Young measures by Ball [3].

Now, since ϱ_n satisfies the approximate equation of continuity (6.2), we may strengthen the convergence in (6.33) to

$$\varrho_n \to \varrho \text{ in } C_{\rm weak}([0,T]; L^\Gamma(\Omega)) \tag{6.34}$$

since the functions

$$t \in [0,T] \mapsto \int_\Omega \varrho_n(t,\cdot)\phi \ {\rm d}x$$

are uniformly Lipschitz continuous for any $\phi \in C_c^\infty(\Omega)$. Consequently, as $L^\Gamma(\Omega) \hookrightarrow\hookrightarrow W^{-1,2}(\Omega)$,

$$\varrho_n \mathbf{u}_n \to \varrho \mathbf{u} \text{ in } \mathcal{D}'((0,T)\times\Omega) \ \Rightarrow \ \overline{\varrho \mathbf{u}} = \varrho \mathbf{u}.$$

In addition, in accordance with (6.27),

$$\nabla_x \varrho_n \to \nabla_x \varrho \text{ weakly in } L^2((0,T)\times\Omega),$$

which, together with (6.34), yields

$$\varrho_n \to \varrho \text{ in } L^2((0,T)\times\Omega). \tag{6.35}$$

At this stage, we may let $n \to \infty$ in the weak formulation (6.5) of the approximate equation of continuity. As, by virtue of (5.6),

$$\Omega \subset \Omega_n, \ |\Omega_n \setminus \Omega| \to 0,$$

and the uniform bounds (6.21)–(6.23), (6.27) hold on Ω_n, we easily deduce

$$\begin{aligned}
&\int_0^T \int_{\Omega_n} \left(\varrho_n \partial_t \varphi + \varrho_n \mathbf{u}_n \cdot \nabla_x \varphi - \varepsilon \nabla_x \varrho_n \cdot \nabla_x \varphi \right) \ {\rm d}x{\rm d}t \\
&= \int_0^T \int_{\Omega} \left(\varrho_n \partial_t \varphi + \varrho_n \mathbf{u}_n \cdot \nabla_x \varphi - \varepsilon \nabla_x \varrho_n \cdot \nabla_x \varphi \right) \ {\rm d}x{\rm d}t \\
&+ \int_0^T \int_{\Omega_n \setminus \Omega} \left(\varrho_n \partial_t \varphi + \varrho_n \mathbf{u}_n \cdot \nabla_x \varphi - \varepsilon \nabla_x \varrho_n \cdot \nabla_x \varphi \right) {\rm d}x{\rm d}t \\
&\to \int_0^T \int_{\Omega} \left(\varrho \partial_t \varphi + \varrho \mathbf{u} \cdot \nabla_x \varphi - \varepsilon \nabla_x \varrho \cdot \nabla_x \varphi \right) \ {\rm d}x{\rm d}t \text{ as } n \to \infty.
\end{aligned}$$

As for the boundary integrals, we have

$$\int_0^T \int_{\partial\Omega_n} \varphi \varrho_B \ [\mathbf{u}_B \cdot \mathbf{n}]_n^- {\rm d}\sigma_x {\rm d}t = \int_0^T \int_{\partial\Omega_n \cap \partial\Omega} \varphi \varrho_B [\mathbf{u}_B \cdot \mathbf{n}]_n^- {\rm d}\sigma_x {\rm d}t$$

$$+\int_0^T \int_{\partial\Omega_n \setminus \partial\Omega} \varphi \varrho_B [\mathbf{u}_B \cdot \mathbf{n}]_n^- \mathrm{d}\sigma_x \mathrm{d}t$$
$$= \int_0^T \int_{\partial\Omega} \varphi \mathbb{1}_{\partial\Omega_n} \varrho_B [\mathbf{u}_B \cdot \mathbf{n}]_n^- \mathrm{d}\sigma_x \mathrm{d}t + \int_0^T \int_{\partial\Omega_n \setminus \partial\Omega} \varphi \varrho_B [\mathbf{u}_B \cdot \mathbf{n}]_n^- \mathrm{d}\sigma_x \mathrm{d}t.$$

By virtue of hypothesis (6.1) on domain/boundary convergence,

$$\int_0^T \int_{\partial\Omega_n \setminus \partial\Omega} \varphi \varrho_B [\mathbf{u}_B \cdot \mathbf{n}]_n^- \mathrm{d}\sigma_x \mathrm{d}t \to 0.$$

By the same token and Lebesgue Convergence Theorem,

$$\int_0^T \int_{\partial\Omega} \varphi \mathbb{1}_{\partial\Omega_n} \varrho_B [\mathbf{u}_B \cdot \mathbf{n}]_n^- \mathrm{d}\sigma_x \mathrm{d}t \to \int_0^T \int_{\partial\Omega} \varphi \varrho_B \, [\mathbf{u}_B \cdot \mathbf{n}]^- \mathrm{d}\sigma_x \mathrm{d}t.$$

Similarly, write

$$\int_0^T \int_{\partial\Omega_n} \varphi \varrho_n \, \left(\mathbf{u}_B \cdot \mathbf{n} - [\mathbf{u}_B \cdot \mathbf{n}]_n^- \right) \mathrm{d}\sigma_x \mathrm{d}t$$
$$= \int_0^T \int_{\partial\Omega_n \setminus \partial\Omega} \varphi \varrho_n \left(\mathbf{u}_B \cdot \mathbf{n} - [\mathbf{u}_B \cdot \mathbf{n}]_n^- \right) \mathrm{d}\sigma_x \, \mathrm{d}t$$
$$+ \int_0^T \int_{\partial\Omega} \varphi \mathbb{1}_{\partial\Omega_n} \varrho_n \left(\mathbf{u}_B \cdot \mathbf{n} - [\mathbf{u}_B \cdot \mathbf{n}]_n^- \right) \mathrm{d}\sigma_x \, \mathrm{d}t.$$

Here, in accordance with the estimate (6.28),

$$\left| \int_0^T \int_{\partial\Omega_n \setminus \partial\Omega} \varphi \varrho_n \left(\mathbf{u}_B \cdot \mathbf{n} - [\mathbf{u}_B \cdot \mathbf{n}]_n^- \right) \mathrm{d}\sigma_x \, \mathrm{d}t \right|$$
$$\leq \left(\int_0^T \int_{\partial\Omega_n \setminus \partial\Omega} \varphi \varrho_n^2 \left(\mathbf{u}_B \cdot \mathbf{n} - [\mathbf{u}_B \cdot \mathbf{n}]_n^- \right) \mathrm{d}\sigma_x \, \mathrm{d}t \right)^{\frac{1}{2}} \times$$
$$\times \sqrt{T} |\partial\Omega_n \setminus \Omega|_{d-1}^{\frac{1}{2}} \|\varphi \mathbf{u}_B\|_{L^\infty((0,T)\times\Omega_n)}^{\frac{1}{2}} \to 0 \text{ as } n \to \infty.$$

Finally, by virtue of (6.35) and the gradient estimate (6.27),

$$\varrho_n \to \varrho \text{ in } L^2((0,T) \times \partial\Omega) \text{ (in the sense of traces);} \tag{6.36}$$

whence

$$\int_0^T \int_{\partial\Omega} \varphi \mathbb{1}_{\partial\Omega_n} \varrho_n \left(\mathbf{u}_B \cdot \mathbf{n} - [\mathbf{u}_B \cdot \mathbf{n}]_n^- \right) \mathrm{d}\sigma_x \, \mathrm{d}t \to \int_0^T \int_{\partial\Omega} \varphi \varrho \, [\mathbf{u}_B \cdot \mathbf{n}]^+ \mathrm{d}\sigma_x \mathrm{d}t.$$

Summing up the previous discussion, we may infer that the limit $(\varrho, \mathbf{u})$ solves the approximate equation of continuity

$$\int_0^T \int_\Omega (\varrho \partial_t \varphi + \varrho \mathbf{u} \cdot \nabla_x \varphi - \varepsilon \nabla_x \varrho \cdot \nabla_x \varphi) \, \mathrm{d}x \mathrm{d}t$$
$$= \int_0^T \int_{\partial\Omega} \varphi \varrho_B [\mathbf{u}_B \cdot \mathbf{n}]^- \, \mathrm{d}\sigma_x \mathrm{d}t + \int_{\partial\Omega} \varphi \varrho [\mathbf{u}_B \cdot \mathbf{n}]^+ \, \mathrm{d}\sigma_x \mathrm{d}t$$
$$- \int_\Omega \varrho_\delta \varphi(0, \cdot) \, \mathrm{d}x \qquad (6.37)$$

for any $\varphi \in C^1_c([0,T) \times \overline{\Omega})$.

6.3.2 Pointwise Convergence of the Temperature and the Limit in the Approximate Entropy Inequality

The following discussion will be repeated, with minor modifications, at each level of the approximation process, in particular for the limits $\varepsilon \to 0$, and $\delta \to 0$. Our goal is to show

$$\vartheta_n(t,x) \to \vartheta(t,x) \text{ for a.a. } (t,x) \in (0,T) \times \Omega \qquad (6.38)$$

up to a suitable subsequence.

First, we need estimates on the time derivative $\partial_t(\varrho_n s_\delta(\varrho_n, \vartheta_n))$ in a suitable (negative) Sobolev space. To this end, we go back to the approximate entropy inequality (6.13). In view of the uniform bounds (6.24), (6.25), we may assume

$$\varrho_n s_\delta(\varrho_n, \vartheta_n) \to \overline{\varrho s_\delta(\varrho, \vartheta)} \text{ weakly in } L^r((0,T) \times \Omega),$$
$$\varrho_n s_\delta(\varrho_n, \vartheta_n) \mathbf{u}_n \to \overline{\varrho s_\delta(\varrho, \vartheta) \mathbf{u}} \text{ weakly in } L^r((0,T) \times \Omega; R^d) \text{ for some } r > 1. \qquad (6.39)$$

The unperturbed entropy flux is estimated as

$$\frac{\kappa(\vartheta_n)}{\vartheta_n} |\nabla_x \vartheta_n| \lesssim \Big(|\nabla_x \log(\vartheta_n)| + \vartheta_n^2 |\nabla_x \vartheta_n| \Big).$$

Its δ-dependent component can be written as

$$\delta\vartheta_n^{\Gamma-1}\nabla_x\vartheta_n = \frac{\delta\Gamma}{2}\vartheta_\delta^{\frac{\Gamma}{2}}\nabla_x\vartheta_n^{\frac{\Gamma}{2}} = \frac{\delta\Gamma}{2}\vartheta_n^{\frac{1}{4}}\vartheta_n^{\frac{\Gamma}{2}-\frac{1}{4}}\nabla_x\vartheta_n^{\frac{\Gamma}{2}};$$

whence controlled by (6.25), whereas the same argument applies to control the component $\delta\vartheta_n^{-2}\nabla_x\vartheta_n$.

Finally, the integrands in the ε-dependent integrals on the right-hand side of (6.13) can be handled as

$$\left\| \nabla_x\varrho_n\cdot\nabla_x\varphi\left(\frac{e_{\rm m}(\varrho_n,\vartheta_n)}{\vartheta_n} - s_{\rm m}(\varrho_n,\vartheta_n) + \frac{p_{\rm m}(\varrho_n,\vartheta_n)}{\varrho_n\vartheta_n}\right)\right\|_{L^1(\Omega_n)}$$
$$\lesssim \|\nabla_x\varphi\|_{L^\infty(\Omega_n;R^d)}\|\nabla_x\varrho_n\|_{L^2(\Omega_n;R^d)}\|\vartheta_n^{-1}\|_{L^3(\Omega_n;R^d)}\left(1+\|\varrho_n^{\frac{2}{3}}\|_{L^6(\Omega_n)}\right).$$

The term

$$\nabla_x\varrho_n\cdot\nabla_x\varphi\,(\delta-\delta\log(\vartheta_n))$$

can be treated in a similar way.

Using the uniform bounds (6.23), (6.24), and (6.27), we may therefore deduce from the approximate entropy inequality (6.13)

$$\int_0^T\int_\Omega \varrho_n s_\delta(\varrho_n,\vartheta_n)\partial_t\varphi\ {\rm d}x{\rm d}t \le \int_0^T\int_\Omega (\mathbf{a}_n\cdot\nabla_x\varphi + b_n\varphi)\ {\rm d}x{\rm d}t$$

for any $\varphi\in C^1_c((0,T)\times\Omega)$, $\varphi\ge 0$, where

$$\|\mathbf{a}_n\|_{L^r((0,T)\times\Omega;R^d)} + \|b_n\|_{L^r(0,T)\times\Omega} \lesssim 1 \text{ for some } r\ge 1.$$

Note that we have used the test function with compact support in Ω to eliminate the boundary integrals.

Thus we may use the Lions–Aubin argument, specifically Lemma 4, to conclude

$$\varrho_n s_\delta(\varrho_n,\vartheta_n)\vartheta_n \to \overline{\varrho s_\delta(\varrho,\vartheta)\vartheta} = \overline{\varrho s_\delta(\varrho,\vartheta)}\vartheta. \tag{6.40}$$

As stated in (6.35), $\{\varrho_n\}_{n=1}^\infty$ converges strongly, and we deduce from (6.40) and monotonicity of the entropy s_δ with respect to ϑ that

$$\overline{\vartheta^3\vartheta} = \overline{\vartheta^3}\ \vartheta,$$

which yields, up to a suitable subsequence, the desired pointwise a.a. convergence claimed in (6.38). Indeed

$$0 = \int_0^T\int_\Omega\left[\overline{\vartheta^3\vartheta} - \overline{\vartheta^3}\vartheta\right]\ {\rm d}x{\rm d}t = \lim_{n\to\infty}\int_0^T\int_\Omega(\vartheta_n^3-\vartheta^3)(\vartheta_n-\vartheta)\ {\rm d}x{\rm d}t$$
$$= \lim_{n\to\infty}\int_0^T\int_\Omega(\vartheta_n-\vartheta)^4\ {\rm d}x{\rm d}t. \tag{6.41}$$

With the strong convergence (6.35), (6.38) at hand, we go back to the approximate entropy inequality (6.13). As $\Omega \subset \Omega_n$, we may replace the non-negative integrals on the right-hand side by their counterparts on Ω:

$$
\begin{aligned}
&\left[\int_{\Omega_n} \varrho_n s_\delta \varphi \, \mathrm{d}x\right]_{t=0}^{t=\tau} \\
&- \int_0^\tau \int_{\Omega_n} \varrho_n s_\delta \partial_t \varphi \, \mathrm{d}x\mathrm{d}t - \int_0^\tau \int_{\Omega_n} \varrho_n s_\delta \mathbf{u}_n \cdot \nabla_x \varphi \, \mathrm{d}x\mathrm{d}t \\
&+ \int_0^\tau \int_{\partial\Omega_n} \varphi \varrho_n s_\delta \left(\mathbf{u}_B \cdot \mathbf{n} - [\mathbf{u}_B \cdot \mathbf{n}]_n^- \right) \mathrm{d}\sigma_x \mathrm{d}t \\
&- \int_0^\tau \int_{\Omega_n} \frac{\mathbf{q}_\delta(\vartheta_n, \nabla_x \vartheta_n)}{\vartheta_n} \cdot \nabla_x \varphi \, \mathrm{d}x\mathrm{d}t \\
&\geq \int_0^\tau \int_\Omega \frac{\varphi}{\vartheta_n} \left(\mathbb{S}_\delta : \nabla_x \mathbf{u}_n - \frac{\mathbf{q} \cdot \nabla_x \vartheta_n}{\vartheta_n} + \frac{\delta}{\vartheta_n^2} \right) \mathrm{d}x\mathrm{d}t \\
&+ \frac{\delta}{2} \int_0^\tau \int_\Omega \frac{\varphi}{\vartheta_n^2} \left(\vartheta_n^\Gamma + \frac{1}{\vartheta_n} \right) |\nabla_x \vartheta_n|^2 \, \mathrm{d}x\mathrm{d}t \\
&+ \frac{\varepsilon}{2} \int_0^\tau \int_\Omega \frac{\varphi}{\vartheta_n} \delta \left(\Gamma \varrho_n^{\Gamma-2} + 2 \right) |\nabla_x \varrho_n|^2 \, \mathrm{d}x\mathrm{d}t - \varepsilon \int_0^\tau \int_{\Omega_n} \varphi \vartheta_n^4 \, \mathrm{d}x\mathrm{d}t \\
&+ \varepsilon \int_0^\tau \int_\Omega \nabla_x \varrho_n \cdot \nabla_x \varphi \left(\frac{e_{\mathrm{m}}(\varrho_n, \vartheta_n)}{\vartheta_n} - s_{\mathrm{m}}(\varrho_n, \vartheta_n) + \frac{p_{\mathrm{m}}(\varrho_n, \vartheta_n)}{\varrho_n \vartheta_n} \right) \mathrm{d}x\mathrm{d}t \\
&+ \varepsilon \int_0^\tau \int_{\Omega_n} \nabla_x \varrho_n \cdot \nabla_x \varphi \left(\delta - \delta \log(\vartheta_n) \right) \mathrm{d}x\mathrm{d}t \\
&+ \int_0^\tau \int_{\partial\Omega_n} \varphi \left[\delta \varrho_B (1 - \log(\vartheta_n)) - \frac{1}{\vartheta_n} f_{i,b} \right] [\mathbf{u}_B \cdot \mathbf{n}]_n^- \mathrm{d}\sigma_x \mathrm{d}t \\
&+ \int_0^\tau \int_{\partial\Omega_n} \varphi \varrho_B \left(\frac{e(\varrho_B, \vartheta_n)}{\vartheta_n} - s(\varrho_B, \vartheta_n) \right) [\mathbf{u}_B \cdot \mathbf{n}]_n^- \mathrm{d}\sigma_x \mathrm{d}t.
\end{aligned}
$$

Moreover, we can perform the limit in all volume integrals using weak lower semi-continuity of the convex functions

$$
|\nabla_x \varrho|^2, \ |\nabla_x \vartheta|^2
$$

obtaining

$$
\begin{aligned}
&- \int_\Omega \varrho_0 s_\delta(\varrho_0, \vartheta_0) \varphi(0, \cdot) \, \mathrm{d}x \\
&- \int_0^T \int_\Omega \varrho s_\delta(\varrho, \vartheta) \partial_t \varphi \, \mathrm{d}x\mathrm{d}t - \int_0^T \int_\Omega \varrho s_\delta(\varrho, \vartheta) \mathbf{u} \cdot \nabla_x \varphi \, \mathrm{d}x\mathrm{d}t \\
&+ \lim_{n \to \infty} \int_0^T \int_{\partial\Omega_n} \varphi \varrho_n s_\delta [\mathbf{u}_B \cdot \mathbf{n}]^+ \left(\mathbf{u}_B \cdot \mathbf{n} - [\mathbf{u}_B \cdot \mathbf{n}]_n^- \right) \mathrm{d}\sigma_x \mathrm{d}t
\end{aligned}
$$

$$
\begin{aligned}
&- \int_0^T \int_\Omega \frac{\mathbf{q}_\delta(\vartheta, \nabla_x \vartheta)}{\vartheta} \cdot \nabla_x \varphi \ \mathrm{d}x\mathrm{d}t \\
&\geq \int_0^T \int_\Omega \frac{\varphi}{\vartheta} \left(\mathbb{S}_\delta(\nabla_x \mathbf{u}) : \nabla_x \mathbf{u} - \frac{\mathbf{q} \cdot \nabla_x \vartheta}{\vartheta} + \frac{\delta}{\vartheta^2} \right) \ \mathrm{d}x\mathrm{d}t \\
&+ \frac{\delta}{2} \int_0^T \int_\Omega \frac{\varphi}{\vartheta^2} \left(\vartheta^\Gamma + \frac{1}{\vartheta} \right) |\nabla_x \vartheta|^2 \ \mathrm{d}x\mathrm{d}t \\
&+ \frac{\varepsilon}{2} \int_0^T \int_\Omega \frac{\varphi}{\vartheta} \delta \left(\Gamma \varrho^{\Gamma - 2} + 2 \right) |\nabla_x \varrho|^2 \ \mathrm{d}x\mathrm{d}t - \varepsilon \int_0^T \int_\Omega \varphi \vartheta^4 \ \mathrm{d}x\mathrm{d}t \\
&+ \varepsilon \int_0^T \int_\Omega \nabla_x \varrho \cdot \nabla_x \varphi \left(\frac{e_{\mathrm{m}}(\varrho, \vartheta)}{\vartheta} - s_{\mathrm{m}}(\varrho, \vartheta) + \frac{p_{\mathrm{m}}(\varrho, \vartheta)}{\varrho \vartheta} \right) \ \mathrm{d}x\mathrm{d}t \\
&+ \varepsilon \int_0^T \int_\Omega \nabla_x \varrho \cdot \nabla_x \varphi \left(\delta - \delta \log(\vartheta) \right) \ \mathrm{d}x\mathrm{d}t \\
&+ \lim_{n \to \infty} \int_0^T \int_{\partial \Omega_n} \varphi \left[\delta \varrho_B (1 - \log(\vartheta_n)) - \frac{1}{\vartheta_n} f_{i,B} \right] [\mathbf{u}_B \cdot \mathbf{n}]_n^- \mathrm{d}\sigma_x \mathrm{d}t \\
&+ \lim_{n \to \infty} \int_0^T \int_{\partial \Omega_n} \varphi \varrho_B \left(\frac{e(\varrho_B, \vartheta_n)}{\vartheta_n} - s(\varrho_B, \vartheta_n) \right) [\mathbf{u}_B \cdot \mathbf{n}]_n^- \mathrm{d}\sigma_x \mathrm{d}t
\end{aligned}
\tag{6.42}
$$

for any $\varphi \in C^1_c([0,T) \times \overline{\Omega}_n)$, $\varphi \geq 0$.

The limit in the boundary integrals is more delicate. First observe, by virtue of the gradient estimate (6.25) and the strong convergence stated in (6.38),

$$
\vartheta_n \to \vartheta \text{ in } L^2((0,T) \times \partial \Omega) \text{ (in the sense of traces).} \tag{6.43}
$$

Next, write

$$
\begin{aligned}
&\int_0^T \int_{\partial \Omega_n} \varphi \varrho_n s_\delta(\varrho_n, \vartheta_n) \ \left(\mathbf{u}_B \cdot \mathbf{n} - [\mathbf{u}_B \cdot \mathbf{n}]_n^- \right) \mathrm{d}\sigma_x \mathrm{d}t \\
&\quad = \int_0^T \int_{\partial \Omega_n} \varphi \varrho_n [s_\delta(\varrho_n, \vartheta_n)]^+ \ \left(\mathbf{u}_B \cdot \mathbf{n} - [\mathbf{u}_B \cdot \mathbf{n}]_n^- \right) \mathrm{d}\sigma_x \mathrm{d}t \\
&\quad + \int_0^T \int_{\partial \Omega_n} \varphi \varrho_n [s_\delta(\varrho_n, \vartheta_n)]^- \ \left(\mathbf{u}_B \cdot \mathbf{n} - [\mathbf{u}_B \cdot \mathbf{n}]_n^- \right) \mathrm{d}\sigma_x \mathrm{d}t,
\end{aligned}
$$

where

$$
\begin{aligned}
&\int_0^T \int_{\partial \Omega_n} \varphi \varrho_n [s_\delta(\varrho_n, \vartheta_n)]^+ \left(\mathbf{u}_B \cdot \mathbf{n} - [\mathbf{u}_B \cdot \mathbf{n}]_n^- \right) \ \mathrm{d}\sigma_x \mathrm{d}t \\
&\quad = \int_0^T \int_{\partial \Omega_n \setminus \partial \Omega} \varphi \varrho_n [s_\delta(\varrho_n, \vartheta_n)]^+ \left(\mathbf{u}_B \cdot \mathbf{n} - [\mathbf{u}_B \cdot \mathbf{n}]_n^- \right) \ \mathrm{d}\sigma_x \mathrm{d}t \\
&\quad + \int_0^T \int_{\partial \Omega} \varphi \mathbb{1}_{\partial \Omega_n} \varrho_n [s_\delta(\varrho_n, \vartheta_n)]^+ \left(\mathbf{u}_B \cdot \mathbf{n} - [\mathbf{u}_B \cdot \mathbf{n}]_n^- \right) \ \mathrm{d}\sigma_x \mathrm{d}t.
\end{aligned}
$$

In accordance with the uniform bounds (6.31), the functions $\varrho_n s_\delta(\varrho_n, \vartheta_n)$ are bounded in

$$L^r\left((0,T) \times \partial\Omega; \left(\mathbf{u}_B \cdot \mathbf{n} - [\mathbf{u}_B \cdot \mathbf{n}]_n^-\right) \mathrm{d}t \otimes \mathrm{d}\sigma_x\right)$$

for some $r > 1$. Consequently, by virtue of hypothesis (6.1) and the pointwise convergence of the traces of ϱ_n, ϑ_n established in (6.36), (6.43),

$$\int_0^T \int_{\partial\Omega_n \setminus \partial\Omega} \varphi \varrho_n [s_\delta(\varrho_n, \vartheta_n)]^+ \left(\mathbf{u}_B \cdot \mathbf{n} - [\mathbf{u}_B \cdot \mathbf{n}]_n^-\right) \, \mathrm{d}\sigma_x \mathrm{d}t \to 0,$$

while

$$\begin{aligned} &\int_0^T \int_{\partial\Omega} \varphi \mathbb{1}_{\partial\Omega_n} \varrho_n [s_\delta(\varrho_n, \vartheta_n)]^+ \left(\mathbf{u}_B \cdot \mathbf{n} - [\mathbf{u}_B \cdot \mathbf{n}]_n^-\right) \, \mathrm{d}\sigma_x \, \mathrm{d}t \\ &\quad \to \int_0^T \int_{\partial\Omega} \varphi \varrho [s_\delta(\varrho, \vartheta)]^+ \, [\mathbf{u}_B \cdot \mathbf{n}]^+ \mathrm{d}\sigma_x \mathrm{d}t \end{aligned} \tag{6.44}$$

as $n \to \infty$ for any bounded test function φ. Finally,

$$\begin{aligned} &\int_0^T \int_{\partial\Omega_n} \varphi \varrho_n [s_\delta(\varrho_n, \vartheta_n)]^- \left(\mathbf{u}_B \cdot \mathbf{n} - [\mathbf{u}_B \cdot \mathbf{n}]_n^-\right) \mathrm{d}\sigma_x \, \mathrm{d}t \\ &\quad \le \int_0^T \int_{\partial\Omega} \varphi \mathbb{1}_{\partial\Omega_n} \varrho_n [s_\delta(\varrho_n, \vartheta_n)]^- \left(\mathbf{u}_B \cdot \mathbf{n} - [\mathbf{u}_B \cdot \mathbf{n}]_n^-\right) \mathrm{d}t. \end{aligned}$$

Consequently, using the bounds (6.31) and the pointwise convergence of the traces, we may apply Fatou's lemma to conclude

$$\begin{aligned} &\limsup_{n \to \infty} \int_0^T \int_{\partial\Omega} \varphi \mathbb{1}_{\partial\Omega_n} \varrho_n [s_\delta(\varrho_n, \vartheta_n)]^- \left(\mathbf{u}_B \cdot \mathbf{n} - [\mathbf{u}_B \cdot \mathbf{n}]_n^-\right) \mathrm{d}t \\ &\quad \le \int_0^T \int_{\partial\Omega} \varphi \varrho [s_\delta(\varrho, \vartheta)]^- \, [\mathbf{u}_B \cdot \mathbf{n}]^+ \mathrm{d}\sigma_x \mathrm{d}t \end{aligned} \tag{6.45}$$

whenever $\varphi \ge 0$.

Now, we rewrite the last two integrals in (6.42) as

$$\begin{aligned} &\int_0^T \int_{\partial\Omega_n} \varphi \left[\delta \varrho_B (1 - \log(\vartheta_n)) - \frac{1}{\vartheta_n} f_{i,B} \right] [\mathbf{u}_B \cdot \mathbf{n}]_n^- \mathrm{d}\sigma_x \mathrm{d}t \\ &+ \int_0^T \int_{\partial\Omega_n} \varphi \varrho_B \left(\frac{e(\varrho_B, \vartheta_n)}{\vartheta_n} - s(\varrho_B, \vartheta_n) \right) [\mathbf{u}_B \cdot \mathbf{n}]_n^- \mathrm{d}\sigma_x \mathrm{d}t \end{aligned}$$

$$
\begin{aligned}
&= - \int_0^T \int_{\partial\Omega_n} \varphi \left(\frac{1}{3}\vartheta_n^3 + \Big(f_{i,B} - \varrho_B e_{\rm el}(\varrho_B) \Big) \frac{1}{\vartheta_n} \right) [\mathbf{u}_B \cdot \mathbf{n}]_n^- {\rm d}\sigma_x {\rm d}t \\
&+ \int_0^T \int_{\partial\Omega_n} \varphi \varrho_B \Big(\delta(1 - \log(\vartheta_n)) - s_{\rm m}(\varrho_B, \vartheta_n) \Big) [\mathbf{u}_B \cdot \mathbf{n}]_n^- {\rm d}\sigma_x {\rm d}t \\
&+ \int_0^T \int_{\partial\Omega_n} \varphi \frac{\varrho_B e_{\rm m}(\varrho_B, \vartheta_n)}{\vartheta_n} [\mathbf{u}_B \cdot \mathbf{n}]_n^- {\rm d}\sigma_x {\rm d}t.
\end{aligned}
$$

For $\varphi \geq 0$, we have

$$
\begin{aligned}
&- \int_0^T \int_{\partial\Omega_n} \varphi \left(\frac{1}{3}\vartheta_n^3 + (f_{i,B} - \varrho_B e_{\rm el}(\varrho_B)) \frac{1}{\vartheta_n} \right) [\mathbf{u}_B \cdot \mathbf{n}]_n^- {\rm d}\sigma_x {\rm d}t \\
&\geq - \int_0^T \int_{\partial\Omega} \varphi \mathbb{1}_{\partial\Omega_n} \left(\frac{1}{3}\vartheta_n^3 + \Big(f_{i,B} - \varrho_B e_{\rm el}(\varrho_B) \Big) \frac{1}{\vartheta_n} \right) [\mathbf{u}_B \cdot \mathbf{n}]_n^- \ dt.
\end{aligned}
$$

Consequently, in accordance with the uniform bounds (6.30), hypothesis (6.1), and the pointwise convergence of the traces, we may apply Fatou's lemma obtaining

$$
\begin{aligned}
&\liminf_{n\to\infty} \int_0^T \int_{\partial\Omega_n} -\varphi \left(\frac{1}{3}\vartheta_n^3 + \Big(f_{i,B} - \varrho_B e_{\rm el}(\varrho_B) \Big) \frac{1}{\vartheta_n} \right) [\mathbf{u}_B \cdot \mathbf{n}]_n^- {\rm d}\sigma_x {\rm d}t \\
&\quad \geq - \int_0^T \int_{\partial\Omega} \varphi \left(\frac{1}{3}\vartheta^3 + \Big(f_{i,B} - \varrho_B e_{\rm el}(\varrho_B) \Big) \frac{1}{\vartheta} \right) \ [\mathbf{u}_B \cdot \mathbf{n}]^- {\rm d}\sigma_x {\rm d}t
\end{aligned}
\tag{6.46}
$$

whenever $\varphi \geq 0$. Next,

$$
\begin{aligned}
&\int_0^T \int_{\partial\Omega_n} \varphi \varrho_B \Big(\delta(1 - \log(\vartheta_n)) - s_{\rm m}(\varrho_B, \vartheta_n) \Big) [\mathbf{u}_B \cdot \mathbf{n}]_n^- {\rm d}\sigma_x {\rm d}t \\
&= \int_0^T \int_{\partial\Omega_n \setminus \partial\Omega} \varphi \varrho_B \Big(\delta(1 - \log(\vartheta_n)) - s_{\rm m}(\varrho_B, \vartheta_n) \Big) [\mathbf{u}_B \cdot \mathbf{n}]_n^- {\rm d}\sigma_x {\rm d}t \\
&+ \int_0^T \int_{\partial\Omega} \varphi \mathbb{1}_{\partial\Omega_n} \varrho_B \Big(\delta(1 - \log(\vartheta_n)) - s_{\rm m}(\varrho_B, \vartheta_n) \Big) [\mathbf{u}_B \cdot \mathbf{n}]_n^- {\rm d}\sigma_x \ {\rm d}t \\
&\to \int_0^T \int_{\partial\Omega} \varphi \varrho_B \Big(\delta(1 - \log(\vartheta)) - s_M(\varrho_B, \vartheta) \Big) \ [\mathbf{u}_B \cdot \mathbf{n}]^- {\rm d}\sigma_x {\rm d}t,
\end{aligned}
\tag{6.47}
$$

where we have used the bounds (6.30) and the Lebesgue Convergence Theorem.

Finally, in accordance with the structural hypotheses (4.12), (4.13),

$$
\varrho e_{\rm m}(\varrho, \vartheta) \lesssim \varrho \vartheta.
$$

Consequently, repeating the above arguments, we may infer that

$$\int_0^T \int_{\partial\Omega_n} \varphi\varrho_B \frac{e_{\rm m}(\varrho_B,\vartheta_n)}{\vartheta_n}[\mathbf{u}_B\cdot\mathbf{n}]_n^- \mathrm{d}\sigma_x \mathrm{d}t \to \int_0^T \int_{\partial\Omega} \varphi\varrho_B \frac{e_{\rm m}(\varrho_B,\vartheta_n)}{\vartheta_n}[\mathbf{u}_B\cdot\mathbf{n}]^- \mathrm{d}\sigma_x \mathrm{d}t \tag{6.48}$$

for any $\varphi \in C^1_c([0,T)\times\Omega)$.

Gathering (6.44)–(6.48), we may evaluate the remaining limits in (6.42) and therefore pass to the limit of the approximate entropy balance:

$$\begin{aligned}
&-\int_\Omega \varrho_0 s_\delta(\varrho_0,\vartheta_0)\varphi(0,\cdot)\,\mathrm{d}x \\
&-\int_0^T\int_\Omega \varrho s_\delta(\varrho,\vartheta)\partial_t\varphi\,\mathrm{d}x\mathrm{d}t - \int_0^T\int_\Omega \varrho s_\delta(\varrho,\vartheta)\mathbf{u}\cdot\nabla_x\varphi\,\mathrm{d}x\mathrm{d}t \\
&+\int_0^T\int_{\partial\Omega}\varphi\varrho s_\delta[\mathbf{u}_B\cdot\mathbf{n}]^+\,\mathrm{d}\sigma_x\mathrm{d}t - \int_0^T\int_\Omega \frac{\mathbf{q}_\delta(\vartheta,\nabla_x\vartheta)}{\vartheta}\cdot\nabla_x\varphi\,\mathrm{d}x\mathrm{d}t \\
&\geq \int_0^T\int_\Omega \frac{\varphi}{\vartheta}\left(\mathbb{S}_\delta(\nabla_x\mathbf{u}):\nabla_x\mathbf{u} - \frac{\mathbf{q}\cdot\nabla_x\vartheta}{\vartheta} + \frac{\delta}{\vartheta^2}\right)\mathrm{d}x\mathrm{d}t \\
&+\frac{\delta}{2}\int_0^T\int_\Omega \frac{\varphi}{\vartheta^2}\left(\vartheta^\Gamma + \frac{1}{\vartheta}\right)|\nabla_x\vartheta|^2\,\mathrm{d}x\mathrm{d}t \\
&+\frac{\varepsilon}{2}\int_0^T\int_\Omega \frac{\varphi}{\vartheta}\delta\left(\Gamma\varrho^{\Gamma-2}+2\right)|\nabla_x\varrho|^2\,\mathrm{d}x\mathrm{d}t - \int_0^T\int_{\Omega_n}\vartheta^4\,\mathrm{d}x\mathrm{d}t \\
&+\varepsilon\int_0^T\int_\Omega \nabla_x\varrho\cdot\nabla_x\varphi\left(\frac{e_{\rm m}(\varrho,\vartheta)}{\vartheta} - s_{\rm m}(\varrho,\vartheta) + \frac{p_{\rm m}(\varrho,\vartheta)}{\varrho\vartheta}\right)\mathrm{d}x\mathrm{d}t \\
&+\varepsilon\int_0^T\int_\Omega \nabla_x\varrho\cdot\nabla_x\varphi\left(\delta - \delta\log(\vartheta)\right)\mathrm{d}x\mathrm{d}t \\
&+\int_0^T\int_{\partial\Omega}\varphi\left[\delta\varrho_B(1-\log(\vartheta)) - \frac{1}{\vartheta}f_{i,B}\right][\mathbf{u}_B\cdot\mathbf{n}]^-\mathrm{d}\sigma_x\mathrm{d}t \\
&+\int_0^T\int_{\partial\Omega}\varphi\varrho_B\left(\frac{e(\varrho_B,\vartheta)}{\vartheta} - s(\varrho_B,\vartheta)\right)[\mathbf{u}_B\cdot\mathbf{n}]^-\mathrm{d}\sigma_x\mathrm{d}t
\end{aligned} \tag{6.49}$$

for any $\varphi \in C^1_c([0,T)\times\overline{\Omega})$, $\varphi \geq 0$.

6.3.3 Momentum Equation

In view of the pointwise convergence of the approximate densities and temperatures stated in (6.35), (6.38), the limit in the approximate momentum equation can be performed as soon as we show

$$\varrho_n \mathbf{u}_n \otimes \mathbf{u}_n \to \overline{\varrho \mathbf{u} \otimes \mathbf{u}} \text{ weakly in } L^r((0,T) \times \Omega; R^{d \times d}) \text{ for some } r > 1,$$
$$\overline{\varrho \mathbf{u} \otimes \mathbf{u}} = \varrho \mathbf{u} \otimes \mathbf{u}, \tag{6.50}$$

and

$$\nabla_x \varrho_n \to \nabla_x \varrho \text{ in } L^2((0,T) \times \Omega; R^d). \tag{6.51}$$

To see (6.50), observe that the time derivative of the function

$$t \mapsto \int_\Omega \varrho_n \mathbf{u}_n(t, \cdot) \cdot \phi \ \mathrm{d}x, \ \phi \in X_n$$

can be computed via the approximate momentum balance (6.3). Applying the Arzelà–Ascoli Theorem, we get

$$\left[t \mapsto \int_\Omega \varrho_n \mathbf{u}_n(t, \cdot) \cdot \phi \ \mathrm{d}x \right] \to \left[t \mapsto \int_\Omega \varrho_n \mathbf{u}_n(t, \cdot) \cdot \phi \ \mathrm{d}x \right] \text{ in } C[0,T]$$

for any $\phi \in X_n$. Seeing that $\cup_{n=1}^\infty X_n$ is dense in $L^r(\Omega; R^d)$ for any finite r, we conclude

$$\varrho_n \mathbf{u}_n \to \varrho \mathbf{u} \text{ in } C_{\mathrm{weak}}([0,T]; L^{\frac{2\Gamma}{\Gamma+1}}(\Omega; R^d)). \tag{6.52}$$

Thus a direct application of the Lions–Aubin argument yields (6.50).

To see (6.51), we evoke the approximate equation of continuity (5.11)–(5.13) multiplied on $\varrho_n \phi$ (cf. (5.35)):

$$\begin{aligned}
&\left[\int_{\Omega_n} \frac{1}{2} \varrho_n^2 \ \mathrm{d}x \right]_{t=0}^{t=T} + \varepsilon \int_0^T \int_{\Omega_n} |\nabla_x \varrho_n|^2 \ \mathrm{d}x\mathrm{d}t \\
&+ \frac{1}{2} \int_{\partial\Omega_n} \varrho_n^2 \ \left(\mathbf{u}_B \cdot \mathbf{n} - [\mathbf{u}_B \cdot \mathbf{n}]_n^- \right) \mathrm{d}\sigma_x \mathrm{d}t \\
&- \frac{1}{2} \int_0^T \int_{\partial\Omega_n} (\varrho_n - \varrho_B)^2 [\mathbf{u}_B \cdot \mathbf{n}]_n^- \mathrm{d}\sigma_x \mathrm{d}t \\
&= - \frac{1}{2} \int_0^T \int_{\Omega_n} \frac{1}{2} \varrho_n^2 \mathrm{div}_x \mathbf{u}_n \ \mathrm{d}x\mathrm{d}t - \frac{1}{2} \int_{\partial\Omega_n} \varrho_B [\mathbf{u}_B \cdot \mathbf{n}]_n^- \mathrm{d}\sigma_x \mathrm{d}t.
\end{aligned} \tag{6.53}$$

A similar relation can be derived also for the limit problem via the abstract integration by parts formula (see Proposition 1, where $V = W^{1,2}(\Omega_n)$, $H = L^2(\Omega_n)$) applied to Eq. (6.37):

$$\begin{aligned}
&\left[\int_{\Omega} \frac{1}{2} \varrho^2 \ \mathrm{d}x \right]_{t=0}^{t=T} + \varepsilon \int_0^T \int_{\Omega} |\nabla_x \varrho|^2 \ \mathrm{d}x\mathrm{d}t + \frac{1}{2} \int_{\partial\Omega} \varrho^2 [\mathbf{u}_B \cdot \mathbf{n}]^+ \mathrm{d}\sigma_x \mathrm{d}t \\
&- \frac{1}{2} \int_0^T \int_{\partial\Omega} (\varrho - \varrho_B)^2 [\mathbf{u}_B \cdot \mathbf{n}]^- \mathrm{d}\sigma_x \mathrm{d}t \\
&= - \frac{1}{2} \int_0^T \int_{\Omega} \frac{1}{2} \varrho^2 \mathrm{div}_x \mathbf{u} \ \mathrm{d}x\mathrm{d}t - \frac{1}{2} \int_{\partial\Omega} \varrho_B [\mathbf{u}_B \cdot \mathbf{n}]^- \mathrm{d}\sigma_x \mathrm{d}t.
\end{aligned} \tag{6.54}$$

Thus letting $n \to \infty$ in (6.53) and using the weak lower semi-continuity of convex functions, we deduce

$$\limsup_{n \to \infty} \int_0^T \int_{\Omega_n} |\nabla_x \varrho_n|^2 \, \mathrm{d}x\mathrm{d}t \le \int_0^T \int_{\Omega} |\nabla_x \varrho|^2 \, \mathrm{d}x\mathrm{d}t, \tag{6.55}$$

which yields (6.51).

Thus the resulting approximate momentum equation reads

$$\begin{aligned} &-\int_{\Omega} \mathbf{m}_0 \cdot \boldsymbol{\varphi} \, \mathrm{d}x \\ &= \int_0^T \int_{\Omega} \Big[\varrho \mathbf{u} \cdot \partial_t \boldsymbol{\varphi} + \varrho \mathbf{u} \otimes \mathbf{u} : \nabla_x \boldsymbol{\varphi} + p_\delta(\varrho, \vartheta) \mathrm{div}_x \boldsymbol{\varphi} - \mathbb{S}_\delta : \nabla_x \boldsymbol{\varphi} \Big] \, \mathrm{d}x \\ &- \varepsilon \int_0^T \int_{\Omega} \nabla_x \varrho \cdot \nabla_x \mathbf{u} \cdot \boldsymbol{\varphi} \, \mathrm{d}x\mathrm{d}t + \int_0^T \int_{\Omega} \varrho \mathbf{g} \cdot \boldsymbol{\varphi} \, \mathrm{d}x\mathrm{d}t \end{aligned} \tag{6.56}$$

for any $\boldsymbol{\varphi} \in C^1_c([0,T); X_n)$ and any $n > 0$. The validity of (6.56) can be easily extended to the class $\boldsymbol{\varphi} \in C^1 \cap C_0([0,T) \times \Omega; R^d)$ via a density argument. Note that the necessary bounds to ensure equi-integrability of the pressure follow from (6.32) whenever $\Gamma > 6$.

6.3.3.1 Total Energy Balance

Our ultimate goal is to let $n \to \infty$ in the total energy balance (6.14). Repeating similar arguments used in the preceding part, it is not difficult to pass to the limit in the volume integrals. As for the surface integrals, the only difficult term reads

$$\begin{aligned} &\int_0^T \psi \int_{\partial\Omega_n} \varrho_n e_\delta \, \left(\mathbf{u}_B \cdot \mathbf{n} - [\mathbf{u}_B \cdot \mathbf{n}]_n^- \right) \mathrm{d}\sigma_x \mathrm{d}t \\ &\quad + \delta \int_0^T \int_{\partial\Omega_n} \left(\frac{1}{\Gamma - 1} \varrho_n^\Gamma + \varrho_n^2 \right) \, \left(\mathbf{u}_B \cdot \mathbf{n} - [\mathbf{u}_B \cdot \mathbf{n}]_n^- \right) \mathrm{d}\sigma_x \mathrm{d}t \\ &\quad \ge \int_0^T \psi \int_{\partial\Omega_n} \varrho_n e_\delta [\mathbf{u}_B \cdot \mathbf{n}]^+ \mathrm{d}\sigma_x \mathrm{d}t \\ &\quad + \delta \int_0^T \int_{\partial\Omega_n} \left(\frac{1}{\Gamma - 1} \varrho_n^\Gamma + \varrho_n^2 \right) [\mathbf{u}_B \cdot \mathbf{n}]^+ \, \mathrm{d}\sigma_x \mathrm{d}t, \ \psi \ge 0, \end{aligned} \tag{6.57}$$

appearing in the left-hand side of the inequality (6.14). Therefore it is enough to control it from below as $\psi \ge 0$. As the integrand being non-negative, we get

$$
\begin{aligned}
&\int_{\partial\Omega_n} \varrho_n e_\delta [\mathbf{u}_B \cdot \mathbf{n}]^+ \mathrm{d}\sigma_x \mathrm{d}t + \delta \int_0^\tau \int_{\partial\Omega_n} \left(\frac{1}{\Gamma - 1} \varrho_n^\Gamma + \varrho_n^2 \right) [\mathbf{u}_B \cdot \mathbf{n}]^+ \ \mathrm{d}\sigma_x \\
&\geq \int_{\partial\Omega} \mathbb{1}_{\partial\Omega_n} \varrho_n e_\delta [\mathbf{u}_B \cdot \mathbf{n}]^+ \mathrm{d}\sigma_x \mathrm{d}t \\
&+ \delta \int_0^\tau \int_{\partial\Omega} \mathbb{1}_{\partial\Omega_n} \left(\frac{1}{\Gamma - 1} \varrho_n^\Gamma + \varrho_n^2 \right) [\mathbf{u}_B \cdot \mathbf{n}]^+ \ \mathrm{d}\sigma_x.
\end{aligned}
$$

As the integrands converge $\mathrm{d}t \otimes \sigma_x$ a.a. in $(0, T) \times \partial\Omega$, the desired conclusion follows from Fatou's lemma.

Accordingly, the limit energy inequality reads

$$
\begin{aligned}
&\left[\int_\Omega \left[\frac{1}{2} \varrho |\mathbf{u} - \mathbf{u}_B|^2 + \delta \left(\frac{1}{\Gamma - 1} \varrho^\Gamma + \varrho^2 \right) + \varrho e_\delta \right] \psi \ \mathrm{d}x \right]_{t=0}^{t=\tau} \\
&- \int_0^\tau \partial_t \psi \int_\Omega \left[\frac{1}{2} \varrho |\mathbf{u} - \mathbf{u}_B|^2 + \delta \left(\frac{1}{\Gamma - 1} \varrho^\Gamma + \varrho^2 \right) + \varrho e_\delta \right] \ \mathrm{d}x\mathrm{d}t \\
&+ \int_0^\tau \psi \int_{\partial\Omega} \varrho e_\delta [\mathbf{u}_B \cdot \mathbf{n}]^+ \mathrm{d}\sigma_x \mathrm{d}t + \delta \int_0^\tau \int_{\partial\Omega} \left(\frac{1}{\Gamma - 1} \varrho^\Gamma + \varrho^2 \right) [\mathbf{u}_B \cdot \mathbf{n}]^+ \ \mathrm{d}\sigma_x \mathrm{d}t \\
&+ \int_0^\tau \psi \int_{\partial\Omega} f_{i,b} [\mathbf{u}_B \cdot \mathbf{n}]^- \mathrm{d}\sigma_x \mathrm{d}t \\
&\leq - \int_0^\tau \psi \int_\Omega [\varrho (\mathbf{u} - \mathbf{u}_B) \otimes (\mathbf{u} - \mathbf{u}_B) + p_\delta \mathbb{I} - \mathbb{S}_\delta (\vartheta, \mathbb{D}_x \mathbf{u})] : \mathbb{D}_x \mathbf{u}_B \ \mathrm{d}x\mathrm{d}t \\
&+ \int_0^\tau \psi \int_\Omega \varrho (\mathbf{g} - \partial_t \mathbf{u}_B + \mathbf{u}_B \cdot \nabla_x \mathbf{u}_B) \cdot (\mathbf{u} - \mathbf{u}_B) \ \mathrm{d}x\mathrm{d}t \\
&+ \int_0^\tau \psi \int_{\Omega_n} \left(\delta \frac{1}{\vartheta^2} - \varepsilon \vartheta^5 \right) \ \mathrm{d}x\mathrm{d}t \\
&- \delta \int_0^\tau \psi \int_{\partial\Omega} \left(\frac{1}{\Gamma - 1} \varrho_B^\Gamma + \varrho_B^2 \right) [\mathbf{u}_B \cdot \mathbf{n}]^- \ \mathrm{d}\sigma_x \mathrm{d}t \\
&+ \varepsilon \int_0^\tau \psi \int_\Omega \nabla_x \varrho \cdot \nabla_x (\mathbf{u} - \mathbf{u}_B) \cdot \mathbf{u}_B \ \mathrm{d}x\mathrm{d}t
\end{aligned}
\tag{6.58}
$$

for any $\psi \in C^1[0, T]$, $\psi \geq 0$.

6.4 Limit System

Having performed the limit $n \to \infty$ in the family of Galerkin/domain approximations, we have shown the existence of an approximate solution $(\varrho, \vartheta, \mathbf{u})$ satisfying the following system of equations on the limit domain Ω:

- **Approximate equation of continuity.**

$$\int_0^T \int_\Omega \left(\varrho \partial_t \varphi + \varrho \mathbf{u} \cdot \nabla_x \varphi - \varepsilon \nabla_x \varrho \cdot \nabla_x \varphi \right) \, dxdt$$
$$= \int_0^T \int_{\partial\Omega} \varphi \varrho_B [\mathbf{u}_B \cdot \mathbf{n}]^- \, d\sigma_x dt + \int_{\partial\Omega} \varphi \varrho [\mathbf{u}_B \cdot \mathbf{n}]^+ \, d\sigma_x dt$$
$$- \int_\Omega \varrho_{0,\delta} \varphi(0, \cdot) \, dx \tag{6.59}$$

for any $\varphi \in C^1_c([0,T) \times \overline{\Omega})$.

- **Approximate momentum balance.**

$$- \int_\Omega \mathbf{m}_0 \cdot \boldsymbol{\varphi}(0, \cdot) \, dx$$
$$= \int_0^T \int_\Omega \Big[\varrho \mathbf{u} \cdot \partial_t \boldsymbol{\varphi} + \varrho \mathbf{u} \otimes \mathbf{u} : \nabla_x \boldsymbol{\varphi} + p_\delta(\varrho, \vartheta) \mathrm{div}_x \boldsymbol{\varphi} - \mathbb{S}_\delta : \nabla_x \boldsymbol{\varphi} \Big] \, dx$$
$$- \varepsilon \int_0^T \int_\Omega \nabla_x \varrho \cdot \nabla_x \mathbf{u} \cdot \boldsymbol{\varphi} \, dxdt + \int_0^T \int_\Omega \varrho \mathbf{g} \cdot \boldsymbol{\varphi} \, dxdt \tag{6.60}$$

for any $\boldsymbol{\varphi} \in C^1 \cap C_0([0,T) \times \Omega; R^d)$.

- **Approximate entropy inequality.**

$$- \int_\Omega \varrho_{0,\delta} s_\delta(\varrho_{0,\delta}, \vartheta_{0,\delta}) \varphi(0, \cdot) \, dx$$
$$- \int_0^T \int_\Omega \varrho s_\delta(\varrho, \vartheta) \partial_t \varphi \, dxdt - \int_0^T \int_\Omega \varrho s_\delta(\varrho, \vartheta) \mathbf{u} \cdot \nabla_x \varphi \, dxdt$$
$$+ \int_0^T \int_{\partial\Omega} \varphi \varrho s_\delta [\mathbf{u}_B \cdot \mathbf{n}]^+ \, d\sigma_x dt - \int_0^T \int_\Omega \frac{\mathbf{q}_\delta(\vartheta, \nabla_x \vartheta)}{\vartheta} \cdot \nabla_x \varphi \, dxdt$$
$$\geq \int_0^T \int_\Omega \frac{\varphi}{\vartheta} \left(\mathbb{S}_\delta(\nabla_x \mathbf{u}) : \nabla_x \mathbf{u} - \frac{\mathbf{q} \cdot \nabla_x \vartheta}{\vartheta} + \frac{\delta}{\vartheta^2} \right) \, dxdt$$
$$+ \frac{\delta}{2} \int_0^T \int_\Omega \frac{\varphi}{\vartheta^2} \left(\vartheta^\Gamma + \frac{1}{\vartheta} \right) |\nabla_x \vartheta|^2 \, dxdt$$
$$+ \frac{\varepsilon}{2} \int_0^T \int_\Omega \frac{\varphi}{\vartheta} \delta \left(\Gamma \varrho^{\Gamma - 2} + 2 \right) |\nabla_x \varrho|^2 \, dxdt - \int_0^\tau \int_{\Omega_n} \vartheta^4 \, dxdt$$

$$
\begin{aligned}
&+\varepsilon \int_0^T \int_\Omega \nabla_x \varrho \cdot \nabla_x \varphi \left(\frac{e_{\rm m}(\varrho,\vartheta)}{\vartheta} - s_{\rm m}(\varrho,\vartheta) + \frac{p_{\rm m}(\varrho,\vartheta)}{\varrho\vartheta} \right) \,{\rm d}x{\rm d}t \\
&+\varepsilon \int_0^T \int_\Omega \nabla_x \varrho \cdot \nabla_x \varphi \left(\delta - \delta \log(\vartheta) \right) \,{\rm d}x{\rm d}t \\
&+\int_0^T \int_{\partial\Omega} \varphi \left[\delta \varrho_B (1 - \log(\vartheta)) - \frac{1}{\vartheta} f_{i,B} \right] [\mathbf{u}_B \cdot \mathbf{n}]^- {\rm d}\sigma_x {\rm d}t \\
&+\int_0^T \int_{\partial\Omega} \varphi \varrho_B \left(\frac{e(\varrho_B,\vartheta)}{\vartheta} - s(\varrho_B,\vartheta) \right) [\mathbf{u}_B \cdot \mathbf{n}]^- {\rm d}\sigma_x {\rm d}t
\end{aligned}
\tag{6.61}
$$

for any $\varphi \in C^1_c([0,T) \times \overline{\Omega})$, $\varphi \geq 0$.

- **Approximate energy balance.**

$$
\begin{aligned}
&\left[\int_\Omega \left[\frac{1}{2} \varrho |\mathbf{u} - \mathbf{u}_B|^2 + \delta \left(\frac{1}{\Gamma - 1} \varrho^\Gamma + \varrho^2 \right) + \varrho e_\delta \right] \psi \,{\rm d}x \right]_{t=0}^{t=\tau} \\
&- \int_0^\tau \partial_t \psi \int_\Omega \left[\frac{1}{2} \varrho |\mathbf{u} - \mathbf{u}_B|^2 + \delta \left(\frac{1}{\Gamma - 1} \varrho^\Gamma + \varrho^2 \right) + \varrho e_\delta \right] \,{\rm d}x{\rm d}t \\
&+ \int_0^\tau \psi \int_{\partial\Omega} \varrho e_\delta [\mathbf{u}_B \cdot \mathbf{n}]^+ {\rm d}\sigma_x {\rm d}t \\
&+ \delta \int_0^\tau \int_{\partial\Omega} \left(\frac{1}{\Gamma - 1} \varrho^\Gamma + \varrho^2 \right) [\mathbf{u}_B \cdot \mathbf{n}]^+ \,{\rm d}\sigma_x {\rm d}t \\
&+ \int_0^\tau \psi \int_{\partial\Omega} f_{i,B} [\mathbf{u}_B \cdot \mathbf{n}]^- {\rm d}\sigma_x {\rm d}t \\
&\leq - \int_0^\tau \psi \int_\Omega \left[\varrho (\mathbf{u} - \mathbf{u}_B) \otimes (\mathbf{u} - \mathbf{u}_B) + p_\delta \mathbb{I} - \mathbb{S}_\delta(\vartheta, \mathbb{D}_x \mathbf{u}) \right] : \mathbb{D}_x \mathbf{u}_B \,{\rm d}x{\rm d}t \\
&+ \int_0^\tau \psi \int_\Omega \varrho (\mathbf{g} - \partial_t \mathbf{u}_B + \mathbf{u}_B \cdot \nabla_x \mathbf{u}_B) \cdot (\mathbf{u} - \mathbf{u}_B) \,{\rm d}x{\rm d}t \\
&+ \int_0^\tau \psi \int_\Omega \left(\delta \frac{1}{\vartheta^2} - \varepsilon \vartheta^5 \right) \,{\rm d}x{\rm d}t \\
&- \delta \int_0^\tau \psi \int_{\partial\Omega} \left(\frac{1}{\Gamma - 1} \varrho_B^\Gamma + \varrho_B^2 \right) [\mathbf{u}_B \cdot \mathbf{n}]^- \,{\rm d}\sigma_x {\rm d}t \\
&+ \varepsilon \int_0^\tau \psi \int_\Omega \nabla_x \varrho \cdot \nabla_x (\mathbf{u} - \mathbf{u}_B) \cdot \mathbf{u}_B \,{\rm d}x{\rm d}t
\end{aligned}
\tag{6.62}
$$

for any $\psi \in C^1[0,T]$, $\psi \geq 0$.

6.4.1 Density Renormalization

For future use, it is convenient to record a *renormalized version* of the approximate equation of continuity (6.59). Multiplying (6.2) on $b'(\varrho_n)$, we obtain

$$\begin{aligned}\partial_t b(\varrho_n) + \mathrm{div}_x(b(\varrho_n)\mathbf{u}_n) + \Big(b'(\varrho_n)\varrho_n - b(\varrho_n)\Big)\mathrm{div}_x\mathbf{u}_n \\ = \varepsilon\Big(b'(\varrho_n)\nabla_x\varrho_n\Big) - \varepsilon b''(\varrho_n)|\nabla_x\varrho_n|^2 \qquad (6.63)\end{aligned}$$

in $(0,T)\times\Omega_n$, with the boundary conditions

$$\varepsilon b'(\varrho_n)\nabla_x\varrho_n\cdot\mathbf{n} = b'(\varrho_n)(\varrho_n - \varrho_B)[\mathbf{u}_B\cdot\mathbf{n}]_n^- \text{ on } (0,T)\times\partial\Omega_n. \qquad (6.64)$$

Multiplying (6.63) on a test function $\varphi\in C_c^1([0,T)\times\overline{\Omega}_n)$ and integrating by parts yield

$$\begin{aligned}&\int_0^T\int_{\Omega_n}\Big[b(\varrho_n)\partial_t\varphi + b(\varrho_n)\mathbf{u}_n\cdot\nabla_x\varphi + \Big(b(\varrho_n) - b'(\varrho_n)\varrho_n\Big)\mathrm{div}_x\mathbf{u}_n\varphi\Big]\ \mathrm{d}x\mathrm{d}t \\ &\quad - \int_0^T\int_{\Omega_n}\Big[\varepsilon b'(\varrho_n)\nabla_x\varrho_n\cdot\nabla_x\varphi + \varepsilon b''(\varrho_n)|\nabla_x\varrho_n|^2\varphi\Big]\ \mathrm{d}x\mathrm{d}t \\ &\quad = -\int_{\Omega_n} b(\varrho_{0,n,\delta})\varphi(0,\cdot)\ \mathrm{d}x \\ &\quad + \int_0^T\int_{\partial\Omega_n}\Big[b(\varrho_n)\mathbf{u}_B\cdot\mathbf{n} - b'(\varrho_n)(\varrho_n-\varrho_B)[\mathbf{u}_B\cdot\mathbf{n}]_n^-\Big]\ \mathrm{d}\sigma_x\mathrm{d}t. \qquad (6.65)\end{aligned}$$

Finally, using the strong convergence of the density and the density gradients established in (6.35) and (6.51), respectively, and the strong convergence of the density traces (6.36), we may let $n\to\infty$ in (6.65) obtaining

renormalized approximate equation of continuity:

$$\begin{aligned}&\int_0^T\int_{\Omega}\Big[b(\varrho)\partial_t\varphi + b(\varrho)\mathbf{u}\cdot\nabla_x\varphi + \Big(b(\varrho) - b'(\varrho)\varrho\Big)\mathrm{div}_x\mathbf{u}\varphi\Big]\ \mathrm{d}x\mathrm{d}t \\ &\quad - \int_0^T\int_{\Omega}\Big[\varepsilon b'(\varrho)\nabla_x\varrho\cdot\nabla_x\varphi + \varepsilon b''(\varrho)|\nabla_x\varrho|^2\varphi\Big]\ \mathrm{d}x\mathrm{d}t \\ &\quad = -\int_{\Omega_n} b(\varrho_{0,\delta})\varphi(0,\cdot)\ \mathrm{d}x \\ &\quad + \int_0^T\int_{\partial\Omega}\Big[b(\varrho)\mathbf{u}_B\cdot\mathbf{n} - b'(\varrho)(\varrho-\varrho_B)[\mathbf{u}_B\cdot\mathbf{n}]^-\Big]\ \mathrm{d}\sigma_x\mathrm{d}t \qquad (6.66)\end{aligned}$$

for any $\varphi \in C^1_c([0,T) \times \overline{\Omega})$ and any $b \in C^2(R)$, $|b''(\varrho)| \lesssim 1$. Note that the strong convergence of the gradients (6.51), together with the bound (6.55), yields

$$\int_0^T \int_{\Omega_n} b''(\varrho_n)|\nabla_x \varrho_n|^2 \varphi \ \mathrm{d}x\mathrm{d}t \to \int_0^T \int_{\Omega} b''(\varrho)|\nabla_x \varrho|^2 \varphi \ \mathrm{d}x.$$

Chapter 7
Vanishing Artificial Diffusion Limit

Our next goal is to perform the vanishing diffusion limit $\varepsilon \to 0$ in the sequence of approximate solutions constructed in Chap. 6. At this stage, the approximate solutions $\{\varrho_\varepsilon, \vartheta_\varepsilon, \mathbf{u}_\varepsilon\}_{\varepsilon>0}$ are already defined on a fixed spatial domain Ω and satisfy the following system of equations/inequalities identified in Chap. 6:

- **Approximate equation of continuity.**

$$\int_0^T \int_\Omega (\varrho_\varepsilon \partial_t \varphi + \varrho_\varepsilon \mathbf{u}_\varepsilon \cdot \nabla_x \varphi - \varepsilon \nabla_x \varrho_\varepsilon \cdot \nabla_x \varphi) \ \mathrm{d}x\mathrm{d}t$$
$$= \int_0^T \int_{\partial\Omega} \varphi \varrho_B [\mathbf{u}_B \cdot \mathbf{n}]^- \ \mathrm{d}\sigma_x \mathrm{d}t + \int_{\partial\Omega} \varphi \varrho_\varepsilon [\mathbf{u}_B \cdot \mathbf{n}]^+ \ \mathrm{d}\sigma_x \mathrm{d}t$$
$$- \int_\Omega \varrho_{0,\delta} \varphi(0,\cdot) \ \mathrm{d}x \tag{7.1}$$

for any $\varphi \in C^1_c([0,T) \times \overline{\Omega})$, together with the associated energy balance

$$\left[\int_\Omega \frac{1}{2} \varrho_\varepsilon^2 \ \mathrm{d}x \right]_{t=0}^{t=T} + \varepsilon \int_0^T \int_\Omega |\nabla_x \varrho_\varepsilon|^2 \ \mathrm{d}x\mathrm{d}t + \frac{1}{2} \int_{\partial\Omega} \varrho^2 [\mathbf{u}_B \cdot \mathbf{n}]^+ \mathrm{d}\sigma_x \mathrm{d}t$$
$$- \frac{1}{2} \int_0^T \int_{\partial\Omega} (\varrho_\varepsilon - \varrho_B)^2 [\mathbf{u}_B \cdot \mathbf{n}]^- \mathrm{d}\sigma_x \mathrm{d}t$$
$$= -\frac{1}{2} \int_0^T \int_\Omega \frac{1}{2} \varrho_\varepsilon^2 \mathrm{div}_x \mathbf{u} \ \mathrm{d}x\mathrm{d}t - \frac{1}{2} \int_{\partial\Omega} \varrho_B [\mathbf{u}_B \cdot \mathbf{n}]^- \mathrm{d}\sigma_x \mathrm{d}t. \tag{7.2}$$

E. Feireisl, A. Novotný, *Mathematics of Open Fluid Systems*, Nečas Center Series, https://doi.org/10.1007/978-3-030-94793-4_7

- **Approximate momentum balance.**

$$
\begin{aligned}
&-\int_{\Omega} \mathbf{m}_0 \cdot \boldsymbol{\varphi}(0,\cdot)\ \mathrm{d}x \\
&= \int_0^T \int_{\Omega} \Big[\varrho_\varepsilon \mathbf{u}_\varepsilon \cdot \partial_t \boldsymbol{\varphi} + \varrho_\varepsilon \mathbf{u}_\varepsilon \otimes \mathbf{u}_\varepsilon : \nabla_x \boldsymbol{\varphi} + p_\delta(\varrho_\varepsilon, \vartheta_\varepsilon) \mathrm{div}_x \boldsymbol{\varphi} \Big]\ \mathrm{d}x\mathrm{d}t \\
&- \int_0^T \int_{\Omega} \mathbb{S}_\delta(\vartheta_\varepsilon, \mathbb{D}_x \mathbf{u}_\varepsilon) : \nabla_x \boldsymbol{\varphi}\ \mathrm{d}x\mathrm{d}t \\
&- \varepsilon \int_0^T \int_{\Omega} \nabla_x \varrho_\varepsilon \cdot \nabla_x \mathbf{u}_\varepsilon \cdot \boldsymbol{\varphi}\ \mathrm{d}x\mathrm{d}t + \int_0^T \int_{\Omega} \varrho_\varepsilon \mathbf{g} \cdot \boldsymbol{\varphi}\ \mathrm{d}x\mathrm{d}t
\end{aligned} \tag{7.3}
$$

for any $\boldsymbol{\varphi} \in C^1_c([0,T) \times \Omega; R^d)$.

- **Approximate entropy inequality.**

$$
\begin{aligned}
&-\int_{\Omega} \varrho_{0,\delta} s_\delta(\varrho_{0,\delta}, \vartheta_{0,\delta}) \varphi(0,\cdot)\ \mathrm{d}x \\
&- \int_0^T \int_{\Omega} \varrho_\varepsilon s_\delta(\varrho_\varepsilon, \vartheta_\varepsilon) \partial_t \varphi\ \mathrm{d}x\mathrm{d}t - \int_0^T \int_{\Omega} \varrho_\varepsilon s_\delta(\varrho_\varepsilon, \vartheta_\varepsilon) \mathbf{u}_\varepsilon \cdot \nabla_x \varphi\ \mathrm{d}x\mathrm{d}t \\
&+ \int_0^T \int_{\partial\Omega} \varphi \varrho_\varepsilon s_\delta [\mathbf{u}_B \cdot \mathbf{n}]^+\ \mathrm{d}\sigma_x \mathrm{d}t - \int_0^T \int_{\Omega} \frac{\mathbf{q}_\delta(\vartheta_\varepsilon, \nabla_x \vartheta_\varepsilon)}{\vartheta_\varepsilon} \cdot \nabla_x \varphi\ \mathrm{d}x\mathrm{d}t \\
&\geq \int_0^T \int_{\Omega} \frac{\varphi}{\vartheta_\varepsilon} \left(\mathbb{S}_\delta(\vartheta_\varepsilon, \nabla_x \mathbf{u}_\varepsilon) : \mathbb{D}_x \mathbf{u}_\varepsilon - \frac{\mathbf{q}_\delta \cdot \nabla_x \vartheta_\varepsilon}{\vartheta_\varepsilon} + \frac{\delta}{\vartheta_\varepsilon^2} \right)\ \mathrm{d}x\mathrm{d}t \\
&+ \frac{\delta}{2} \int_0^T \int_{\Omega} \frac{\varphi}{\vartheta_c^2} \left(\vartheta_\varepsilon^\Gamma + \frac{1}{\vartheta_\varepsilon} \right) |\nabla_x \vartheta_\varepsilon|^2\ \mathrm{d}x\mathrm{d}t \\
&+ \frac{\varepsilon}{2} \int_0^T \int_{\Omega} \frac{\varphi}{\vartheta_\varepsilon} \delta \left(\Gamma \varrho_\varepsilon^{\Gamma-2} + 2 \right) |\nabla_x \varrho_\varepsilon|^2\ \mathrm{d}x\mathrm{d}t - \int_0^T \int_{\Omega_n} \vartheta_\varepsilon^4\ \mathrm{d}x\mathrm{d}t \\
&+ \varepsilon \int_0^T \int_{\Omega} \nabla_x \varrho_\varepsilon \cdot \nabla_x \varphi \left(\frac{e_{\mathrm{m}}(\varrho_\varepsilon, \vartheta_\varepsilon)}{\vartheta_\varepsilon} - s_{\mathrm{m}}(\varrho_\varepsilon, \vartheta_\varepsilon) + \frac{p_{\mathrm{m}}(\varrho_\varepsilon, \vartheta_\varepsilon)}{\varrho_\varepsilon \vartheta_\varepsilon} \right)\ \mathrm{d}x\mathrm{d}t \\
&+ \varepsilon \int_0^T \int_{\Omega} \nabla_x \varrho_\varepsilon \cdot \nabla_x \varphi \left(\delta - \delta \log(\vartheta_\varepsilon) \right)\ \mathrm{d}x\mathrm{d}t \\
&+ \int_0^T \int_{\partial\Omega} \varphi \left[\delta \varrho_B (1 - \log(\vartheta_\varepsilon)) - \frac{1}{\vartheta_\varepsilon} f_{i,B} \right] [\mathbf{u}_B \cdot \mathbf{n}]^- \mathrm{d}\sigma_x \mathrm{d}t \\
&+ \int_0^T \int_{\partial\Omega} \varphi \varrho_B \left(\frac{e(\varrho_B, \vartheta_\varepsilon)}{\vartheta_\varepsilon} - s(\varrho_B, \vartheta_\varepsilon) \right) [\mathbf{u}_B \cdot \mathbf{n}]^- \mathrm{d}\sigma_x \mathrm{d}t
\end{aligned} \tag{7.4}
$$

for any $\varphi \in C^1_c([0,T) \times \overline{\Omega})$, $\varphi \geq 0$.

- **Approximate energy balance.**

$$
\begin{aligned}
&\left[\int_\Omega \left[\frac{1}{2}\varrho_\varepsilon|\mathbf{u}_\varepsilon - \mathbf{u}_B|^2 + \delta\left(\frac{1}{\Gamma-1}\varrho_\varepsilon^\Gamma + \varrho_\varepsilon^2\right) + \varrho_\varepsilon e_\delta\right] \psi \ \mathrm{d}x\right]_{t=0}^{t=\tau} \\
&- \int_0^\tau \partial_t\psi \int_\Omega \left[\frac{1}{2}\varrho_\varepsilon|\mathbf{u}_\varepsilon - \mathbf{u}_B|^2 + \delta\left(\frac{1}{\Gamma-1}\varrho_\varepsilon^\Gamma + \varrho_\varepsilon^2\right) + \varrho_\varepsilon e_\delta\right] \ \mathrm{d}x\mathrm{d}t \\
&+ \int_0^\tau \psi \int_{\partial\Omega} \varrho_\varepsilon e_\delta [\mathbf{u}_B\cdot\mathbf{n}]^+ \mathrm{d}\sigma_x \mathrm{d}t \\
&+ \delta \int_0^\tau \int_{\partial\Omega} \left(\frac{1}{\Gamma-1}\varrho_\varepsilon^\Gamma + \varrho_\varepsilon^2\right) [\mathbf{u}_B\cdot\mathbf{n}]^+ \ \mathrm{d}\sigma_x\mathrm{d}t \\
&+ \int_0^\tau \psi \int_{\partial\Omega} f_{i,B}[\mathbf{u}_B\cdot\mathbf{n}]^- \mathrm{d}\sigma_x\mathrm{d}t \\
&\leq - \int_0^\tau \psi \int_\Omega [\varrho_\varepsilon(\mathbf{u}_\varepsilon - \mathbf{u}_B)\otimes(\mathbf{u}_\varepsilon - \mathbf{u}_B) \ \mathrm{d}x \\
&\quad + p_\delta \mathbb{I} - \mathbb{S}_\delta(\vartheta_\varepsilon, \mathbb{D}_x\mathbf{u}_\varepsilon)] : \mathbb{D}_x\mathbf{u}_B \mathrm{d}t \\
&+ \int_0^\tau \psi \int_\Omega \varrho_\varepsilon(\mathbf{g} - \partial_t\mathbf{u}_B + \mathbf{u}_B\cdot\nabla_x\mathbf{u}_B)\cdot(\mathbf{u}_\varepsilon - \mathbf{u}_B) \ \mathrm{d}x\mathrm{d}t \\
&+ \int_0^\tau \psi \int_\Omega \left(\delta\frac{1}{\vartheta_\varepsilon^2} - \varepsilon\vartheta_\varepsilon^5\right) \ \mathrm{d}x\mathrm{d}t \\
&- \delta \int_0^\tau \psi \int_{\partial\Omega} \left(\frac{1}{\Gamma-1}\varrho_B^\Gamma + \varrho_B^2\right)[\mathbf{u}_B\cdot\mathbf{n}]^- \ \mathrm{d}\sigma_x\mathrm{d}t \\
&+ \varepsilon \int_0^\tau \psi \int_\Omega \nabla_x\varrho_\varepsilon\cdot\nabla_x(\mathbf{u}_\varepsilon - \mathbf{u}_B)\cdot\mathbf{u}_B \ \mathrm{d}x\mathrm{d}t
\end{aligned} \tag{7.5}
$$

for any $\psi \in C_c^1[0,T)$, $\psi \geq 0$.

Finally, we recall the renormalized approximate equation of continuity (6.66),

$$
\begin{aligned}
&\int_0^T \int_\Omega \left[b(\varrho_\varepsilon)\partial_t\varphi + b(\varrho_\varepsilon)\mathbf{u}_\varepsilon\cdot\nabla_x\varphi + \Big(b(\varrho_\varepsilon) - b'(\varrho_\varepsilon)\varrho_\varepsilon\Big)\mathrm{div}_x\mathbf{u}_\varepsilon\varphi\right] \ \mathrm{d}x\mathrm{d}t \\
&\quad - \int_0^T \int_\Omega \left[\varepsilon b'(\varrho_\varepsilon)\nabla_x\varrho_\varepsilon\cdot\nabla_x\varphi + \varepsilon b''(\varrho_\varepsilon)|\nabla_x\varrho_\varepsilon|^2\varphi\right] \ \mathrm{d}x\mathrm{d}t \\
&\quad = - \int_\Omega b(\varrho_{0,\delta})\varphi(0,\cdot) \ \mathrm{d}x \\
&\quad + \int_0^T \int_{\partial\Omega} \varphi\left[b(\varrho_\varepsilon)\mathbf{u}_B\cdot\mathbf{n} - b'(\varrho_\varepsilon)(\varrho_\varepsilon - \varrho_B)[\mathbf{u}_B\cdot\mathbf{n}]^-\right]\mathrm{d}\sigma_x\mathrm{d}t
\end{aligned} \tag{7.6}
$$

for any $\varphi \in C^1_c([0,T) \times \overline{\Omega})$ and any $b \in C^2(R)$, $|b''(\varrho)| \lesssim 1$.

Remark 11. The reader will have noticed that (7.6) yields both (7.1) and (7.2) for a conveniently chosen b and φ.

7.1 Uniform Bounds

The uniform bounds are obtained in the same way as in Chap. 6 resulting from the "ballistic energy" inequality (6.20). Of course, certain estimates depend on the vanishing parameter ε. The list of relevant estimates is therefore a copy of (6.21)–(6.31):

$$\sup_{\tau\in[0,T]} \|\varrho_\varepsilon(\tau,\cdot)\|_{L^{\frac{5}{3}}(\Omega)} + \sup_{\tau\in[0,T]} \|\vartheta_\varepsilon(\tau,\cdot)\|_{L^4(\Omega)} \lesssim 1, \tag{7.7}$$

$$\sup_{\tau\in[0,T]} \|\sqrt{\varrho_\varepsilon}\mathbf{u}_\varepsilon(\tau,\cdot)\|_{L^2(\Omega;R^d)} \lesssim 1, \tag{7.8}$$

$$\sup_{\tau\in[0,T]} \delta \int_\Omega \varrho_\varepsilon^\Gamma(\tau,\cdot)\, \mathrm{d}x \lesssim 1, \tag{7.9}$$

$$\delta \int_0^T \int_\Omega \frac{1}{\vartheta_\varepsilon^3}\, \mathrm{d}x\mathrm{d}t + \varepsilon \int_0^T \int_\Omega \vartheta_\varepsilon^5\, \mathrm{d}x\mathrm{d}t \lesssim 1, \tag{7.10}$$

$$\delta \int_0^T \int_\Omega \left[|\nabla_x \mathbf{u}_\varepsilon|^2 + \left(\frac{1}{\vartheta_\varepsilon^3} + \vartheta_\varepsilon^{\Gamma-2} |\nabla_x \vartheta_\varepsilon|^2 \right) \right] \mathrm{d}x\mathrm{d}t \lesssim 1, \tag{7.11}$$

$$\varepsilon\delta \int_0^T \int_\Omega \frac{1}{\vartheta_\varepsilon} \left(\varrho_\varepsilon^{\Gamma-2} + 1 \right) |\nabla_x \varrho_\varepsilon|^2\, \mathrm{d}x\mathrm{d}t \lesssim 1 \tag{7.12}$$

$$\varepsilon \int_0^T \int_\Omega |\nabla_x \varrho_\varepsilon|^2\, \mathrm{d}x\mathrm{d}t \lesssim 1 \tag{7.13}$$

$$-\int_0^T \int_{\partial\Omega} \varrho_\varepsilon^2\, [\mathbf{u}_B \cdot \mathbf{n}]^-\, \mathrm{d}\sigma_x \mathrm{d}t + \int_0^T \int_{\partial\Omega} \varrho_\varepsilon^2\, [\mathbf{u}_B \cdot \mathbf{n}]^+ \mathrm{d}\sigma_x \mathrm{d}t \le c(\delta), \tag{7.14}$$

$$-\int_0^T \int_{\partial\Omega} \left(\frac{1}{\vartheta_\varepsilon} + \vartheta_\varepsilon^3 \right) [\mathbf{u}_B \cdot \mathbf{n}]^- \mathrm{d}\sigma_x \mathrm{d}t \lesssim 1, \tag{7.15}$$

$$\int_0^T \int_{\partial\Omega} \Big(\varrho_\varepsilon e_\delta(\varrho_\varepsilon, \vartheta_\varepsilon) - \varrho_\varepsilon s_\delta(\varrho_\varepsilon, \vartheta_\varepsilon) \Big) [\mathbf{u}_B \cdot \mathbf{n}]^+ \mathrm{d}\sigma_x \mathrm{d}t \lesssim 1. \tag{7.16}$$

7.1.1 Pressure Estimates

Unlike in Chap. 6, where the bound (6.32) guaranteed equi-integrability of the augmented pressure p_δ, the present estimates (7.7)–(7.16) are not sufficient to render the pressure term $p_\delta(\varrho_\varepsilon, \vartheta_\varepsilon)$ bounded in a reflexive space $L^r((0,T) \times \Omega)$, $r > 1$. To solve the problem, consider the quantity

$$\varphi(t,x) = \phi \nabla_x \Delta_x^{-1}[\phi b(\varrho_\varepsilon)], \quad \phi \in C_c^\infty(\Omega)$$

as a test function in the approximate momentum balance (7.3), where Δ_x^{-1} is the inverse of the Laplace operator on the space R^d defined via the convolution with the Poisson kernel. After a routine manipulation we arrive at

$$\begin{aligned}
&\int_0^T \int_\Omega p_\delta(\varrho_\varepsilon, \vartheta_\varepsilon) \left[\phi^2 b(\varrho_\varepsilon) + \nabla_x \phi \cdot \nabla_x \Delta_x^{-1}[\phi b(\varrho_\varepsilon)]\right] \, \mathrm{d}x\mathrm{d}t \\
&\quad = \left[\int_\Omega \phi \varrho_\varepsilon \mathbf{u}_\varepsilon \cdot \nabla_x \Delta_x^{-1}[\phi b(\varrho_\varepsilon)] \, \mathrm{d}x\right]_{t=0}^{t=T} \\
&\quad - \int_0^T \int_\Omega \phi \varrho_\varepsilon \mathbf{u}_\varepsilon \cdot \partial_t \Big(\nabla_x \Delta_x^{-1}[\phi b(\varrho_\varepsilon)]\Big) \, \mathrm{d}x\mathrm{d}t \\
&\quad - \int_0^T \int_\Omega \phi \varrho_\varepsilon \mathbf{u}_\varepsilon \otimes \mathbf{u}_\varepsilon : \nabla_x^2 \Delta_x^{-1}[\phi b(\varrho_\varepsilon)] \, \mathrm{d}x\mathrm{d}t \\
&\quad - \int_0^T \int_\Omega \varrho_\varepsilon \mathbf{u}_\varepsilon \otimes \mathbf{u}_\varepsilon \cdot \nabla_x \phi \cdot \nabla_x \Delta_x^{-1}[\phi b(\varrho_\varepsilon)] \, \mathrm{d}x\mathrm{d}t \\
&\quad + \int_0^T \int_\Omega \phi \mathbb{S}_\delta(\vartheta_\varepsilon, \mathbb{D}_x \mathbf{u}_\varepsilon) : \nabla_x^2 \Delta_x^{-1}[\phi b(\varrho_\varepsilon)] \, \mathrm{d}x\mathrm{d}t \\
&\quad + \int_0^T \int_\Omega \mathbb{S}_\delta(\vartheta_\varepsilon, \mathbb{D}_x \mathbf{u}_\varepsilon) \cdot \nabla_x \phi \cdot \nabla_x \Delta_x^{-1}[\phi b(\varrho_\varepsilon)] \, \mathrm{d}x\mathrm{d}t \\
&\quad - \int_0^T \int_\Omega \phi \varrho_\varepsilon \mathbf{g} \cdot \nabla_x \Delta_x^{-1}[\phi b(\varrho_\varepsilon)] \, \mathrm{d}x\mathrm{d}t \\
&\quad + \varepsilon \int_0^T \int_\Omega \phi \nabla_x \varrho_\varepsilon \cdot \nabla_x \mathbf{u}_\varepsilon \cdot \nabla_x \Delta_x^{-1}[\phi b(\varrho_\varepsilon)] \, \mathrm{d}x\mathrm{d}t. \qquad (7.17)
\end{aligned}$$

Our aim is to show that all integrals on the right-hand side of (7.17) are bounded in terms of the uniform bounds (7.7)–(7.16). To see this, we have to evaluate the time derivative

$$\partial_t \Big(\nabla_x \Delta_x^{-1}[\phi b(\varrho_\varepsilon)]\Big) = \nabla_x \Delta_x^{-1}[\phi \partial_t b(\varrho_\varepsilon)].$$

For $b(\varrho_\varepsilon) = \varrho_\varepsilon$ we have

$$\phi \partial_t \varrho_\varepsilon = -\phi \mathrm{div}_x(\varrho_\varepsilon \mathbf{u}_\varepsilon) + \varepsilon \phi \Delta_x \varrho_\varepsilon \text{ in } \mathcal{D}'((0,T) \times \Omega).$$

As ϕ is compactly supported, the boundary conditions become irrelevant. Consequently,

$$\begin{aligned}
\nabla_x \Delta_x^{-1}[\phi \partial_t \varrho_\varepsilon] &= -\nabla_x \Delta_x^{-1} \mathrm{div}_x(\phi \varrho_\varepsilon \mathbf{u}_\varepsilon) + \nabla_x \Delta_x^{-1}[\varrho_\varepsilon \mathbf{u}_\varepsilon \cdot \nabla_x \phi] \\
&\quad + \varepsilon \nabla_x \Delta_x^{-1} \mathrm{div}_x[\phi \nabla_x \varrho_\varepsilon] - \varepsilon \nabla_x \Delta_x^{-1}[\nabla_x \phi \nabla_x \varrho_\varepsilon].
\end{aligned}$$

In view of the uniform bound (7.13) and the smoothing properties of the operator Δ_x^{-1}, we get

$$\left\| \varepsilon \nabla_x \Delta_x^{-1} \mathrm{div}_x [\phi \nabla_x \varrho_\varepsilon] - \varepsilon \nabla_x \Delta_x^{-1} [\nabla_x \phi \nabla_x \varrho_\varepsilon] \right\|_{L^2((0,T)\times\Omega;R^d)} \leq c(\phi)\sqrt{\varepsilon}, \tag{7.18}$$

and, similarly,

$$\begin{aligned} &\left\| -\nabla_x \Delta_x^{-1} \mathrm{div}_x (\phi \varrho_\varepsilon \mathbf{u}_\varepsilon) + \nabla_x \Delta_x^{-1} [\varrho_\varepsilon \mathbf{u}_\varepsilon \cdot \nabla_x \phi] \right\|_{L^r(\Omega;R^d)} \\ &\leq c(r,\phi) \|\varrho_\varepsilon \mathbf{u}_\varepsilon\|_{L^r(\Omega;R^d)} \text{ for any } 1 < r < \infty. \end{aligned} \tag{7.19}$$

Now, going back to (7.17) we first observe that

$$\left\| \nabla_x \Delta_x^{-1} [\phi \varrho_\varepsilon] \right\|_{L^\infty((0,T)\times\Omega;R^d)} \leq c(\phi) \text{ provided } \Gamma > d \text{ in (7.9)} \tag{7.20}$$

and

$$\operatorname*{ess\ sup}_{\tau\in(0,T)} \|\nabla_x^2 \Delta_x^{-1} [\phi \varrho_\varepsilon]\|_{L^\Gamma(\Omega;R^{d\times d})} \leq c(\phi). \tag{7.21}$$

As $\Gamma > 6$, the estimates (7.18)–(7.21), together with the energy bounds (7.7)–(7.16) render the right-hand side of (7.17) bounded uniformly for $\varepsilon \to 0$. Keeping in mind that (7.18)–(7.21) depend on the compactly supported function ϕ, we deduce *local* pressure estimates,

$$\int_0^T \int_K p_\delta(\varrho_\varepsilon, \vartheta_\varepsilon)\varrho_\varepsilon \ \mathrm{d}x \ \mathrm{d}t \leq c(K) \text{ for any compact } K \subset \Omega. \tag{7.22}$$

To extend (7.22) up to the boundary, we use the same strategy, this time with a test function

$$\varphi(t,x) = \mathcal{B}[\Phi], \ \Phi \in L^q(\Omega), \ \int_\Omega \Phi \ \mathrm{d}x = 0,$$

where $\mathcal{B}$ is the Bogovskii operator introduced in Sect. 0.8. After a straightforward manipulation, we obtain

$$\begin{aligned} \int_0^T \int_\Omega \Phi p_\delta \ \mathrm{d}x\mathrm{d}t &= \left[\int_\Omega \varrho_\delta \mathbf{u}_\delta \cdot \mathcal{B}[\Phi] \ \mathrm{d}x \right]_{t=0}^{t=T} \\ &- \int_0^T \int_\Omega \varrho_\delta \mathbf{u}_\delta \otimes \mathbf{u}_\delta : \nabla_x \mathcal{B}[\Phi] \ \mathrm{d}x\mathrm{d}t + \int_0^T \int_\Omega \mathbb{S}_\delta : \nabla_x \mathcal{B}[\Phi] \ \mathrm{d}x\mathrm{d}t \\ &- \int_0^T \int_\Omega \varrho_\delta \mathbf{g} \cdot \mathcal{B}[\Phi] \ \mathrm{d}x\mathrm{d}t. \end{aligned} \tag{7.23}$$

As stated in Theorem 2, the operator $\mathcal{B}$ shares the smoothing properties with $\nabla_x \Delta_x^{-1}$ considered in the previous step. Using the available uniform bounds we may therefore check that the integral on the left-hand side of (7.23) is bounded. Choosing a suitable Φ singular near the boundary, we may combine

(7.23) with the already existing bound (7.22) to conclude

$$\int_0^T \int_\Omega p_\delta(\varrho_\varepsilon, \vartheta_\varepsilon) \mathrm{dist}^{-\omega}[x, \partial\Omega] \ \mathrm{d}x \ \mathrm{d}t \lesssim 1 \text{ for some } \omega > 0. \tag{7.24}$$

The estimates (7.22) and (7.24) imply equi-integrability of the pressure sequence $\{p_\delta(\varrho_\varepsilon, \vartheta_\varepsilon)\}_{\varepsilon>0}$.

7.2 Convergence $\varepsilon \to 0$

We have collected all available bounds to perform the limit $\varepsilon \to 0$ in the family of approximate equations. Similarly to Chap. 6, we may extract suitable subsequences,

$$\begin{aligned} &\varrho_\varepsilon \to \varrho \text{ in } C_{\mathrm{weak}}([0,T]; L^\Gamma(\Omega)), \\ &\vartheta_\varepsilon \to \vartheta \text{ weakly-(*) in } L^\infty(0,T; L^4(\Omega)), \\ &\mathbf{u}_\varepsilon \to \mathbf{u} \text{ weakly in } L^2(0,T; W^{1,2}(\Omega; R^d)). \end{aligned} \tag{7.25}$$

7.2.1 Approximate Equation of Continuity

Exactly as in Sect. 6.3.1, we show

$$\varrho_\varepsilon \mathbf{u}_\varepsilon \to \varrho \mathbf{u} \text{ weakly-(*) in } L^\infty(0,T; L^{\frac{2\Gamma}{\Gamma+1}}(\Omega; R^d)).$$

Moreover, as the parameter $\delta > 0$ is fixed at this stage, we can use the boundary estimates (7.4) to extract another subsequence such that

$$\varrho_\varepsilon \to \varrho \text{ weakly in } L^2((0,T) \times \partial\Omega; [\mathbf{u}_B \cdot \mathbf{n}]^+ (\mathrm{d}\sigma_x \otimes \mathrm{d}t)). \tag{7.26}$$

Finally, in view of (7.13),

$$\varepsilon \nabla_x \varrho_\varepsilon \to 0 \text{ in } L^2((0,T) \times \Omega; R^d).$$

Letting $\varepsilon \to 0$ in the approximate equation of continuity (7.1), we conclude

$$\begin{aligned} &\int_0^T \int_\Omega \Big(\varrho \partial_t \varphi + \varrho \mathbf{u} \cdot \nabla_x \varphi \Big) \ \mathrm{d}x\mathrm{d}t \\ &= \int_0^T \int_{\partial\Omega} \varphi \varrho_B [\mathbf{u}_B \cdot \mathbf{n}]^- \ \mathrm{d}\sigma_x \mathrm{d}t + \int_{\partial\Omega} \varphi \varrho [\mathbf{u}_B \cdot \mathbf{n}]^+ \ \mathrm{d}\sigma_x \mathrm{d}t \\ &- \int_\Omega \varrho_{0,\delta} \varphi(0, \cdot) \ \mathrm{d}x \end{aligned} \tag{7.27}$$

for any $\varphi \in C^1_c([0,T) \times \overline{\Omega})$.

7.2.2 Pointwise Convergence of the Temperature

Our next goal is to show, again for a suitable subsequence,

$$\vartheta_\varepsilon(t,x) \to \vartheta(t,x) \text{ for a.a. } (t,x) \in (0,T) \times \Omega. \tag{7.28}$$

The arguments are exactly the same as in Sect. 6.3.2 up to the identity (6.40)

$$\varrho_\varepsilon s_\delta(\varrho_\varepsilon,\vartheta_\varepsilon)\vartheta_\varepsilon \to \overline{\varrho s_\delta(\varrho,\vartheta)\vartheta} = \overline{\varrho s_\delta(\varrho,\vartheta)}\vartheta. \tag{7.29}$$

Now, seeing that

$$\overline{\varrho s_\delta(\varrho,\vartheta)\vartheta} = \overline{\varrho s_m(\varrho,\vartheta)\vartheta} + \delta\overline{\varrho\log(\vartheta)\vartheta} + \frac{4a}{3}\overline{\vartheta^3\vartheta},$$
$$\overline{\varrho s_\delta(\varrho,\vartheta)}\vartheta = \overline{\varrho s_m(\varrho,\vartheta)}\vartheta + \delta\overline{\varrho\log(\vartheta)}\vartheta + \frac{4a}{3}\overline{\vartheta^3}\vartheta$$

we have to show

$$\overline{\varrho s_m(\varrho,\vartheta)\vartheta} \geq \overline{\varrho s_m(\varrho,\vartheta)}\vartheta,\ \overline{\varrho\log(\vartheta)\vartheta} \geq \overline{\varrho\log(\vartheta)}\vartheta \tag{7.30}$$

to obtain the desired conclusion

$$\overline{\vartheta^3\vartheta} = \overline{\vartheta^3}\vartheta.$$

Unfortunately, in contrast with Chap. 6, the pointwise convergence of approximate densities is not available at this stage of the proof, and a refined argument must be used to show (7.30).

As the approximate densities satisfy the renormalized equation (7.6), we may use the available uniform bounds to deduce

$$b(\varrho_\varepsilon) \to \overline{b(\varrho)} \text{ in } C_{\text{weak}}([0,T];L^q(\Omega)),\ 1 < q < \infty$$

for any bounded b passing to a suitable subsequence as the case may be. Moreover, by virtue of the gradient bound (7.11), we may assume

$$g(\vartheta_\varepsilon) \to \overline{g(\vartheta_\varepsilon)} \text{ weakly in } L^2(0,T;W^{1,2}(\Omega))$$

for any globally Lipschitz g. We conclude that

$$\overline{b(\varrho)g(\vartheta)} = \overline{b(\varrho)}\ \overline{b(\vartheta)} \text{ for any } b,g \in C_c(R). \tag{7.31}$$

Next, as the entropy s_δ is increasing in ϑ,

$$\int_0^T\int_\Omega \Big(\varrho_\varepsilon s_\delta(\varrho_\varepsilon,\vartheta_\varepsilon) - \varrho_\varepsilon s_\delta(\varrho_\varepsilon,\vartheta)\Big)\Big(\vartheta_\varepsilon - \vartheta\Big)\ \mathrm{d}x\mathrm{d}t \geq 0,$$

where, in accordance with (7.29),

$$\lim_{\varepsilon\to 0}\int_\Omega \Big(\varrho_\varepsilon s_\delta(\varrho_\varepsilon,\vartheta_\varepsilon) - \varrho_\varepsilon s_\delta(\varrho_\varepsilon,\vartheta)\Big)\Big(\vartheta_\varepsilon - \vartheta\Big)\ \mathrm{d}x\mathrm{d}t$$
$$= \lim_{\varepsilon\to 0}\int_0^T\int_\Omega \varrho_\varepsilon s_\delta(\varrho_\varepsilon,\vartheta)(\vartheta-\vartheta_\varepsilon)\ \mathrm{d}x\mathrm{d}t.$$

We claim that (7.31) yields

$$\lim_{\varepsilon\to 0}\int_0^T\int_\Omega \varrho_\varepsilon s_\delta(\varrho_\varepsilon,\vartheta)(\vartheta-\vartheta_\varepsilon)\ \mathrm{d}x\mathrm{d}t = 0.$$

Indeed this follows for the observation that $(t,x,\varrho)\mapsto s_\delta(\varrho,\vartheta(t,x))$ can be seen as a Caratheodory function of ϱ and ϑ; whence the desired conclusion follows from (7.31) and the Fundamental Theorem of the theory of Young measures, see Ball [3] or Pedregal [79, Chapter 6, Theorem 6.2]. A detailed proof of this statement can be found in [43, Chapter 3, Section 3.6.2].

Thus we have shown

$$\int_0^T\int_\Omega \Big(\varrho_\varepsilon s_\delta(\varrho_\varepsilon,\vartheta_\varepsilon) - \varrho_\varepsilon s_\delta(\varrho_\varepsilon,\vartheta)\Big)\Big(\vartheta_\varepsilon - \vartheta\Big)\ \mathrm{d}x\mathrm{d}t \to 0 \text{ as } \varepsilon\to 0,$$

yielding

$$\overline{\vartheta^3\vartheta} = \overline{\vartheta^3}\vartheta$$

which implies (7.28) up to extracting another subsequence, cf. (6.41).

7.2.3 Pointwise Convergence of the Density

Our ultimate goal before performing the limit in the remaining field equations is to establish pointwise convergence of the approximate densities. This must be done both in the interior of Ω and on the outflow component of the boundary.

7.2.3.1 Lions Identity

The crucial tool for showing compactness of the family of approximate densities is an intimate relation between the limit $\overline{\varrho \mathrm{div}_x \mathbf{u}}$ and the pressure commonly called *Lions identity.* This is the fundamental idea behind the theory of weak solutions of the compressible Navier–Stokes system developed by P.-L. Lions [68].

We start with the pressure identity (7.17) evaluated for $b(\varrho_\varepsilon) = \varrho_\varepsilon$:

$$\int_0^T\int_\Omega p_\delta(\varrho_\varepsilon,\vartheta_\varepsilon)\left[\phi^2\varrho_\varepsilon + \nabla_x\phi\cdot\nabla_x\Delta_x^{-1}[\phi\varrho_\varepsilon]\right]\ \mathrm{d}x\mathrm{d}t$$

$$
\begin{aligned}
&= \left[\int_\Omega \phi \varrho_\varepsilon \mathbf{u}_\varepsilon \cdot \nabla_x \Delta_x^{-1}[\phi \varrho_\varepsilon] \, \mathrm{d}x \right]_{t=0}^{t=T} \\
&- \int_0^T \int_\Omega \phi \varrho_\varepsilon \mathbf{u}_\varepsilon \cdot \partial_t \Big(\nabla_x \Delta_x^{-1}[\phi \varrho_\varepsilon] \Big) \, \mathrm{d}x\mathrm{d}t \\
&- \int_0^T \int_\Omega \phi \varrho_\varepsilon \mathbf{u}_\varepsilon \otimes \mathbf{u}_\varepsilon : \nabla_x^2 \Delta_x^{-1}[\phi \varrho_\varepsilon] \, \mathrm{d}x\mathrm{d}t \\
&- \int_0^T \int_\Omega \varrho_\varepsilon \mathbf{u}_\varepsilon \otimes \mathbf{u}_\varepsilon \cdot \nabla_x \phi \cdot \nabla_x \Delta_x^{-1}[\phi \varrho_\varepsilon] \, \mathrm{d}x\mathrm{d}t \\
&+ \int_0^T \int_\Omega \phi \mathbb{S}_\delta(\vartheta_\varepsilon, \mathbb{D}_x \mathbf{u}_\varepsilon) : \nabla_x^2 \Delta_x^{-1}[\phi \varrho_\varepsilon] \, \mathrm{d}x\mathrm{d}t \\
&+ \int_0^T \int_\Omega \mathbb{S}_\delta(\vartheta_\varepsilon, \mathbb{D}_x \mathbf{u}_\varepsilon) \cdot \nabla_x \phi \cdot \nabla_x \Delta_x^{-1}[\phi \varrho_\varepsilon] \, \mathrm{d}x\mathrm{d}t \\
&- \int_0^T \int_\Omega \phi \varrho_\varepsilon \mathbf{g} \cdot \nabla_x \Delta_x^{-1}[\phi \varrho_\varepsilon] \, \mathrm{d}x\mathrm{d}t \\
&+ \varepsilon \int_0^T \int_\Omega \phi \nabla_x \varrho_\varepsilon \cdot \nabla_x \mathbf{u}_\varepsilon \cdot \nabla_x \Delta_x^{-1}[\phi \varrho_\varepsilon] \, \mathrm{d}x\mathrm{d}t, \qquad (7.32)
\end{aligned}
$$

with

$$
\begin{aligned}
\nabla_x \Delta_x^{-1}[\phi \partial_t \varrho_\varepsilon] &= -\nabla_x \Delta_x^{-1} \mathrm{div}_x(\phi \varrho_\varepsilon \mathbf{u}_\varepsilon) + \nabla_x \Delta_x^{-1}[\varrho_\varepsilon \mathbf{u}_\varepsilon \cdot \nabla_x \phi] \\
&+ \varepsilon \nabla_x \Delta_x^{-1} \mathrm{div}_x[\phi \nabla_x \varrho_\varepsilon] - \varepsilon \nabla_x \Delta_x^{-1}[\nabla_x \phi \nabla_x \varrho_\varepsilon]
\end{aligned}
$$

for $\phi \in C_c^1(\Omega)$.

Using the uniform bounds (7.7)–(7.16), the pressure estimates (7.23) and (7.24), the strong convergence of the temperature (7.28), and the smoothing properties of the operator Δ_x^{-1}, we let $\varepsilon \to 0$ in (7.32) obtaining

$$
\begin{aligned}
&\lim_{\varepsilon \to 0} \int_0^T \int_\Omega \phi^2 p_\delta(\varrho_\varepsilon, \vartheta_\varepsilon) \varrho_\varepsilon \, \mathrm{d}x\mathrm{d}t + \int_0^T \int_\Omega \overline{p_\delta(\varrho, \vartheta)} \nabla_x \phi \cdot \nabla_x \Delta_x^{-1}[\phi \varrho] \, \mathrm{d}x\mathrm{d}t \\
&= \left[\int_\Omega \phi \varrho \mathbf{u} \cdot \nabla_x \Delta_x^{-1}[\phi \varrho] \, \mathrm{d}x \right]_{t=0}^{t=T} \\
&- \int_0^T \int_\Omega \phi \varrho \mathbf{u} \cdot \nabla_x \Delta_x^{-1}[\varrho \mathbf{u} \cdot \nabla_x \phi] \, \mathrm{d}x\mathrm{d}t \\
&+ \lim_{\varepsilon \to 0} \int_0^T \int_\Omega \phi \varrho_\varepsilon \mathbf{u}_\varepsilon \cdot \nabla_x \Delta_x^{-1} \mathrm{div}_x(\phi \varrho_\varepsilon \mathbf{u}_\varepsilon) \, \mathrm{d}x\mathrm{d}t \\
&- \lim_{\varepsilon \to 0} \int_0^T \int_\Omega \phi \varrho_\varepsilon \mathbf{u}_\varepsilon \otimes \mathbf{u}_\varepsilon : \nabla_x^2 \Delta_x^{-1}[\phi \varrho_\varepsilon] \, \mathrm{d}x\mathrm{d}t \\
&- \int_0^T \int_\Omega \varrho \mathbf{u} \otimes \mathbf{u} \cdot \nabla_x \phi \cdot \nabla_x \Delta_x^{-1}[\phi \varrho] \, \mathrm{d}x\mathrm{d}t
\end{aligned}
$$

$$+ \lim_{\varepsilon \to 0} \int_0^T \int_\Omega \phi \mathbb{S}_\delta(\vartheta_\varepsilon, \mathbb{D}_x \mathbf{u}_\varepsilon) : \nabla_x^2 \Delta_x^{-1}[\phi \varrho_\varepsilon] \, dx dt$$
$$+ \int_0^T \int_\Omega \mathbb{S}_\delta(\vartheta, \mathbb{D}_x \mathbf{u}) \cdot \nabla_x \phi \cdot \nabla_x \Delta_x^{-1}[\phi \varrho] \, dx dt$$
$$- \int_0^T \int_\Omega \phi \varrho \mathbf{g} \cdot \nabla_x \Delta_x^{-1}[\phi \varrho] \, dx dt. \tag{7.33}$$

The next step is letting $\varepsilon \to 0$ in the approximate momentum equation (7.3):

$$- \int_\Omega \mathbf{m}_0 \cdot \boldsymbol{\varphi}(0, \cdot) \, dx$$
$$= \int_0^T \int_\Omega \left[\varrho \mathbf{u} \cdot \partial_t \boldsymbol{\varphi} + \varrho \mathbf{u} \otimes \mathbf{u} : \nabla_x \boldsymbol{\varphi} + \overline{p_\delta(\varrho, \vartheta)} \mathrm{div}_x \boldsymbol{\varphi} \right] \, dx dt$$
$$- \int_0^T \int_\Omega \mathbb{S}_\delta(\vartheta, \mathbb{D}_x \mathbf{u}) : \nabla_x \boldsymbol{\varphi} \, dx dt + \int_0^T \int_\Omega \varrho \mathbf{g} \cdot \boldsymbol{\varphi} \, dx dt \tag{7.34}$$

for any $\boldsymbol{\varphi} \in C_c^1([0,T] \times \Omega; R^d)$.

As we already know the limit $(\varrho, \mathbf{u})$ satisfies the equation of continuity, we can use

$$\boldsymbol{\varphi}(t,x) = \phi \nabla_x \Delta_x^{-1}[\phi b(\varrho_\varepsilon)], \quad \phi \in C_c^\infty(\Omega)$$

as a test function in (7.34):

$$\int_0^T \int_\Omega \phi^2 \overline{p_\delta(\varrho, \vartheta)} \varrho \, dx dt + \int_0^T \int_\Omega \overline{p_\delta(\varrho, \vartheta)} \nabla_x \phi \cdot \nabla_x \Delta_x^{-1}[\phi \varrho] \, dx dt$$
$$= \left[\int_\Omega \phi \varrho \mathbf{u} \cdot \nabla_x \Delta_x^{-1}[\phi \varrho] \, dx \right]_{t=0}^{t=T}$$
$$- \int_0^T \int_\Omega \phi \varrho \mathbf{u} \cdot \nabla_x \Delta_x^{-1}[\varrho \mathbf{u} \cdot \nabla_x \phi] \, dx dt$$
$$+ \int_0^T \int_\Omega \phi \varrho \mathbf{u} \cdot \nabla_x \Delta_x^{-1} \mathrm{div}_x(\phi \varrho \mathbf{u}) \, dx dt$$
$$- \int_0^T \int_\Omega \phi \varrho \mathbf{u} \otimes \mathbf{u} : \nabla_x^2 \Delta_x^{-1}[\phi \varrho] \, dx dt$$
$$- \int_0^T \int_\Omega \varrho \mathbf{u} \otimes \mathbf{u} \cdot \nabla_x \phi \cdot \nabla_x \Delta_x^{-1}[\phi \varrho] \, dx dt$$
$$+ \int_0^T \int_\Omega \phi \mathbb{S}_\delta(\vartheta, \mathbb{D}_x \mathbf{u}) : \nabla_x^2 \Delta_x^{-1}[\phi \varrho] \, dx dt$$
$$+ \int_0^T \int_\Omega \mathbb{S}_\delta(\vartheta, \mathbb{D}_x \mathbf{u}) \cdot \nabla_x \phi \cdot \nabla_x \Delta_x^{-1}[\phi \varrho] \, dx dt$$

$$- \int_0^T \int_\Omega \phi\varrho\mathbf{g} \cdot \nabla_x \Delta_x^{-1}[\phi\varrho] \, \mathrm{d}x\mathrm{d}t. \tag{7.35}$$

Putting together (7.33) and (7.35) we conclude

$$\begin{aligned}
\lim_{\varepsilon\to 0} \int_0^T \int_\Omega \phi^2 p_\delta(\varrho_\varepsilon, \vartheta_\varepsilon)\varrho_\varepsilon \, \mathrm{d}x\mathrm{d}t - \int_0^T \int_\Omega \phi^2 \overline{p_\delta(\varrho,\vartheta)\varrho} \, \mathrm{d}x\mathrm{d}t \\
= \lim_{\varepsilon\to 0} \int_0^T \int_\Omega \phi\varrho_\varepsilon \mathbf{u}_\varepsilon \cdot \nabla_x \Delta_x^{-1} \mathrm{div}_x(\phi\varrho_\varepsilon\mathbf{u}_\varepsilon) \, \mathrm{d}x\mathrm{d}t \\
- \int_0^T \int_\Omega \phi\varrho\mathbf{u} \cdot \nabla_x \Delta_x^{-1} \mathrm{div}_x(\phi\varrho\mathbf{u}) \, \mathrm{d}x\mathrm{d}t \\
- \lim_{\varepsilon\to 0} \int_0^T \int_\Omega \phi\varrho_\varepsilon \mathbf{u}_\varepsilon \otimes \mathbf{u}_\varepsilon : \nabla_x^2 \Delta_x^{-1}[\phi\varrho_\varepsilon] \, \mathrm{d}x\mathrm{d}t \\
+ \int_0^T \int_\Omega \phi\varrho\mathbf{u} \otimes \mathbf{u} : \nabla_x^2 \Delta_x^{-1}[\phi\varrho] \, \mathrm{d}x\mathrm{d}t \\
+ \lim_{\varepsilon\to 0} \int_0^T \int_\Omega \phi\mathbb{S}_\delta(\vartheta_\varepsilon, \mathbb{D}_x\mathbf{u}_\varepsilon) : \nabla_x^2 \Delta_x^{-1}[\phi\varrho_\varepsilon] \, \mathrm{d}x\mathrm{d}t \\
- \int_0^T \int_\Omega \phi\mathbb{S}_\delta(\vartheta, \mathbb{D}_x\mathbf{u}) : \nabla_x^2 \Delta_x^{-1}[\phi\varrho] \, \mathrm{d}x\mathrm{d}t.
\end{aligned} \tag{7.36}$$

Now, we use Compactness Lemma (Lemma 2) to show

$$\begin{aligned}
\lim_{\varepsilon\to 0} \int_0^T \int_\Omega \phi\varrho_\varepsilon \mathbf{u}_\varepsilon \cdot \nabla_x \Delta_x^{-1} \mathrm{div}_x(\phi\varrho_\varepsilon\mathbf{u}_\varepsilon) \, \mathrm{d}x\mathrm{d}t \\
- \lim_{\varepsilon\to 0} \int_0^T \int_\Omega \phi\varrho_\varepsilon \mathbf{u}_\varepsilon \otimes \mathbf{u}_\varepsilon : \nabla_x^2 \Delta_x^{-1}[\phi\varrho_\varepsilon] \, \mathrm{d}x\mathrm{d}t \\
= \lim_{\varepsilon\to 0} \int_0^T \int_{R^d} \phi\mathbf{u}_\varepsilon \cdot \Big[\varrho_\varepsilon \nabla_x \Delta_x^{-1} \mathrm{div}_x(\phi\varrho_\varepsilon\mathbf{u}_\varepsilon) - \varrho_\varepsilon\mathbf{u}_\varepsilon \cdot \nabla_x \Delta_x^{-1}[\phi\varrho_\varepsilon]\Big] \mathrm{d}x\mathrm{d}t \\
= \int_0^T \int_{R^d} \phi\mathbf{u} \cdot \Big[\varrho \nabla_x \Delta_x^{-1} \mathrm{div}_x(\phi\varrho\mathbf{u}) - \varrho\mathbf{u} \cdot \nabla_x \Delta_x^{-1}[\phi\varrho]\Big] \mathrm{d}x\mathrm{d}t.
\end{aligned} \tag{7.37}$$

Indeed, as

$$\varrho_\varepsilon \to \varrho \text{ in } C_{\mathrm{weak}}([0,T]; L^\Gamma(\Omega)), \;\; \varrho_\varepsilon\mathbf{u}_\varepsilon \to \varrho\mathbf{u} \text{ in } C_{\mathrm{weak}}([0,T]; L^{\frac{2\Gamma}{\Gamma+1}}(\Omega; R^d)),$$

Lemma 2 yields

$$\begin{aligned}&\varrho_\varepsilon \nabla_x \Delta_x^{-1} \mathrm{div}_x(\phi \varrho_\varepsilon \mathbf{u}_\varepsilon) - \varrho_\varepsilon \mathbf{u}_\varepsilon \cdot \nabla_x \Delta_x^{-1}[\phi \varrho_\varepsilon] \\ &\to \varrho \nabla_x \Delta_x^{-1} \mathrm{div}_x(\phi \varrho \mathbf{u}) - \varrho \mathbf{u} \cdot \nabla_x \Delta_x^{-1}[\phi \varrho] \text{ in } L^q(\Omega),\ q = \frac{2\Gamma}{2 + \Gamma + 1}\end{aligned}$$

for *any* $t \in [0, T]$, while

$$\mathbf{u}_\varepsilon \to \mathbf{u} \text{ weakly in } L^2(0, T; W^{1,2}(\Omega; R^d)),$$

which yields the desired conclusion.

Consequently, relation (7.36) reduces to

$$\begin{aligned}\lim_{\varepsilon \to 0} \int_0^T \int_\Omega \phi^2 p_\delta(\varrho_\varepsilon, \vartheta_\varepsilon) \varrho_\varepsilon \ \mathrm{d}x\mathrm{d}t &- \int_0^T \int_\Omega \phi^2 \overline{p_\delta(\varrho, \vartheta)} \varrho \ \mathrm{d}x\mathrm{d}t \\ &= \lim_{\varepsilon \to 0} \int_0^T \int_\Omega \phi \mathbb{S}_\delta(\vartheta_\varepsilon, \mathbb{D}_x \mathbf{u}_\varepsilon) : \nabla_x^2 \Delta_x^{-1}[\phi \varrho_\varepsilon] \ \mathrm{d}x\mathrm{d}t \\ &- \int_0^T \int_\Omega \phi \mathbb{S}_\delta(\vartheta, \mathbb{D}_x \mathbf{u}) : \nabla_x^2 \Delta_x^{-1}[\phi \varrho] \ \mathrm{d}x\mathrm{d}t. \qquad (7.38)\end{aligned}$$

Finally, write

$$\begin{aligned}&\int_\Omega \phi \mathbb{S}_\delta(\vartheta_\varepsilon, \mathbb{D}_x \mathbf{u}_\varepsilon) : \nabla_x^2 \Delta_x^{-1}[\phi \varrho_\varepsilon] \ \mathrm{d}x - \int_\Omega \phi \mathbb{S}_\delta(\vartheta, \mathbb{D}_x \mathbf{u}) : \nabla_x^2 \Delta_x^{-1}[\phi \varrho] \ \mathrm{d}x \\ &= \int_{R^d} \Big[\phi \mathbb{S}_\delta(\vartheta_\varepsilon, \mathbb{D}_x \mathbf{u}_\varepsilon) : \nabla_x^2 \Delta_x^{-1}[\phi \varrho_\varepsilon] - \phi \mathbb{S}_\delta(\vartheta, \mathbb{D}_x \mathbf{u}) : \nabla_x^2 \Delta_x^{-1}[\phi \varrho] \Big] \mathrm{d}x \\ &= \int_{R^d} \Big[\phi \varrho_\varepsilon \nabla_x^2 \Delta_x^{-1} : [\phi \mathbb{S}_\delta(\vartheta_\varepsilon, \mathbb{D}_x \mathbf{u}_\varepsilon)] - \phi \varrho \nabla_x^2 \Delta_x^{-1} : [\phi \mathbb{S}_\delta(\vartheta, \mathbb{D}_x \mathbf{u})] \Big] \mathrm{d}x.\end{aligned}$$

Furthermore,

$$\begin{aligned}\nabla_x^2 \Delta_x^{-1} : \Big[\phi \mathbb{S}(\vartheta_\varepsilon, \nabla_x \mathbf{u}_\varepsilon) \Big] &= \phi \left(\frac{4}{3} \mu(\vartheta_\varepsilon) + \eta(\vartheta_\varepsilon) \right) \mathrm{div}_x \mathbf{u}_\varepsilon \\ &+ \nabla_x^2 \Delta_x^{-1} : \Big[\phi \mathbb{S}(\vartheta_\varepsilon, \nabla_x \mathbf{u}_\varepsilon \Big] - \phi \left(\frac{4}{3} \mu(\vartheta_\varepsilon) + \eta(\vartheta_\varepsilon) \right) \mathrm{div}_x \mathbf{u}_\varepsilon,\end{aligned}$$

and, similarly,

$$\begin{aligned}\nabla_x^2 \Delta_x^{-1} : \Big[\phi \mathbb{S}(\vartheta, \nabla_x \mathbf{u}) \Big] &= \phi \left(\frac{4}{3} \mu(\vartheta) + \eta(\vartheta) \right) \mathrm{div}_x \mathbf{u} \\ &+ \nabla_x^2 \Delta_x^{-1} : \Big[\phi \mathbb{S}(\vartheta, \nabla_x \mathbf{u}) \Big] - \phi \left(\frac{4}{3} \mu(\vartheta) + \eta(\vartheta) \right) \mathrm{div}_x \mathbf{u}.\end{aligned}$$

Seeing that

$$\varrho_\varepsilon \to \varrho \text{ in } C_{\mathrm{weak}}([0, T]; L^\Gamma(\Omega)),$$

we may use Commutator Lemma (Lemma 3) to conclude

$$\lim_{\varepsilon \to 0} \int_0^T \int_\Omega \phi \mathbb{S}_\delta(\vartheta_\varepsilon, \mathbb{D}_x \mathbf{u}_\varepsilon) : \nabla_x^2 \Delta_x^{-1}[\phi \varrho_\varepsilon] \, \mathrm{d}x\mathrm{d}t$$
$$- \int_0^T \int_\Omega \phi \mathbb{S}_\delta(\vartheta, \mathbb{D}_x \mathbf{u}) : \nabla_x^2 \Delta_x^{-1}[\phi \varrho] \, \mathrm{d}x\mathrm{d}t$$
$$= \int_0^T \int_\Omega \phi^2 \left(\frac{4}{3}\mu(\vartheta) + \eta(\vartheta) \right) \left(\overline{\varrho \mathrm{div}_x \mathbf{u}} - \varrho \mathrm{div}_x \mathbf{u} \right) \mathrm{d}x\mathrm{d}t.$$

Going back to (7.38) we obtain

the **effective viscous flux identity (Lions' identity):**

$$\lim_{\varepsilon \to 0} \int_0^T \int_\Omega \phi^2 \Big[p_\delta(\varrho_\varepsilon, \vartheta_\varepsilon)\varrho_\varepsilon - \overline{p_\delta(\varrho, \vartheta)\varrho} \Big] \, \mathrm{d}x\mathrm{d}t$$
$$= \int_0^T \int_\Omega \phi^2 \left(\frac{4}{3}\mu(\vartheta) + \eta(\vartheta) \right) \left(\overline{\varrho \mathrm{div}_x \mathbf{u}} - \varrho \mathrm{div}_x \mathbf{u} \right) \mathrm{d}x\mathrm{d}t \qquad (7.39)$$

for any $\phi \in C_c(\Omega)$.

7.2.3.2 Oscillation Defect

As the pressure p_δ is non-decreasing in ϱ and we have already established pointwise a.a. convergence of the approximate temperatures,

$$\lim_{\varepsilon \to 0} \int_0^T \int_\Omega \phi^2 \Big[p_\delta(\varrho_\varepsilon, \vartheta_\varepsilon)\varrho_\varepsilon - \overline{p_\delta(\varrho, \vartheta)\varrho} \Big] \, \mathrm{d}x\mathrm{d}t \geq 0$$

for any $\phi \in C_c(\Omega)$. Consequently, the effective viscous flux identity (7.39) yields

$$\overline{\varrho \mathrm{div}_x \mathbf{u}} \geq \varrho \mathrm{div}_x \mathbf{u} \text{ a.a. in } (0, T) \times \Omega. \qquad (7.40)$$

Now, we use the fact that the approximate densities satisfy the renormalized equation (7.6). Keeping the boundary bounds (7.14) in mind, we may perform the limit $\varepsilon \to 0$ in (7.6) for a *convex* function b:

$$\int_0^T \int_\Omega \left[\overline{b(\varrho)} \partial_t \psi + \overline{\Big(b(\varrho) - b'(\varrho)\varrho \Big) \mathrm{div}_x \mathbf{u} \psi} \right] \mathrm{d}x\mathrm{d}t$$
$$\geq -\psi(0) \int_\Omega b(\varrho_{0,\delta}) \, \mathrm{d}x$$

$$+\int_0^T \psi \int_{\partial\Omega} \left[\overline{b(\varrho)}\mathbf{u}_B \cdot \mathbf{n} - \overline{b'(\varrho)(\varrho - \varrho_B)}[\mathbf{u}_B \cdot \mathbf{n}]^- \right] \mathrm{d}\sigma_x \mathrm{d}t \tag{7.41}$$

for any $\psi \in C^1_c[0,T)$, $\psi \geq 0$. Here the boundary values are generated by the weak limit of $\{\varrho_\varepsilon\}_{\varepsilon>0}$ in $L^2((0,T)\times\partial\Omega; |\mathbf{u}_B\cdot\mathbf{n}|(\mathrm{d}t \otimes \mathrm{d}\sigma_x))$.

As we have already shown, the limit $(\varrho, \mathbf{u})$ satisfies the equation of continuity (7.27). The crucial fact, elaborated and proved in Appendix, Theorem 19 and Remark 21, asserts that the renormalized version of (7.27) is satisfied as well. Specifically,

$$\begin{aligned}
&\int_0^\tau \int_\Omega \left[b(\varrho)\partial_t\varphi + b(\varrho)\mathbf{u}\cdot\nabla_x\varphi + \Big(b(\varrho) - b'(\varrho)\varrho\Big)\mathrm{div}_x\mathbf{u}\varphi \right] \,\mathrm{d}x\mathrm{d}t \\
&\quad = \left[\int_\Omega b(\varrho)\varphi \,\mathrm{d}x \right]_{t=0}^{t=\tau} \\
&\quad + \int_0^\tau \int_{\partial\Omega} \varphi b(\varrho_B)\, [\mathbf{u}_B\cdot\mathbf{n}]^- \mathrm{d}\sigma_x\mathrm{d}t + \int_0^\tau \int_{\partial\Omega} \varphi b(\varrho)\, [\mathbf{u}_B\cdot\mathbf{n}]^+ \mathrm{d}\sigma_x\mathrm{d}t
\end{aligned} \tag{7.42}$$

holds for any $0 < \tau \leq T$, any $\varphi \in C^1([0,T]\times\overline{\Omega})$. We emphasize that validity of (7.42) depends in a crucial way on certain compatibility of the boundary data with the geometry of the spatial domain Ω formulated as hypothesis (B.1) in Appendix, specifically,

$$[0,T]\times\partial\Omega = \Gamma^0_{\mathrm{in}} \cup \Gamma^0_{\mathrm{out}} \cup \Gamma^0_{\mathrm{wall}} \cup \Gamma_{\mathrm{sing}},$$

where Γ^0_{in}, Γ^0_{out}, Γ^0_{wall} are open in $[0,T]\times\partial\Omega$,

$$\begin{aligned}
\Gamma^0_{\mathrm{in}} &\subset \Big\{ (t,x) \ \Big|\ \mathbf{n}(x) \text{ exists}, \ \mathbf{u}_B(t,x)\cdot\mathbf{n}(x) < 0 \Big\}, \\
\Gamma^0_{\mathrm{out}} &\subset \Big\{ (t,x) \ \Big|\ \mathbf{n}(x) \text{ exists}, \ \mathbf{u}_B(t,x)\cdot\mathbf{n}(x) > 0 \Big\}, \\
\Gamma^0_{\mathrm{wall}} &\subset \Big\{ (t,x) \ \Big|\ \mathbf{n}(x) \text{ exists}, \ \mathbf{u}_B(t,x)\cdot\mathbf{n}(x) = 0 \Big\},
\end{aligned}$$

$$\text{and } \Gamma_{\mathrm{sing}} \subset [0,T]\times\partial\Omega \text{ compact}, \ |\Gamma_{\mathrm{sing}}|_{\mathrm{d}t\otimes\mathrm{d}\sigma_x} = 0. \tag{7.43}$$

Taking $\varphi = \psi(t)$ in (7.42) and subtracting the resulting expression from (7.41), we get

$$\begin{aligned}
&\int_0^T \int_\Omega \left[\Big(\overline{b(\varrho)} - b(\varrho)\Big)\partial_t\psi + \overline{\Big(b(\varrho) - b'(\varrho)\varrho\Big)\mathrm{div}_x\mathbf{u}\psi} \right] \,\mathrm{d}x\mathrm{d}t \\
&\quad - \int_0^T \int_\Omega \Big(b(\varrho) - b'(\varrho)\varrho\Big)\mathrm{div}_x\mathbf{u}\psi \,\mathrm{d}x\mathrm{d}t \\
&\quad \geq \int_0^T \psi \int_{\partial\Omega} \left[\overline{b(\varrho)} - \overline{b'(\varrho)(\varrho - \varrho_B)} - b(\varrho_B) \right] \,[\mathbf{u}_B\cdot\mathbf{n}]^- \mathrm{d}\sigma_x\mathrm{d}t \\
&\quad + \int_0^T \psi \int_{\partial\Omega} \Big(\overline{b(\varrho)} - b(\varrho)\Big)\,[\mathbf{u}_B\cdot\mathbf{n}]^+ \mathrm{d}\sigma_x\mathrm{d}t.
\end{aligned} \tag{7.44}$$

As b is convex,

$$\left[\overline{b(\varrho)} - \overline{b'(\varrho)(\varrho - \varrho_B)} - b(\varrho_B)\right][\mathbf{u}_B \cdot \mathbf{n}]^- \geq 0,$$

and (7.44) gives rise to

$$\int_\Omega \left(\overline{b(\varrho)} - b(\varrho)\right)(\tau, \cdot)\ \mathrm{d}x + \int_0^\tau \int_{\partial\Omega} \left(\overline{b(\varrho)} - b(\varrho)\right) [\mathbf{u}_B \cdot \mathbf{n}]^+ \mathrm{d}\sigma_x \mathrm{d}t$$
$$\leq \int_0^\tau \int_\Omega \left[\overline{\left(b(\varrho) - b'(\varrho)\varrho\right)\mathrm{div}_x \mathbf{u}} - \left(b(\varrho) - b'(\varrho)\varrho\right)\mathrm{div}_x \mathbf{u}\right]\ \mathrm{d}x\mathrm{d}t \quad (7.45)$$

for any $0 \leq \tau \leq T$. Finally, for $b(\varrho) = \varrho \log(\varrho)$, the inequality (7.40) yields

$$\int_\Omega \left(\overline{\varrho \log(\varrho)} - \varrho \log(\varrho)\right)(\tau, \cdot)\ \mathrm{d}x$$
$$+ \int_0^\tau \int_{\partial\Omega} \left(\overline{\varrho \log(\varrho)} - \varrho \log(\varrho)\right)\ [\mathbf{u}_B \cdot \mathbf{n}]^+ \mathrm{d}\sigma_x \mathrm{d}t$$
$$\leq \int_0^\tau \int_\Omega \left[\varrho \mathrm{div}_x \mathbf{u} - \overline{\varrho \mathrm{div}_x \mathbf{u}}\right]\ \mathrm{d}x\mathrm{d}t \leq 0. \quad (7.46)$$

As $\varrho \mapsto \varrho \log(\varrho)$ is strictly convex, relation (7.46) yields the desired conclusion

$$\varrho_\varepsilon \to \varrho \text{ a.a. in } (0,T) \times \Omega,$$
$$\varrho_\varepsilon \to \varrho \text{ a.a. in } (0,T) \times \partial\Omega \text{ with respect to the measure } [\mathbf{u}_B \cdot \mathbf{n}]^+(\mathrm{d}t \otimes \mathrm{d}\sigma_x), \quad (7.47)$$

up to a suitable subsequence.

7.3 Limit System

Before passing to the limit $\varepsilon \to 0$ in the approximate system (7.1)–(7.5), first observe that, by interpolation,

$$\int_0^T \int_{\partial\Omega} \|\vartheta_\varepsilon - \vartheta\|^2\ \mathrm{d}\sigma_x \mathrm{d}t \lesssim \int_0^T \|\vartheta_\varepsilon - \vartheta\|^2_{W^{\alpha,2}(\Omega)} \mathrm{d}t$$
$$\leq \int_0^T \|\vartheta_\varepsilon - \vartheta\|^{2\alpha}_{W^{1,2}(\Omega)} \|\vartheta_\varepsilon - \vartheta\|^{2(1-\alpha)}_{L^2(\Omega)} \mathrm{d}t$$

for any $\frac{1}{2} < \alpha \leq 1$. Consequently, in accordance with the strong convergence of $\{\vartheta_\varepsilon\}_{\varepsilon>0}$ established in (7.28),

$$\vartheta_\varepsilon \to \vartheta \text{ in } L^2((0,T) \times \partial\Omega), \quad (7.48)$$

With the uniform bounds (7.7)–(7.16), the pressure bounds (7.22), (7.24), and the pointwise convergence established in (7.28), (7.47), and (7.48) at hand, it is a routine matter to perform the limit $\varepsilon \to 0$ in the approximate system (7.1)–(7.6) combining the Lebesgue Convergence Theorem, Fatou's lemma, and the weak lower semi-continuity of convex functions. The resulting system reads:

- **Equation of continuity.**

$$\begin{aligned}\int_0^T \int_\Omega \Big(\varrho \partial_t \varphi + \varrho \mathbf{u} \cdot \nabla_x \varphi\Big) \ \mathrm{d}x\mathrm{d}t \\ = \int_0^T \int_{\partial\Omega} \varphi \varrho_B [\mathbf{u}_B \cdot \mathbf{n}]^- \ \mathrm{d}\sigma_x \mathrm{d}t + \int_{\partial\Omega} \varphi \varrho [\mathbf{u}_B \cdot \mathbf{n}]^+ \ \mathrm{d}\sigma_x \mathrm{d}t \\ - \int_\Omega \varrho_{0,\delta} \varphi(0,\cdot) \ \mathrm{d}x \qquad (7.49)\end{aligned}$$

for any $\varphi \in C^1_c([0,T) \times \overline{\Omega})$, and its

renormalized version:

$$\begin{aligned}\int_0^T \int_\Omega \Big[b(\varrho)\partial_t \varphi + b(\varrho)\mathbf{u} \cdot \nabla_x \varphi + \Big(b(\varrho) - b'(\varrho)\varrho\Big) \mathrm{div}_x \mathbf{u} \varphi\Big] \ \mathrm{d}x\mathrm{d}t \\ = - \int_\Omega b(\varrho_{0,\delta}) \varphi(0,\cdot) \ \mathrm{d}x \\ + \int_0^T \int_{\partial\Omega} \varphi b(\varrho_B) \ [\mathbf{u}_B \cdot \mathbf{n}]^- \mathrm{d}\sigma_x \mathrm{d}t + \int_0^T \int_{\partial\Omega} \varphi b(\varrho) \ [\mathbf{u}_B \cdot \mathbf{n}]^+ \mathrm{d}\sigma_x \mathrm{d}t \\ (7.50)\end{aligned}$$

for any $\varphi \in C^1_c([0,T) \times \overline{\Omega})$ and any $b \in C^1(R)$, $b' \in C_c(R)$.

- **Momentum equation.**

$$\begin{aligned}- \int_\Omega \mathbf{m}_0 \cdot \boldsymbol{\varphi}(0,\cdot) \ \mathrm{d}x \\ = \int_0^T \int_\Omega \Big[\varrho \mathbf{u} \cdot \partial_t \boldsymbol{\varphi} + \varrho \mathbf{u} \otimes \mathbf{u} : \nabla_x \boldsymbol{\varphi} + p_\delta(\varrho, \vartheta) \mathrm{div}_x \boldsymbol{\varphi}\Big] \ \mathrm{d}x\mathrm{d}t \\ - \int_0^T \int_\Omega \mathbb{S}_\delta(\vartheta, \mathbb{D}_x \mathbf{u}) : \nabla_x \boldsymbol{\varphi} \ \mathrm{d}x\mathrm{d}t + \int_0^T \int_\Omega \varrho \mathbf{g} \cdot \boldsymbol{\varphi} \ \mathrm{d}x\mathrm{d}t \qquad (7.51)\end{aligned}$$

for any $\boldsymbol{\varphi} \in C^1_c([0,T] \times \Omega; R^d)$.

- **Entropy balance (inequality).**

$$
\begin{aligned}
&-\int_\Omega \varrho_{0,\delta} s_\delta(\varrho_{0,\delta}, \vartheta_{0,\delta})\varphi(0,\cdot)\ \mathrm{d}x \\
&-\int_0^T \int_\Omega \varrho s_\delta(\varrho,\vartheta)\partial_t \varphi\ \mathrm{d}x\mathrm{d}t - \int_0^T\int_\Omega \varrho s_\delta(\varrho,\vartheta)\mathbf{u}\cdot\nabla_x\varphi\ \mathrm{d}x\mathrm{d}t \\
&+\int_0^T\int_{\partial\Omega} \varphi\varrho s_\delta[\mathbf{u}_B\cdot\mathbf{n}]^+\ \mathrm{d}\sigma_x\mathrm{d}t - \int_0^T\int_\Omega \frac{\mathbf{q}_\delta(\vartheta,\nabla_x\vartheta)}{\vartheta}\cdot\nabla_x\varphi\ \mathrm{d}x\mathrm{d}t \\
&\geq \int_0^T\int_\Omega \frac{\varphi}{\vartheta}\left(\mathbb{S}_\delta(\vartheta,\nabla_x\mathbf{u}):\mathbb{D}_x\mathbf{u} - \frac{\mathbf{q}_\delta\cdot\nabla_x\vartheta}{\vartheta} + \frac{\delta}{\vartheta^2}\right)\ \mathrm{d}x\mathrm{d}t \\
&+\frac{\delta}{2}\int_0^T\int_\Omega \frac{\varphi}{\vartheta^2}\left(\vartheta^\Gamma + \frac{1}{\vartheta}\right)|\nabla_x\vartheta|^2\ \mathrm{d}x\mathrm{d}t - \int_0^\tau\int_{\Omega_n}\vartheta^4\ \mathrm{d}x\mathrm{d}t \\
&+\int_0^T\int_{\partial\Omega}\varphi\left[\delta\varrho_B(1-\log(\vartheta)) - \frac{1}{\vartheta_\varepsilon}f_{i,B}\right][\mathbf{u}_B\cdot\mathbf{n}]^-\mathrm{d}\sigma_x\mathrm{d}t \\
&+\int_0^T\int_{\partial\Omega}\varphi\varrho_B\left(\frac{e(\varrho_B,\vartheta)}{\vartheta} - s(\varrho_B,\vartheta)\right)[\mathbf{u}_B\cdot\mathbf{n}]^-\mathrm{d}\sigma_x\mathrm{d}t
\end{aligned}
\tag{7.52}
$$

for any $\varphi \in C^1_c([0,T)\times\overline{\Omega})$, $\varphi \geq 0$.

- **Energy balance (inequality).**

$$
\begin{aligned}
&\left[\int_\Omega \left[\frac{1}{2}\varrho|\mathbf{u}-\mathbf{u}_B|^2 + \delta\left(\frac{1}{\Gamma-1}\varrho^\Gamma + \varrho^2\right) + \varrho e_\delta(\varrho,\vartheta)\right]\psi\ \mathrm{d}x\right]_{t=0}^{t=\tau} \\
&-\int_0^\tau \partial_t\psi\int_\Omega\left[\frac{1}{2}\varrho|\mathbf{u}-\mathbf{u}_B|^2 + \delta\left(\frac{1}{\Gamma-1}\varrho^\Gamma + \varrho^2\right) + \varrho e_\delta\right]\ \mathrm{d}x\mathrm{d}t \\
&+\int_0^\tau\psi\int_{\partial\Omega}\varrho e_\delta[\mathbf{u}_B\cdot\mathbf{n}]^+\mathrm{d}\sigma_x\mathrm{d}t \\
&+\delta\int_0^\tau\int_{\partial\Omega}\left(\frac{1}{\Gamma-1}\varrho^\Gamma + \varrho^2\right)[\mathbf{u}_B\cdot\mathbf{n}]^+\ \mathrm{d}\sigma_x\mathrm{d}t \\
&+\int_0^\tau\psi\int_{\partial\Omega} f_{i,B}[\mathbf{u}_B\cdot\mathbf{n}]^-\mathrm{d}\sigma_x\mathrm{d}t \\
&\leq -\int_0^\tau\psi\int_\Omega\left[\varrho(\mathbf{u}-\mathbf{u}_B)\otimes(\mathbf{u}-\mathbf{u}_B) + p_\delta\mathbb{I} - \mathbb{S}_\delta\right]:\nabla_x\mathbf{u}_B\ \mathrm{d}x\mathrm{d}t
\end{aligned}
$$

$$
\begin{aligned}
&+\frac{1}{2}\int_0^\tau \psi \int_\Omega \varrho \mathbf{u}\cdot\nabla_x|\mathbf{u}_B|^2 \ \mathrm{d}x\mathrm{d}t\\
&+\int_0^\tau \psi \int_\Omega \varrho(\mathbf{g}-\partial_t\mathbf{u}_B+\mathbf{u}_B\cdot\nabla_x\mathbf{u}_B)\cdot(\mathbf{u}-\mathbf{u}_B)\ \mathrm{d}x\mathrm{d}t\\
&+\int_0^\tau \psi \int_\Omega \frac{\delta}{\vartheta^2}\ \mathrm{d}x\mathrm{d}t-\delta\int_0^\tau \psi\int_{\partial\Omega}\left(\frac{1}{\Gamma-1}\varrho_B^\Gamma+\varrho_B^2\right)[\mathbf{u}_B\cdot\mathbf{n}]^-\ \mathrm{d}\sigma_x\mathrm{d}t
\end{aligned}
\tag{7.53}
$$

for any $\psi \in C^1[0,T]$, $\psi \geq 0$.

Chapter 8
Vanishing Artificial Pressure Limit

The ultimate step of the global existence proof consists in performing the limit $\delta \to 0$ in the following system of integral identities:

- **Equation of continuity.**

$$
\begin{aligned}
&\int_0^T \int_\Omega \left[\varrho_\delta \partial_t \varphi + \varrho_\delta \mathbf{u}_\delta \cdot \nabla_x \varphi \right] \, \mathrm{d}x \\
&\quad = \int_0^T \int_{\partial\Omega} \varphi \varrho_B \, [\mathbf{u}_B \cdot \mathbf{n}]^- \mathrm{d}\sigma_x \mathrm{d}t + \int_0^T \int_{\partial\Omega} \varphi \varrho_\delta \, [\mathbf{u}_B \cdot \mathbf{n}]^+ \mathrm{d}\sigma_x \mathrm{d}t \\
&\quad - \int_\Omega \varrho_{0,\delta} \varphi(0,\cdot) \, \mathrm{d}x
\end{aligned}
\tag{8.1}
$$

  for any test function $\varphi \in C^1_c([0,T) \times \overline{\Omega})$.
- **Momentum equation**

$$
\begin{aligned}
&\int_0^T \int_\Omega \left[\varrho_\delta \mathbf{u}_\delta \cdot \partial_t \boldsymbol{\varphi} + \varrho_\delta \mathbf{u}_\delta \otimes \mathbf{u}_\delta : \nabla_x \boldsymbol{\varphi} + p_\delta \mathrm{div}_x \boldsymbol{\varphi} - \mathbb{S}_\delta : \nabla_x \boldsymbol{\varphi} \right] \, \mathrm{d}x \\
&\quad + \int_0^T \int_\Omega \varrho_\delta \mathbf{g} \cdot \boldsymbol{\varphi} \, \mathrm{d}x\mathrm{d}t = - \int_\Omega \mathbf{m}_0 \cdot \boldsymbol{\varphi}(0,\cdot) \, \mathrm{d}x
\end{aligned}
\tag{8.2}
$$

  for any $\boldsymbol{\varphi} \in C^1_c[0,T) \times \Omega; R^d)$.

E. Feireisl, A. Novotný, *Mathematics of Open Fluid Systems*, Nečas Center Series, https://doi.org/10.1007/978-3-030-94793-4_8

- **Entropy inequality**

$$
\begin{aligned}
&-\int_{\Omega} \varrho_{0,\delta} s_\delta(\varrho_{0,\delta}, \vartheta_{0,\delta}) \varphi(0,\cdot)\ \mathrm{d}x \\
&-\int_0^T \int_{\Omega} \varrho_\delta s_\delta \partial_t \varphi\ \mathrm{d}x\mathrm{d}t - \int_0^T \int_{\Omega} \varrho_\delta s_\delta \mathbf{u}_\delta \cdot \nabla_x \varphi\ \mathrm{d}x\mathrm{d}t \\
&+\int_0^T \int_{\partial\Omega} \varphi \varrho_\delta s_\delta\ [\mathbf{u}_B \cdot \mathbf{n}]^+ \mathrm{d}\sigma_x \mathrm{d}t - \int_0^T \int_{\Omega} \frac{\mathbf{q}_\delta}{\vartheta_\delta} \cdot \nabla_x \varphi\ \mathrm{d}x\mathrm{d}t \\
&\geq \int_0^T \int_{\Omega} \frac{\varphi}{\vartheta_\delta} \left(\mathbb{S}_\delta : \nabla_x \mathbf{u}_\delta - \frac{\mathbf{q}_\delta \cdot \nabla_x \vartheta_\delta}{\vartheta_\delta} + \frac{\delta}{\vartheta_\delta^2} \right) \mathrm{d}x\mathrm{d}t \\
&+\int_0^T \int_{\partial\Omega} \varphi \left[\delta \varrho_B (1 - \log(\vartheta_\delta)) - \frac{1}{\vartheta_\delta} f_{i,B} \right] [\mathbf{u}_B \cdot \mathbf{n}]^- \mathrm{d}\sigma_x \mathrm{d}t \\
&+\int_0^T \int_{\partial\Omega} \varphi \left(\frac{e(\varrho_B, \vartheta_\delta)}{\vartheta_\delta} - s(\varrho_B, \vartheta_\delta) \right) \varrho_B\ [\mathbf{u}_B \cdot \mathbf{n}]^- \mathrm{d}\sigma_x \mathrm{d}t
\end{aligned}
\tag{8.3}
$$

for any $\varphi \in C^1([0,T) \times \overline{\Omega})$, $\varphi \geq 0$.

- **Total energy balance.**

$$
\begin{aligned}
&\left[\int_{\Omega} \left[\frac{1}{2} \varrho_\delta |\mathbf{u}_\delta - \mathbf{u}_B|^2 + \delta \left(\frac{1}{\Gamma - 1} \varrho_\delta^\Gamma + \varrho_\delta^2 \right) + \varrho_\delta e_\delta \right] \psi\ \mathrm{d}x \right]_{t=0}^{t=\tau} \\
&-\int_0^\tau \partial_t \psi \int_{\Omega} \left[\frac{1}{2} \varrho_\delta |\mathbf{u}_\delta - \mathbf{u}_B|^2 + \delta \left(\frac{1}{\Gamma - 1} \varrho_\delta^\Gamma + \varrho_\delta^2 \right) + \varrho_\delta e_\delta \right] \mathrm{d}x\mathrm{d}t \\
&+\int_0^\tau \psi \int_{\partial\Omega} \varrho_\delta e_\delta\ [\mathbf{u}_B \cdot \mathbf{n}]^+ \mathrm{d}\sigma_x \mathrm{d}t \\
&+\delta \int_0^\tau \psi \int_{\partial\Omega} \left(\frac{1}{\Gamma - 1} \varrho_\delta^\Gamma + \varrho_\delta^2 \right) [\mathbf{u}_B \cdot \mathbf{n}]^+ \mathrm{d}\sigma_x \mathrm{d}t \\
&+\int_0^T \psi \int_{\partial\Omega} f_{i,B}\ [\mathbf{u}_B \cdot \mathbf{n}]^- \mathrm{d}\sigma_x \mathrm{d}t \\
&\leq -\int_0^\tau \psi \int_{\Omega} [\varrho_\delta (\mathbf{u}_\delta - \mathbf{u}_B) \otimes (\mathbf{u}_\delta - \mathbf{u}_B) + p_\delta \mathbb{I} - \mathbb{S}_\delta] : \mathbb{D}_x \mathbf{u}_B\ \mathrm{d}x\mathrm{d}t \\
&+\int_0^\tau \psi \int_{\Omega} \varrho_\delta (\mathbf{g} - \partial_t \mathbf{u}_B + \mathbf{u}_B \cdot \nabla_x \mathbf{u}_B) \cdot (\mathbf{u}_\delta - \mathbf{u}_B)\ \mathrm{d}x\mathrm{d}t \\
&+\delta \int_0^\tau \psi \int_{\Omega} \frac{1}{\vartheta_\delta^2}\ \mathrm{d}x\mathrm{d}t \\
&-\delta \int_0^\tau \psi \int_{\partial\Omega} \left(\frac{1}{\Gamma - 1} \varrho_B^\Gamma + \varrho_B^2 \right) [\mathbf{u}_B \cdot \mathbf{n}]^- \mathrm{d}\sigma_x \mathrm{d}t
\end{aligned}
\tag{8.4}
$$

for any $\psi \in C^1[0,T]$, $\psi \geq 0$.

In addition, we also know that ϱ_δ, $\mathbf{u}_\delta$ solve the renormalized equation of continuity

$$\int_0^T \int_\Omega \Big[b(\varrho_\delta)\partial_t\varphi + b(\varrho_\delta)\mathbf{u}_\delta \cdot \nabla_x\varphi + \Big(b(\varrho_\delta) - b'(\varrho_\delta)\varrho_\delta \Big) \mathrm{div}_x \mathbf{u}_\delta \varphi \Big] \ \mathrm{d}x\mathrm{d}t$$
$$= - \int_\Omega b(\varrho_{0,\delta})\varphi(0,\cdot) \ \mathrm{d}x$$
$$+ \int_0^T \int_{\partial\Omega} \varphi b(\varrho_B) \ [\mathbf{u}_B \cdot \mathbf{n}]^- \mathrm{d}\sigma_x \mathrm{d}t + \int_0^T \int_{\partial\Omega} \varphi b(\varrho_\delta) \ [\mathbf{u}_B \cdot \mathbf{n}]^+ \mathrm{d}\sigma_x \mathrm{d}t \tag{8.5}$$

for any $\varphi \in C^1_c([0,T) \times \overline{\Omega})$ and any $b \in C^1[0,\infty)$, $b' \in C_c[0,\infty)$.

Similarly to the preceding chapters, our aim is to perform the limit $\delta \to 0$ to obtain a weak solution $(\varrho, \mathbf{u}, \vartheta)$ of the Navier–Stokes–Fourier system. In contrast with the previous part, this last step differs substantially for the real gas EOS specified through (4.8)–(4.14), and the hard sphere gas EOS introduced in (4.53)–(4.55). Accordingly, we handle the two cases separately.

8.1 Vanishing Artificial Pressure—Real Gas EOS

8.1.1 Energy Estimates

We start with the uniform bounds that can be deduced from the approximate energy balance (8.4) combined with the entropy inequality (8.3). Similarly to the preceding part, we consider $\psi = 1$ and $\varphi = \psi(t)$ as test functions in (8.4), (8.3), respectively. Subtracting the resulting integral identities, we obtain

$$\int_\Omega \left[\frac{1}{2}\varrho_\delta|\mathbf{u}_\delta - \mathbf{u}_B|^2 + \varrho_\delta e_\delta - \varrho_\delta s_\delta + \delta\left(\frac{1}{\Gamma - 1}\varrho_\delta^\Gamma + \varrho_\delta^2 \right) \right] (\tau,\cdot) \ \mathrm{d}x$$
$$+ \int_0^\tau \int_{\partial\Omega} (\varrho_\delta e_\delta - \varrho_\delta s_\delta) \ [\mathbf{u}_B \cdot \mathbf{n}]^+ \mathrm{d}\sigma_x \mathrm{d}t$$
$$+ \int_0^\tau \int_{\partial\Omega} \left(1 - \frac{1}{\vartheta_\delta} \right) f_{i,B} \ [\mathbf{u}_B \cdot \mathbf{n}]^- \mathrm{d}\sigma_x \mathrm{d}t$$
$$+ \int_0^\tau \int_\Omega \frac{1}{\vartheta_\delta} \left(\mathbb{S}_\delta(\vartheta_\delta, \mathbb{D}_x\mathbf{u}_\delta) : \mathbb{D}_x\mathbf{u}_\delta - \frac{\mathbf{q}_\delta(\vartheta_\delta, \nabla_x\vartheta_\delta) \cdot \nabla_x\vartheta_\delta}{\vartheta_\delta} \right) \ \mathrm{d}x\mathrm{d}t$$
$$\le \int_\Omega \left[\frac{1}{2}\frac{|\mathbf{m}_0|^2}{\varrho_{0,\delta}} - \mathbf{m}_0 \cdot \mathbf{u}_B + \frac{1}{2}\varrho_{0,\delta}|\mathbf{u}_B|^2 \right] \ \mathrm{d}x$$

$$
\begin{aligned}
&+\int_{\Omega}\left[\delta\left(\frac{1}{\Gamma-1}\varrho_{0,\delta}^{\Gamma}+\varrho_{0,\delta}^{2}\right)+\varrho_{0,\delta}e_{\delta}(\varrho_{0,\delta},\vartheta_{0,\delta})-\varrho_{0,\delta}s_{\delta}(\varrho_{0,\delta},\vartheta_{0,\delta})\right]\,\mathrm{d}x\\
&-\int_0^{\tau}\int_{\Omega}\left[\varrho_{\delta}(\mathbf{u}_{\delta}-\mathbf{u}_B)\otimes(\mathbf{u}_{\delta}-\mathbf{u}_B)+p_{\delta}\mathbb{I}-\mathbb{S}_{\delta}(\vartheta_{\delta},\mathbb{D}_x\mathbf{u}_{\delta})\right]:\mathbb{D}_x\mathbf{u}_B\,\mathrm{d}x\mathrm{d}t\\
&+\int_0^{\tau}\int_{\Omega}\varrho_{\delta}(\mathbf{g}-\partial_t\mathbf{u}_B-\mathbf{u}_B\cdot\nabla_x\mathbf{u}_B)\cdot(\mathbf{u}_{\delta}-\mathbf{u}_B)\,\mathrm{d}x\mathrm{d}t\\
&+\delta\int_0^{\tau}\int_{\Omega}\left(\frac{1}{\vartheta_{\delta}^2}-\frac{1}{\vartheta_{\delta}^3}\right)\,\mathrm{d}x\mathrm{d}t\\
&-\delta\int_0^{\tau}\int_{\partial\Omega}\left(\frac{1}{\Gamma-1}\varrho_B^{\Gamma}+\varrho_B^2\right)[\mathbf{u}_B\cdot\mathbf{n}]^-\,\mathrm{d}\sigma_x\mathrm{d}t\\
&-\delta\int_0^{\tau}\int_{\partial\Omega}\varrho_B(1-\log(\vartheta_{\delta}))[\mathbf{u}_B\cdot\mathbf{n}]^-\mathrm{d}\sigma_x\mathrm{d}t\\
&-\int_0^{\tau}\int_{\partial\Omega}\left(\frac{e(\varrho_B,\vartheta_{\delta})}{\vartheta_{\delta}}-s(\varrho_B,\vartheta_{\delta})\right)\varrho_B[\mathbf{u}_B\cdot\mathbf{n}]^-\mathrm{d}\sigma_x\mathrm{d}t
\end{aligned}
\tag{8.6}
$$

for a.a. $\tau\in(0,T)$.

Our goal is to apply Gronwall's argument to the inequality (8.6). First observe that

$$
\begin{aligned}
&\delta\int_0^{\tau}\int_{\Omega}\left(\frac{1}{\vartheta_{\delta}^2}-\frac{1}{\vartheta_{\delta}^3}\right)\,\mathrm{d}x\mathrm{d}t\\
&\quad-\delta\int_0^{\tau}\int_{\partial\Omega}\left(\frac{1}{\Gamma-1}\varrho_B^{\Gamma}+\varrho_B^2\right)[\mathbf{u}_B\cdot\mathbf{n}]^-\,\mathrm{d}\sigma_x\mathrm{d}t\lesssim\delta.
\end{aligned}
\tag{8.7}
$$

Next, using the constitutive restrictions for real gas EOS, specifically (4.15), we get

$$
\begin{aligned}
&-\int_0^{\tau}\int_{\partial\Omega}\left(\frac{e(\varrho_B,\vartheta_{\delta})}{\vartheta_{\delta}}-s(\varrho_B,\vartheta_{\delta})\right)\varrho_B[\mathbf{u}_B\cdot\mathbf{n}]^-\mathrm{d}\sigma_x\mathrm{d}t\\
&\qquad\lesssim\int_0^{\tau}\int_{\partial\Omega}\left(\frac{a}{3}\vartheta_{\delta}^3+[\log(\vartheta_{\delta})]^-+1\right)[\mathbf{u}_B\cdot\mathbf{n}]^-\mathrm{d}\sigma_x\mathrm{d}t\\
&\qquad\lesssim\left(1+\int_0^{\tau}\int_{\partial\Omega}[\log(\vartheta_{\delta})]^-[\mathbf{u}_B\cdot\mathbf{n}]^-\mathrm{d}\sigma_x\mathrm{d}t\right).
\end{aligned}
\tag{8.8}
$$

Similarly,

$$
\begin{aligned}
&-\delta\int_0^{\tau}\int_{\partial\Omega}\varrho_B(1-\log(\vartheta_{\delta}))[\mathbf{u}_B\cdot\mathbf{n}]^-\mathrm{d}\sigma_x\mathrm{d}t\\
&\qquad\lesssim\delta\left(1+\int_0^{\tau}\int_{\partial\Omega}[\log(\vartheta_{\delta})]^-[\mathbf{u}_B\cdot\mathbf{n}]^-\mathrm{d}\sigma_x\mathrm{d}t\right).
\end{aligned}
\tag{8.9}
$$

Consequently, the integrals on the right-hand side of (8.8), (8.9) are controlled by the terms on the left-hand side of (8.6), specifically,

$$- \int_0^\tau \int_{\partial\Omega} \frac{1}{\vartheta_\delta} f_{i,B} [\mathbf{u}_B \cdot \mathbf{n}]^- \mathrm{d}\sigma_x \mathrm{d}t,$$

as long as $f_{i,B}$ is bounded below away from zero as in (5.48). Thus the inequality (8.6) gives rise to

$$\begin{aligned}
&\int_\Omega \left[\frac{1}{2} \varrho_\delta |\mathbf{u}_\delta - \mathbf{u}_B|^2 + \varrho_\delta e_\delta - \varrho_\delta s_\delta + \delta \left(\frac{1}{\Gamma - 1} \varrho_\delta^\Gamma + \varrho_\delta^2 \right) \right] (\tau, \cdot) \, \mathrm{d}x \\
&+ \int_0^\tau \int_{\partial\Omega} (\varrho_\delta e_\delta - \varrho_\delta s_\delta) [\mathbf{u}_B \cdot \mathbf{n}]^+ \mathrm{d}\sigma_x \mathrm{d}t - \frac{1}{2} \int_0^\tau \int_{\partial\Omega} \frac{1}{\vartheta_\delta} f_{i,B} [\mathbf{u}_B \cdot \mathbf{n}]^- \mathrm{d}\sigma_x \mathrm{d}t \\
&+ \int_0^\tau \int_\Omega \frac{1}{\vartheta_\delta} \left(\mathbb{S}_\delta : \nabla_x \mathbf{u}_\delta - \frac{\mathbf{q}_\delta \cdot \nabla_x \vartheta_\delta}{\vartheta_\delta} \right) \, \mathrm{d}x \mathrm{d}t \\
&\lesssim \int_\Omega \left[\frac{1}{2} \frac{|\mathbf{m}_0|^2}{\varrho_{0,\delta}} - \mathbf{m}_0 \cdot \mathbf{u}_B + \frac{1}{2} \varrho_{0,\delta} |\mathbf{u}_B|^2 \right] \, \mathrm{d}x \\
&+ \int_\Omega \left[\delta \left(\frac{1}{\Gamma - 1} \varrho_{0,\delta}^\Gamma + \varrho_{0,\delta}^2 \right) + \varrho_{0,\delta} e_\delta(\varrho_{0,\delta}, \vartheta_{0,\delta}) - \varrho_{0,\delta} s_\delta(\varrho_{0,\delta}, \vartheta_{0,\delta}) \right] \, \mathrm{d}x \\
&- \int_0^\tau \int_\Omega [\varrho_\delta (\mathbf{u}_\delta - \mathbf{u}_B) \otimes (\mathbf{u}_\delta - \mathbf{u}_B) + p_\delta \mathbb{I}] : \nabla_x \mathbf{u}_B \, \mathrm{d}x \mathrm{d}t \\
&+ \int_0^\tau \int_\Omega \mathbb{S}_\delta(\vartheta_\delta, \nabla_x \mathbf{u}_\delta) : \nabla_x \mathbf{u}_B \, \mathrm{d}x \mathrm{d}t \\
&+ \int_0^\tau \int_\Omega \varrho_\delta (\mathbf{g} - \partial_t \mathbf{u}_B + \mathbf{u}_B \cdot \nabla_x \mathbf{u}_B) \cdot (\mathbf{u}_\delta - \mathbf{u}_B) \, \mathrm{d}x \mathrm{d}t + 1
\end{aligned} \tag{8.10}$$

for a.a. $\tau \in (0, T)$.

Finally, we have

$$\begin{aligned}
&\int_\Omega \mathbb{S}_\delta(\vartheta_\delta, \nabla_x \mathbf{u}_\delta) : \nabla_x \mathbf{u}_B \, \mathrm{d}x \\
&= \int_\Omega (\mu(\vartheta_\delta) + \delta \vartheta_\delta) \left(\nabla_x \mathbf{u}_\delta + \nabla_x \mathbf{u}_\delta^t - \frac{2}{d} \mathrm{div}_x \mathbf{u}_\delta \mathbb{I} \right) : \nabla_x \mathbf{u}_B \, \mathrm{d}x \\
&+ \int_\Omega \eta(\vartheta_\delta) \mathrm{div}_x \mathbf{u}_\delta \mathrm{div}_x \mathbf{u}_B \, \mathrm{d}x \le \omega \int_\Omega \frac{1}{\vartheta_\delta} \mathbb{S}_\delta : \nabla_x \mathbf{u}_\delta \, \mathrm{d}x + c_1(\omega, \mathbf{u}_B) \int_\Omega \vartheta_\delta^2 \, \mathrm{d}x \\
&\le \omega \int_\Omega \frac{1}{\vartheta_\delta} \mathbb{S}_\delta : \mathbb{D}_x \mathbf{u}_\delta \, \mathrm{d}x + c_2(\omega, \mathbf{u}_B) \left(1 + \int_\Omega \varrho_\delta e_\delta \, \mathrm{d}x \right)
\end{aligned}$$

for any $\omega > 0$. As the norm of the initial/boundary data is controlled uniformly for $\delta \to 0$, we may infer that all integrals on the right-hand side of (8.10) can be "absorbed" by the left-hand side via Gronwall's argument.

Summing up (8.8)–(8.10), we obtain the following bounds:

$$\operatorname*{ess\,sup}_{\tau \in (0,T)} \int_\Omega \left[\frac{1}{2} \varrho_\delta |\mathbf{u}_\delta - \mathbf{u}_B|^2 + \varrho_\delta e_\delta - \varrho_\delta s_\delta \right] \, \mathrm{d}x \mathrm{d}t \lesssim 1$$

$$
\begin{aligned}
&\operatorname*{ess\ sup}_{\tau\in(0,T)} \delta \int_\Omega \left(\frac{1}{\Gamma - 1} \varrho_\delta^\Gamma + \varrho_\delta^2 \right) (\tau, \cdot) \, \mathrm{d}x \lesssim 1, \\
&- \int_0^T \int_{\partial\Omega} \left(\vartheta_\delta^3 + \frac{1}{\vartheta_\delta} \right) [\mathbf{u}_B \cdot \mathbf{n}]^- \mathrm{d}\sigma_x \mathrm{d}t \lesssim 1, \\
&\int_0^T \int_{\partial\Omega} (\varrho_\delta e_\delta - \varrho_\delta s_\delta) [\mathbf{u}_B \cdot \mathbf{n}]^+ \mathrm{d}\sigma_x \mathrm{d}t \lesssim 1, \\
&\int_0^T \int_\Omega \frac{1}{\vartheta_\delta} \left(\mathbb{S}_\delta(\vartheta_\delta, \mathbb{D}_x \mathbf{u}_\delta) : \mathbb{D}_x \mathbf{u}_\delta - \frac{\mathbf{q}_\delta(\vartheta_\delta, \nabla_x \vartheta_\delta) \cdot \nabla_x \vartheta_\delta}{\vartheta_\delta} \right) \, \mathrm{d}x\mathrm{d}t \lesssim 1
\end{aligned}
\tag{8.11}
$$

uniformly for $\delta \to 0$. In view of the structural hypotheses (4.2)–(4.20), the above bounds give rise to the following estimates:

$$
\operatorname*{ess\ sup}_{\tau\in(0,T)} \| \sqrt{\varrho_\delta} \mathbf{u}_\delta(\tau, \cdot) \|_{L^2(\Omega; R^d)} \lesssim 1; \tag{8.12}
$$

$$
\operatorname*{ess\ sup}_{\tau\in(0,T)} \| \varrho_\delta(\tau, \cdot) \|_{L^{\frac{5}{3}}(\Omega)} \lesssim 1; \tag{8.13}
$$

$$
\operatorname*{ess\ sup}_{\tau\in(0,T)} \| \varrho_\delta(\tau, \cdot) \|_{L^\Gamma(\Omega)} \lesssim \left(\delta^\Gamma \right)^{-1}; \tag{8.14}
$$

$$
\operatorname*{ess\ sup}_{\tau\in(0,T)} \| \vartheta_\delta(\tau, \cdot) \|_{L^4(\Omega)} \lesssim 1; \tag{8.15}
$$

and

$$
\int_0^T \int_\Omega \left(|\nabla_x \log(\vartheta_\delta)|^2 + |\nabla_x \vartheta_\delta^{\frac{3}{2}}|^2 \right) \, \mathrm{d}x\mathrm{d}t \lesssim 1. \tag{8.16}
$$

In addition, we have

$$
|\nabla_x \mathbf{u}_\delta + \nabla_x \mathbf{u}_\delta - \frac{2}{d} \mathrm{div}_x \mathbf{u}_\delta \mathbb{I}| = \vartheta_\delta^{\frac{1-\Lambda}{2}} \vartheta_\delta^{\frac{\Lambda-1}{2}} |\nabla_x \mathbf{u}_\delta + \nabla_x \mathbf{u}_\delta - \frac{2}{d} \mathrm{div}_x \mathbf{u}_\delta \mathbb{I}|,
$$

where, by virtue of (8.11) and hypothesis (4.18),

$$
\left\| \vartheta_\delta^{\frac{\Lambda-1}{2}} (\nabla_x \mathbf{u}_\delta + \nabla_x \mathbf{u}_\delta - \frac{2}{d} \mathrm{div}_x \mathbf{u}_\delta \mathbb{I}) \right\|_{L^2((0,T)\times\Omega; R^{d\times d})} \lesssim 1,
$$

and, by virtue of (8.15),

$$
\operatorname*{ess\ sup}_{\tau\in(0,T)} \| \vartheta_\delta^{\frac{1-\Lambda}{2}} \|_{L^s(\Omega)} \lesssim 1, \ s = \frac{8}{1-\Lambda}.
$$

Consequently, by interpolation,

$$
\left\| \nabla_x \mathbf{u}_\delta + \nabla_x \mathbf{u}_\delta - \frac{2}{d} \mathrm{div}_x \mathbf{u}_\delta \mathbb{I} \right\|_{L^2(0,T; L^{\frac{8}{5-\Lambda}}(\Omega; R^{d\times d}))} \lesssim 1.
$$

Finally, using the Korn–Poincaré inequality, we conclude

$$\int_0^T \int_\Omega \|\mathbf{u}_\delta\|^2_{W^{1,q}(\Omega;R^d)} \, \mathrm{d}x\mathrm{d}t \lesssim 1, \; q = \frac{8}{5-\Lambda}. \tag{8.17}$$

Now, similarly, we combine the temperature estimates (8.15), (8.16) with the Sobolev embedding $W^{1,2} \hookrightarrow L^6$, $(d = 3)$ to obtain

$$\|\vartheta_\delta\|_{L^3(0,T;L^9(\Omega))} \lesssim 1. \tag{8.18}$$

Moreover, following the previous arguments leading to (8.17) with (8.15) replaced by (8.18), we get

$$\int_0^T \int_\Omega \|\mathbf{u}_\delta\|^r_{W^{1,s}(\Omega;R^d)} \, \mathrm{d}x\mathrm{d}t \lesssim 1, \; r = \frac{6}{4-\Lambda}, \quad s = \frac{18}{10-\Lambda}. \tag{8.19}$$

Finally, seeing that

$$\frac{8}{5-\Lambda} \geq \frac{16}{9}$$

we may infer that

$$\int_0^T \int_\Omega |\mathbb{S}(\vartheta_\delta, \mathbb{D}_x \mathbf{u}_\delta)|^q \, \mathrm{d}x\mathrm{d}t \lesssim 1 \text{ for some } q > 1. \tag{8.20}$$

Remark 12. The above estimates seem optimal for $d = 3$ and can be possibly improved if $d = 2$. The interested reader may work out the details.

8.1.2 Pressure Estimates

The (internal) pressure estimates are obtained, similarly to Sect. 7.1.1, by testing the approximate momentum equation (8.2) on

$$\varphi(t,x) = \phi \nabla_x \Delta_x^{-1}[\phi b(\varrho_\delta)], \quad \phi \in C_c^\infty(\Omega).$$

The resulting expression reads

$$\int_0^T \int_\Omega p_\delta \left[\phi^2 b(\varrho_\delta) + \nabla_x \phi \cdot \nabla_x \Delta_x^{-1}[\phi b(\varrho_\delta)]\right] \, \mathrm{d}x\mathrm{d}t$$

$$= \left[\int_\Omega \phi \varrho_\delta \mathbf{u}_\delta \cdot \nabla_x \Delta_x^{-1}[\phi b(\varrho_\delta)] \, \mathrm{d}x\right]_{t=0}^{t=T}$$

$$
\begin{aligned}
&-\int_0^T \int_\Omega \phi \varrho_\delta \mathbf{u}_\delta \cdot \partial_t \Big(\nabla_x \Delta_x^{-1} [\phi b(\varrho_\delta)] \Big) \ \mathrm{d}x\mathrm{d}t \\
&-\int_0^T \int_\Omega \phi \varrho_\delta \mathbf{u}_\delta \otimes \mathbf{u}_\delta : \nabla_x^2 \Delta_x^{-1} [\phi b(\varrho_\delta)] \ \mathrm{d}x\mathrm{d}t \\
&-\int_0^T \int_\Omega \varrho_\delta \mathbf{u}_\delta \otimes \mathbf{u}_\delta \cdot \nabla_x \phi \cdot \nabla_x \Delta_x^{-1} [\phi b(\varrho_\delta)] \ \mathrm{d}x\mathrm{d}t \\
&+\int_0^T \int_\Omega \phi \mathbb{S}_\delta : \nabla_x^2 \Delta_x^{-1} [\phi b(\varrho_\delta)] \ \mathrm{d}x\mathrm{d}t \\
&+\int_0^T \int_\Omega \mathbb{S}_\delta \cdot \nabla_x \phi \cdot \nabla_x \Delta_x^{-1} [\phi b(\varrho_\delta)] \ \mathrm{d}x\mathrm{d}t \\
&-\int_0^T \int_\Omega \phi \varrho_\delta \mathbf{g} \cdot \nabla_x \Delta_x^{-1} [\phi b(\varrho_\delta)] \ \mathrm{d}x\mathrm{d}t, \qquad (8.21)
\end{aligned}
$$

where, by virtue of the renormalized equation of continuity (8.5),

$$
\begin{aligned}
&\partial_t \Big(\nabla_x \Delta_x^{-1} [\phi b(\varrho_\delta)] \Big) \\
&= -\nabla_x \Delta_x^{-1} \Big[\phi \mathrm{div}_x (b(\varrho_\delta) \mathbf{u}_\delta) \Big] + \nabla_x \Delta_x^{-1} \Big[\phi (b(\varrho_\delta) - b'(\varrho_\delta) \varrho_\delta) \mathrm{div}_x \mathbf{u}_\delta \Big]. \qquad (8.22)
\end{aligned}
$$

Note that ϕ is compactly supported in Ω so that the boundary behaviour of ϱ_δ, $\mathbf{u}_\delta$ is irrelevant.

Next, we repeat the same procedure with the test function

$$
\boldsymbol{\varphi}(t,x) = \mathcal{B}[\Phi], \ \Phi \in L^q(\Omega), \ \int_\Omega \Phi \ \mathrm{d}x = 0,
$$

where $\mathcal{B}$ is the Bogovskii operator, see Sect. 0.8. After a straightforward manipulation, we obtain

$$
\begin{aligned}
&\int_0^T \int_\Omega \Phi p_\delta(\varrho_\delta, \vartheta_\delta) \ \mathrm{d}x\mathrm{d}t = \left[\int_\Omega \varrho_\delta \mathbf{u}_\delta \cdot \mathcal{B}[\Phi] \ \mathrm{d}x \right]_{t=0}^{t=T} \\
&\quad - \int_0^T \int_\Omega \varrho_\delta \mathbf{u}_\delta \otimes \mathbf{u}_\delta : \nabla_x \mathcal{B}[\Phi] \ \mathrm{d}x\mathrm{d}t + \int_0^T \int_\Omega \mathbb{S}_\delta : \nabla_x \mathcal{B}[\Phi] \ \mathrm{d}x\mathrm{d}t \qquad (8.23) \\
&\quad - \int_0^T \int_\Omega \varrho_\delta \mathbf{g} \cdot \mathcal{B}[\Phi] \ \mathrm{d}x\mathrm{d}t.
\end{aligned}
$$

In view of the uniform bounds (8.12)–(8.20), we deduce from (8.21), (8.22) the *interior* pressure estimates

$$
\int_0^T \int_K \left(p(\varrho_\delta, \vartheta_\delta) + \delta \varrho_\delta^\Gamma \right) \varrho_\delta^\omega \mathrm{d}x\mathrm{d}t \le c(K) \text{ for some } \omega > 0 \qquad (8.24)
$$

for any compact $K \subset \Omega$. In addition, (8.23) together with (8.24) yields the *boundary* pressure estimates

$$\int_0^T \int_\Omega \left(p(\varrho_\delta, \vartheta_\delta) + \delta \varrho_\delta^\Gamma \right) \mathrm{dist}^{-\omega}[x, \partial\Omega] \, \mathrm{d}x\mathrm{d}t \lesssim 1 \text{ for some } \omega > 0, \tag{8.25}$$

cf. Sect. 7.1.1 for details.

8.1.3 Positivity of the Absolute Temperature

Up to this moment, positivity of the (absolute) temperature ϑ has been guaranteed by integrability of the regularizing term $\delta \frac{1}{\vartheta_\delta^2}$. From the point of view of physics but also for purely mathematical reasons, it is desirable to keep ϑ positive in the asymptotic limit. To this end, we establish the bound

$$\int_0^T \int_\Omega \left(|\log(\vartheta_\delta)|^2 + |\nabla_x \log(\vartheta_\delta)|^2 \right) \, \mathrm{d}x\mathrm{d}t \lesssim 1 \tag{8.26}$$

that implies, in particular, $\vartheta_\delta > 0$ a.a. As the corresponding estimate of the gradient part has been obtained in (8.16), we have to show that ϑ_δ cannot vanish asymptotically as $\delta \to 0$ on a "large" subset of $(0,T) \times \Omega$. This problem is intimately related to the (hypothetical) possibility of the total vacuum ($\varrho \equiv 0$) appearing in a finite time. To avoid this rather unphysical situation, we suppose the inflow part of the boundary is always active, more specifically,

$$\sup_{\tau \in (0,T)} \int_{\partial\Omega} [\mathbf{u}_B(\tau, \cdot) \cdot \mathbf{n}]^- \mathrm{d}\sigma_x < 0. \tag{8.27}$$

In the opposite case, we may suppose purely tangential motion along the boundary:

$$\mathbf{u}_B \cdot \mathbf{n}|_{\partial\Omega} = 0. \tag{8.28}$$

If (8.28) holds, the system is mechanically isolated without mass transport through the boundary, and the analysis, including the existence theory, is essentially the same as in the monograph [43, Chapter 2, Section 2.2.4]. We therefore focus on (8.27).

Remark 13. The condition (8.28) is quite restrictive and not really necessary. A close inspection of the proof in [43, Chapter 2, Section 2.2.4] reveals that it is enough to make sure that the total mass of the fluid remains positive. Since

$$\frac{\mathrm{d}}{\mathrm{d}t}\int_{\Omega}\varrho \ \mathrm{d}x \geq -\int_{\partial\Omega}\varrho \ [\mathbf{u}_B\cdot\mathbf{n}]^{+}\mathrm{d}\sigma_x, \ \int_{\Omega}\varrho(0,\cdot) \ \mathrm{d}x = \int_{\Omega}\varrho_0 \ \mathrm{d}x,$$

we can replace (8.28) by a milder condition on integrability of

$$[\mathbf{u}_B\cdot\mathbf{n}]^{+}|_{\partial\Omega},$$

or even less a restrictive one as soon as $\int_{\Omega}\varrho \ \mathrm{d}x$ remains positive.

If (8.27) holds true, the desired estimate (8.26) can be deduced from (8.11), (8.16), and the Poincaré inequality. Indeed it is enough to consider only $[\log(\vartheta_\delta)]^{-}$ as the $[\log(\vartheta_\delta)]^{+}$ is controlled by (8.15). Next, we have

$$\begin{aligned}
0 < & \inf_{\tau\in(0,T)}\int_{\partial\Omega}\left|[\mathbf{u}_B(\tau,\cdot)\cdot\mathbf{n}]^{-}\right| \mathrm{d}\sigma_x \mathrm{d}t \\
\leq & \int_{\partial\Omega,|[\mathbf{u}_B(t,\cdot)\cdot\mathbf{n}]^{-}|<\omega}\left|[\mathbf{u}_B(t,\cdot)\cdot\mathbf{n}]^{-}\right| \mathrm{d}\sigma_x \\
& + \int_{\partial\Omega,|[\mathbf{u}_B(t,\cdot)\cdot\mathbf{n}]^{-}|\geq\omega}\left|[\mathbf{u}_B(t,\cdot)\cdot\mathbf{n}]^{-}\right| \mathrm{d}\sigma_x \leq \omega|\partial\Omega| \\
& + \left|\{[\mathbf{u}_B(t,\cdot)\cdot\mathbf{n}]^{-}\geq\omega\}\right| \sup_{\partial\Omega}|\mathbf{u}_B|.
\end{aligned}$$

Consequently, there exists $\omega > 0$ such that

$$\inf_{\tau\in(0,T)}\left|\{[\mathbf{u}_B(\tau,\cdot)\cdot\mathbf{n}]^{-}\geq\omega\}\right|_{\mathrm{d}\sigma_x} > 0. \tag{8.29}$$

Thus we may use the Poincaré inequality to conclude

$$\begin{aligned}
& \int_0^T\int_{\Omega}|[\log(\vartheta_\delta)]^{-}|^2 \ \mathrm{d}x\mathrm{d}t \\
& \quad \lesssim \int_0^T\int_{\Omega}|\nabla_x[\log(\vartheta_\delta)]^{-}|^2 \ \mathrm{d}x\mathrm{d}t - \int_0^T\int_{\partial\Omega}\frac{1}{\vartheta_\delta}[\mathbf{u}_B\cdot\mathbf{n}]^{-}\mathrm{d}\sigma_x\mathrm{d}t,
\end{aligned}$$

which yields (8.26).

8.1.4 Convergence $\delta \to 0$

Having collected all available bounds, we are ready to perform the limit $\delta \to 0$. When using embedding theorems of Sobolev type, we focus on the case $d = 3$ as the same obviously holds for $d = 2$.

In view of (8.12) and (8.15), we may assume

$$\varrho_\delta \to \varrho \text{ weakly-(*) in } L^\infty(0,T; L^{\frac{5}{3}}(\Omega)),$$
$$\vartheta_\delta \to \vartheta \text{ weakly-(*) in } L^\infty(0,T; L^4(\Omega))$$

passing to a suitable subsequence if necessary. Similarly, by virtue of (8.17) and (8.19),

$$\mathbf{u}_\delta \to \mathbf{u} \text{ weakly in } L^2(0,T; W^{1,q}(\Omega; R^d)) \cap L^r(0,T; W^{1,s}(\Omega; R^d)),$$
$$q = \frac{8}{5-\Lambda},\ r = \frac{6}{4-\Lambda},\ s = \frac{18}{10-\Lambda}. \tag{8.30}$$

Next, as ϱ_δ, $\mathbf{u}_\delta$ satisfy the equation of continuity (8.1), we get

$$\varrho_\delta \to \varrho \text{ in } C_{\rm weak}([0,T]; L^{\frac{5}{3}}(\Omega)). \tag{8.31}$$

The space $L^{\frac{5}{3}}(\Omega)$ is *compactly* embedded in the dual $W^{-1,p'}$, $p' = \frac{8}{3+\Lambda}$, $\Lambda \in [\frac{1}{2}, 1]$, whence (8.30) and (8.31) give rise to

$$\varrho_\delta \mathbf{u}_\delta \to \varrho \mathbf{u} \text{ weakly-(*) in } L^\infty(0,T; L^{\frac{5}{4}}(\Omega)).$$

Finally, the boundary estimates (8.11) imply, again up to a suitable subsequence,

$$\varrho_\delta \to \varrho \text{ weakly in } L^{\frac{5}{3}}\Big((0,T) \times \partial\Omega; [\mathbf{u}_B \cdot \mathbf{n}]^+ ({\rm d}t \otimes {\rm d}\sigma_x)\Big).$$

Accordingly, we may complete the limit process in the equation of continuity (8.1) obtaining the desired conclusion

$$\int_0^T \int_\Omega \Big[\varrho \partial_t \varphi + \varrho \mathbf{u} \cdot \nabla_x \varphi\Big] \ {\rm d}x = - \int_\Omega \varrho_0 \varphi(0,\cdot) \ {\rm d}x$$
$$+ \int_0^T \int_{\partial\Omega} \varphi \varrho_B [\mathbf{u}_B \cdot \mathbf{n}]^- \ {\rm d}\sigma_x {\rm d}t + \int_0^T \int_{\partial\Omega} \varphi \varrho [\mathbf{u}_B \cdot \mathbf{n}]^+ \ {\rm d}\sigma_x {\rm d}t \tag{8.32}$$

for any test function $\varphi \in C^1_c([0,T) \times \overline{\Omega})$ as long as

$$\varrho_{0,\delta} \to \varrho_0 \text{ in } L^{\frac{5}{3}}(\Omega). \tag{8.33}$$

Similarly, we let $\delta \to 0$ in the renormalized equation of continuity (8.5) to obtain

$$\int_0^T \int_\Omega \left[\overline{b(\varrho)} \partial_t \varphi + \overline{b(\varrho)} \mathbf{u} \cdot \nabla_x \varphi + \overline{\left(b(\varrho) - b'(\varrho)\varrho \right) \mathrm{div}_x \mathbf{u} \varphi} \right] \, \mathrm{d}x$$
$$= - \int_\Omega b(\varrho_0) \varphi(0, \cdot) \, \mathrm{d}x$$
$$+ \int_0^T \int_{\partial\Omega} \varphi b(\varrho_B) [\mathbf{u}_B \cdot \mathbf{n}]^- \, \mathrm{d}\sigma_x \mathrm{d}t + \int_0^T \int_{\partial\Omega} \varphi \overline{b(\varrho)} \, [\mathbf{u}_B \cdot \mathbf{n}]^+ \mathrm{d}\sigma_x \mathrm{d}t \tag{8.34}$$

for any $\varphi \in C^1_c([0,T) \times \overline{\Omega})$ and any $b \in C^1[0,\infty)$, $b' \in C_c[0,\infty)$.

8.1.5 Limit in the Momentum Equation

Next, we perform the limit in the approximate momentum balance (8.2). First, as

$$W^{1,s} \text{ is compactly embedded in } L^5(\Omega), \ s = \frac{18}{10 - \Lambda} \ (d = 3),$$

we get

$$\varrho_\delta \mathbf{u}_\delta \to \varrho \mathbf{u} \text{ in } C_{\mathrm{weak}}([0,T]; L^{\frac{5}{4}}(\Omega; R^d)),$$
$$\varrho_\delta \mathbf{u}_\delta \otimes \mathbf{u}_\delta \to \varrho \mathbf{u} \otimes \mathbf{u} \text{ weakly in } L^r((0,T) \times \Omega; R^{d \times d}) \text{ for some } r > 1. \tag{8.35}$$

Second, the pressure estimates established in (8.24), (8.25) imply

$$\delta \left(\varrho_\delta^\Gamma + \varrho_\delta^2 \right) \to 0 \text{ in } L^1((0,T) \times \Omega).$$

Consequently, letting $\delta \to 0$ in (8.2), we may infer that

$$\int_0^T \int_\Omega \left[\varrho \mathbf{u} \cdot \partial_t \boldsymbol{\varphi} + \varrho \mathbf{u} \otimes \mathbf{u} : \nabla_x \boldsymbol{\varphi} + \overline{p(\varrho, \vartheta)} \mathrm{div}_x \boldsymbol{\varphi} - \overline{\mathbb{S}(\vartheta, \nabla_x \mathbf{u})} : \nabla_x \boldsymbol{\varphi} \right] \, \mathrm{d}x$$
$$+ \int_0^T \int_\Omega \varrho \mathbf{g} \cdot \boldsymbol{\varphi} \, \mathrm{d}x\mathrm{d}t = - \int_\Omega \mathbf{m}_0 \cdot \boldsymbol{\varphi}(0, \cdot) \, \mathrm{d}x \tag{8.36}$$

for any $\varphi \in C^1_c([0,T) \times \Omega; R^d)$, where

$$\overline{p(\varrho,\vartheta)} = \overline{p_{\rm el}(\varrho)} + \overline{p_{\rm m}(\varrho,\vartheta)} + \frac{a}{3}\overline{\vartheta^4},$$

$$\overline{\mathbb{S}(\vartheta,\nabla_x \mathbf{u})} = \overline{\mu(\vartheta)\left(\nabla_x \mathbf{u} + \nabla_x^t \mathbf{u} - \frac{2}{d}{\rm div}_x \mathbf{u}\mathbb{I}\right)} + \overline{\eta(\vartheta){\rm div}_x \mathbf{u}\mathbb{I}}$$

denote the weak (L^1) limits of the corresponding compositions.

8.1.6 Limit in the Energy Balance

Our next goal is to perform the limit in the approximate total energy balance (8.4). Neglecting non-negative integrals on the left-hand side and taking (8.7) into account, we deduce

$$\begin{aligned}
&\left[\int_\Omega \left[\frac{1}{2}\varrho|\mathbf{u}-\mathbf{u}_B|^2 + \overline{\varrho e(\varrho,\vartheta)}\right]\psi \ {\rm d}x\right]_{t=0}^{t=\tau} \\
&- \int_0^\tau \partial_t\psi \int_\Omega \left[\frac{1}{2}\varrho|\mathbf{u}-\mathbf{u}_B|^2 + \overline{\varrho e(\varrho,\vartheta)}\right] \ {\rm d}x{\rm d}t + \int_0^\tau \int_{\partial\Omega} \psi f_{i,B}[\mathbf{u}_B\cdot\mathbf{n}]^- {\rm d}\sigma_x {\rm d}t \\
&+ \liminf_{\delta\to 0} \int_0^\tau \int_{\partial\Omega} \psi\varrho_\delta e_\delta [\mathbf{u}_B\cdot\mathbf{n}]^+ {\rm d}\sigma_x {\rm d}t \\
&\le - \int_0^\tau \psi \int_\Omega \left[\varrho(\mathbf{u}-\mathbf{u}_B)\otimes(\mathbf{u}-\mathbf{u}_B) + \overline{p(\varrho,\vartheta)}\mathbb{I} - \overline{\mathbb{S}(\vartheta,\mathbb{D}_x\mathbf{u})\mathbb{I}}\right] : \mathbb{D}_x\mathbf{u}_B \ {\rm d}x{\rm d}t \\
&+ \int_0^\tau \psi \int_\Omega \varrho(\mathbf{g} - \partial_t\mathbf{u}_B + \mathbf{u}_B\cdot\nabla_x\mathbf{u}_B)\cdot(\mathbf{u}-\mathbf{u}_B) \ {\rm d}x{\rm d}t
\end{aligned} \tag{8.37}$$

for any $\psi \in C^1[0,T]$, $\psi \ge 0$.

Obviously,

$$\int_0^\tau \int_{\partial\Omega} \psi\varrho_\delta e_\delta(\varrho_\delta,\vartheta_\delta) \ [\mathbf{u}_B\cdot\mathbf{n}]^+ {\rm d}\sigma_x {\rm d}t \ge \int_0^\tau \int_{\partial\Omega} \psi\varrho_\delta e(\varrho_\delta,\vartheta_\delta) \ [\mathbf{u}_B\cdot\mathbf{n}]^+ {\rm d}\sigma_x {\rm d}t$$

and, consequently, by Fatou's lemma,

$$\liminf_{\delta\to 0} \int_0^\tau \int_{\partial\Omega} \psi\varrho_\delta e_\delta [\mathbf{u}_B\cdot\mathbf{n}]^+ {\rm d}\sigma_x {\rm d}t \ge \int_{\partial\Omega} \psi e(\varrho,\vartheta) \ [\mathbf{u}_B\cdot\mathbf{n}]^+ {\rm d}\sigma_x, \ \psi \ge 0$$

as soon as we establish pointwise convergence of boundary densities and temperatures. This issue is addressed in the remaining part of this chapter.

8.1.7 Pointwise Convergence of the Temperature

We establish the pointwise (a.a.) convergence of the temperature,

$$\vartheta_\delta(t,x) \to \vartheta(t,x) \text{ for a.a. } (t,x) \in (0,T)\times\Omega, \tag{8.38}$$

modulo a suitable subsequence.

The proof follows arguments of Sect. 7.2.2. First, we perform the limit in the approximate entropy inequality (8.3).

In view of the uniform bound (8.11), we have

$$\varrho_\delta s(\varrho_\delta, \vartheta_\delta) \text{ bounded in } L^\infty(0,T;L^1(\Omega))$$

uniformly for $\delta \to 0$. In addition, by virtue of (8.13), (8.15), and (8.26),

$$\varrho_\delta s(\varrho_\delta, \vartheta_\delta) \text{ is bounded in } L^2(0,T;L^r(\Omega)) \text{ for some } r > 1.$$

Consequently, passing to a suitable subsequence, we may assume that

$$\varrho_\delta s(\varrho_\delta, \vartheta_\delta) \to \overline{\varrho s(\varrho,\vartheta)} \text{ weakly in } L^p((0,T)\times\Omega) \text{ for some } p > 1. \tag{8.39}$$

Next, making use of the constitutive restrictions imposed on s, we deduce

$$|\varrho_\delta s(\varrho_\delta, \vartheta_\delta)\mathbf{u}_\delta| \lesssim \Big(|\mathbf{u}_\delta||\vartheta_\delta|^3 + \varrho_\delta|\log(\varrho_\delta)||\mathbf{u}_\delta| + |\mathbf{u}_\delta| + \varrho_\delta|\log(\vartheta_\delta)||\mathbf{u}_\delta| \Big),$$

where the energy bounds (8.13), (8.15), and (8.17) yield

$$\Big(|\mathbf{u}_\delta||\vartheta_\delta|^3 + \varrho_\delta|\log(\varrho_\delta)||\mathbf{u}_\delta| + |\mathbf{u}_\delta| + \varrho_\delta|\log(\vartheta_\delta)||\mathbf{u}_\delta| \Big) \text{ bounded in } L^r((0,T)\times\Omega)$$

for some $r > 1$, uniformly for $\delta \to 0$. Indeed possibly the most difficult term can be treated as

$$\|\varrho_\delta \log(\vartheta_\delta)\mathbf{u}_\delta\|_{L^{\frac{30}{29}}(\Omega;R^d)} \le \|\varrho_\delta \mathbf{u}_\delta\|_{L^{\frac{5}{4}}(\Omega;R^d)} \|\log(\vartheta_\delta)\|_{L^6(\Omega)},$$

whence (8.12) and (8.13), together with (8.26) and the Sobolev embedding $W^{1,2} \hookrightarrow L^6 (d=3)$, yield the desired estimate. Thus we conclude

$$\varrho_\delta s(\varrho_\delta, \vartheta_\delta)\mathbf{u}_\delta \to \overline{\varrho s(\varrho,\vartheta)\mathbf{u}} \text{ weakly in } L^r((0,T)\times\Omega;R^d) \text{ for some } r > 1. \tag{8.40}$$

As for the entropy flux $\mathbf{q}_\delta$, we have

$$\frac{\kappa(\vartheta_\delta)}{\vartheta_\delta}|\nabla_x \vartheta_\delta| \lesssim \left(|\nabla_x \log(\vartheta_\delta)| + \vartheta_\delta^{\frac{3}{2}} \left|\nabla_x \vartheta_\delta^{\frac{3}{2}}\right| \right).$$

Consequently, the bounds (8.15) and (8.16) yield

$$\frac{\kappa(\vartheta_\delta)}{\vartheta_\delta}\nabla_x\vartheta_\delta \text{ bounded in } L^r((0,T)\times\Omega;R^d) \text{ for some } r>1 \tag{8.41}$$

uniformly for $\delta \to 0$.

Finally, using the energy bounds (8.11), we may control the δ-dependent integrals. We have

$$\delta\int_0^T \left(\|\vartheta_\delta^{\frac{\Gamma}{2}}\|^2_{W^{1,2}(\Omega)} + \|\vartheta_\delta^{-\frac{1}{2}}\|^2_{W^{1,2}(\Omega)} \right) \mathrm{d}t \lesssim 1. \tag{8.42}$$

Thus writing

$$\delta\vartheta_\delta^{\Gamma-1}\nabla_x\vartheta_\delta = \frac{\delta\Gamma}{2}\vartheta_\delta^{\frac{\Gamma}{2}}\nabla_x\vartheta_\delta^{\frac{\Gamma}{2}} = \frac{\delta\Gamma}{2}\vartheta_\delta^{\frac{1}{4}}\vartheta_\delta^{\frac{\Gamma}{2}-\frac{1}{4}}\nabla_x\vartheta_\delta^{\frac{\Gamma}{2}}$$

we may use (8.15) and (8.42) to conclude

$$\delta\vartheta_\delta^{\Gamma-1}\nabla_x\vartheta_\delta \to 0 \text{ in } L^r((0,T)\times\Omega;R^d) \text{ for some } r>1. \tag{8.43}$$

Moreover, by similar arguments,

$$\frac{\delta}{\vartheta_\delta^2}\nabla_x\vartheta_\delta \to 0 \text{ in } L^r((0,T)\times\Omega;R^d) \text{ for some } r>1. \tag{8.44}$$

At this stage, we are ready to let $\delta \to 0$ in the entropy balance (8.3) and to show the claimed pointwise convergence of the approximate temperatures. First, we deduce from (8.3)

$$\begin{aligned} -\int_0^T\int_\Omega \varrho_\delta s_\delta \partial_t\varphi \, \mathrm{d}x\mathrm{d}t - \int_0^T\int_\Omega \varrho_\delta s_\delta \mathbf{u}_\delta\cdot\nabla_x\varphi \, \mathrm{d}x\mathrm{d}t \\ - \int_0^T\int_\Omega \frac{\mathbf{q}_\delta}{\vartheta}\cdot\nabla_x\varphi \, \mathrm{d}x\mathrm{d}t \geq 0 \end{aligned} \tag{8.45}$$

for any $\varphi \in C^1_c((0,T)\times\Omega)$, $\varphi \geq 0$. Seeing that

$$\delta\varrho_\delta\log(\vartheta_\delta) \to 0 \text{ in } L^2((0,T)\times\Omega)$$

we may use the Lions–Aubin argument (Lemma 4) to deduce

$$\overline{\varrho s(\varrho,\vartheta)g(\vartheta)} = \overline{\varrho s(\varrho,\vartheta)}\;\overline{g(\vartheta)} \tag{8.46}$$

for any $g \in C^1_c(R)$. Note that the identity (8.46) can be extended to any continuous non-decreasing function g as soon as the quantities remain integrable. In particular,

$$\overline{\varrho s(\varrho,\vartheta)\vartheta} = \overline{\varrho s(\varrho,\vartheta)}\;\vartheta. \tag{8.47}$$

Next, we use the renormalized equation (8.5) and apply the same argument to obtain

$$\overline{b(\varrho)g(\vartheta)} = \overline{b(\varrho)}\ \overline{g(\vartheta)} \tag{8.48}$$

for any bounded b.

Finally, write

$$\varrho_\delta s(\varrho_\delta, \vartheta_\delta) = \varrho_\delta s_{\rm m}(\varrho_\delta, \vartheta_\delta) + \frac{4a}{3}\vartheta_\delta^3.$$

As $s_{\rm m}$ is a non-decreasing function of ϑ, we have

$$\Big(\varrho_\delta s_{\rm m}(\varrho_\delta, \vartheta_\delta) - \varrho_\delta s_{\rm m}(\varrho_\delta, \vartheta)\Big)(\vartheta_\delta - \vartheta) \geq 0.$$

Now, exactly as in Sect. 7.2.2, it follows from (8.48) that

$$\varrho_\delta s_{\rm m}(\varrho_\delta, \vartheta)(\vartheta_\delta - \vartheta) \to 0 \text{ weakly in } L^1((0,T) \times \Omega),$$

whence

$$\overline{\varrho s_{\rm m}(\varrho, \vartheta)\vartheta} \geq \overline{\varrho s_{\rm m}(\varrho, \vartheta)}\vartheta. \tag{8.49}$$

By the same token,

$$\overline{\vartheta^3\vartheta} \geq \overline{\vartheta^3}\vartheta. \tag{8.50}$$

Combining (8.47), (8.49), and (8.50), we get

$$\overline{\vartheta^3\vartheta} = \overline{\vartheta^3}\vartheta,$$

which yields, modulo a suitable subsequence, the strong convergence claimed in (8.38).

We are ready to let $\delta \to 0$ in the approximate entropy inequality (8.3). Let us start with showing pointwise convergence of the temperature trace that follows from the pointwise convergence in the interior stated in (8.38) and boundedness of the temperature gradients. Indeed, the energy bounds imply

$$\operatorname*{ess\ sup}_{t\in(0,T)} \|\vartheta_\delta\|_{L^4(\Omega)} + \int_0^T \|\nabla_x\vartheta_\delta\|^2_{L^2(\Omega)} \lesssim 1.$$

Moreover, as shown in (8.38),

$$\vartheta_\delta \to \vartheta \text{ in, say, } L^2((0,T) \times \Omega).$$

Thus, by interpolation,

$$\begin{aligned}\int_0^T \int_{\partial\Omega} \|\vartheta_\delta - \vartheta\|^2 \ {\rm d}\sigma_x {\rm d}t &\lesssim \int_0^T \|\vartheta_\delta - \vartheta\|^2_{W^{\alpha,2}(\Omega)} {\rm d}t \\ &\leq \int_0^T \|\vartheta_\delta - \vartheta\|^{2\alpha}_{W^{1,2}(\Omega)} \|\vartheta_\delta - \vartheta\|^{2(1-\alpha)}_{L^2(\Omega)} {\rm d}t\end{aligned}$$

for any $\frac{1}{2} < \alpha \leq 1$. Consequently,

$$\vartheta_\delta \to \vartheta \text{ in, say, } L^2((0,T) \times \partial\Omega). \tag{8.51}$$

Now, let us examine the boundary integrals in the approximate entropy balance (8.3). First,

$$\int_0^T \int_{\partial\Omega} \varphi \varrho_\delta s_\delta \ [\mathbf{u}_B \cdot \mathbf{n}]^+ \mathrm{d}\sigma_x \mathrm{d}t$$
$$= \delta \int_0^T \int_{\partial\Omega} \varphi \varrho_\delta \log(\vartheta_\delta) \ [\mathbf{u}_B \cdot \mathbf{n}]^+ \mathrm{d}\sigma_x \mathrm{d}t + \int_0^T \int_{\partial\Omega} \varphi S_\delta \ [\mathbf{u}_B \cdot \mathbf{n}]^+ \mathrm{d}\sigma_x \mathrm{d}t,$$

where, by virtue of (8.4) and (8.26),

$$\delta \int_0^T \int_{\partial\Omega} \varphi \varrho_\delta \log(\vartheta_\delta) \ [\mathbf{u}_B \cdot \mathbf{n}]^+ \mathrm{d}\sigma_x \mathrm{d}t$$
$$\leq \delta \int_0^T \int_{\partial\Omega} \varphi \varrho_\delta \log^+(\vartheta_\delta) \ [\mathbf{u}_B \cdot \mathbf{n}]^+ \mathrm{d}\sigma_x \mathrm{d}t \to 0.$$

Second, focusing on the inflow boundary, we have to handle integral

$$\int_0^T \int_{\partial\Omega} \varphi \delta \varrho_B (1 - \log(\vartheta_\delta)) \ [\mathbf{u}_B \cdot \mathbf{n}]^- \mathrm{d}\sigma_x \mathrm{d}t$$
$$+ \int_0^T \int_{\partial\Omega} \varphi \left[-\frac{1}{\vartheta_\delta} f_{i,B} + \varrho_B \left(\frac{e(\varrho_B, \vartheta_\delta)}{\vartheta_\delta} - s(\varrho_B, \vartheta_\delta) \right) \right] \ [\mathbf{u}_B \cdot \mathbf{n}]^- \mathrm{d}\sigma_x \mathrm{d}t.$$

By virtue of the uniform bound (8.26), we have, as a consequence of the trace theorem,

$$\delta \int_0^T \int_{\Gamma_{\rm in}} \varphi \varrho_B (1 - \log(\vartheta_\delta)) \mathbf{u}_B \cdot \mathbf{n} \ \mathrm{d}\sigma_x \mathrm{d}t \to 0 \text{ as } \delta \to 0.$$

Next, by virtue of the strong convergence of the temperature traces established in (8.51), the limit in the integral

$$\int_0^T \int_{\partial\Omega} \varphi \left[-\frac{1}{\vartheta_\delta} f_{i,b} + \varrho_B \left(\frac{e(\varrho_B, \vartheta_\delta)}{\vartheta_\delta} - s(\varrho_B, \vartheta_\delta) \right) \right] \ [\mathbf{u}_B \cdot \mathbf{n}]^- \mathrm{d}\sigma_x \mathrm{d}t$$

can be performed via Fatou's lemma.

Letting $\delta \to 0$ in the approximate entropy balance and exploiting the pointwise convergence of the temperature, we obtain

$$
\begin{aligned}
&- \int_\Omega \varrho_0 s(\varrho_0, \vartheta_0)\varphi(0,\cdot)\ \mathrm{d}x - \int_0^T \int_\Omega \overline{\varrho s}\partial_t \varphi\ \mathrm{d}x\mathrm{d}t \\
&- \int_0^T \int_\Omega \overline{\varrho s}\mathbf{u}\cdot\nabla_x\varphi\ \mathrm{d}x\mathrm{d}t + \limsup_{\delta\to 0}\int_0^T \int_{\partial\Omega} \varphi\varrho_\delta s(\varrho_\delta,\vartheta_\delta)\ [\mathbf{u}_B\cdot\mathbf{n}]^+\mathrm{d}\sigma_x\mathrm{d}t \\
&- \int_0^T \int_\Omega \frac{\mathbf{q}(\vartheta,\nabla_x\vartheta)}{\vartheta}\cdot\nabla_x\varphi\ \mathrm{d}x\mathrm{d}t \\
&\geq \int_0^T \int_\Omega \frac{\varphi}{\vartheta}\left(\mathbb{S}(\vartheta,\mathbb{D}_x\mathbf{u}):\mathbb{D}_x\mathbf{u} - \frac{\mathbf{q}(\vartheta,\nabla_x\vartheta)\cdot\nabla_x\vartheta}{\vartheta}\right)\ \mathrm{d}x\mathrm{d}t \\
&- \int_0^T \int_{\partial\Omega} \varphi\frac{1}{\vartheta}f_{i,B}\ [\mathbf{u}_B\cdot\mathbf{n}]^-\mathrm{d}\sigma_x\mathrm{d}t \\
&+ \int_0^T \int_{\partial\Omega} \varphi\varrho_B\left(\frac{e(\varrho_B,\vartheta)}{\vartheta} - s(\varrho_B,\vartheta)\right)\ [\mathbf{u}_B\cdot\mathbf{n}]^-\mathrm{d}\sigma_x\mathrm{d}t
\end{aligned}
\tag{8.52}
$$

for any $\varphi \in C^1([0,T)\times\overline{\Omega})$, $\varphi \geq 0$.

8.1.8 Pointwise Convergence of the Density

To complete the proof of existence, we have to remove the bars in the renormalized equation of continuity (8.34), the momentum equation (8.36), the total energy balance (8.37), and the entropy inequality (8.52). Moreover, we have to perform the limit in the outflow integrals (8.37) and (8.52). To this end, we show the pointwise convergence of approximate densities both in the interior and on the boundary of Ω.

In view of (8.38), this amounts to showing strong convergence of the approximate densities:

$$
\begin{aligned}
&\varrho_\delta \to \varrho \text{ a.a. in } (0,T)\times\Omega, \\
&\varrho_\delta \to \varrho \text{ a.a. in } (0,T)\times\partial\Omega \\
&\quad \text{with respect to the measure } [\mathbf{u}_B\cdot\mathbf{n}]^+(\mathrm{d}t\otimes\mathrm{d}\sigma_x).
\end{aligned}
\tag{8.53}
$$

Indeed taking for a moment the convergence (8.53) for granted, we have, by Fatou's lemma,

$$
\liminf_{\delta\to 0}\int_0^T \psi \int_{\partial\Omega} \varrho_\delta e_\delta[\mathbf{u}_B\cdot\mathbf{n}]^+\mathrm{d}\sigma_x\mathrm{d}t \geq \int_0^T \psi\int_{\partial\Omega}\varrho e(\varrho,\vartheta)\ [\mathbf{u}_B\cdot\mathbf{n}]^+\mathrm{d}\sigma_x\mathrm{d}t,
$$

$\psi \geq 0$, in the energy balance (8.37), and

$$\limsup_{\delta\to 0}\int_0^T\int_{\partial\Omega}\varphi\varrho_\delta s(\varrho_\delta,\vartheta_\delta)\ [\mathbf{u}_B\cdot\mathbf{n}]^+\mathrm{d}\sigma_x \le \int_0^T\int_{\partial\Omega}\varphi\varrho s(\varrho,\vartheta)\ [\mathbf{u}_B\cdot\mathbf{n}]^+\mathrm{d}\sigma_x, \tag{8.54}$$

$\varphi\ge 0$ in the entropy inequality (8.52). In order to see (8.54), we decompose

$$\varrho_\delta s(\varrho_\delta,\vartheta_\delta)=\varrho_\delta[s(\varrho_\delta,\vartheta_\delta)]^+ + \varrho_\delta[s(\varrho_\delta,\vartheta_\delta)]^-,$$

where, in accordance with the structural restrictions (4.15),

$$\begin{aligned}\varrho_\delta[s(\varrho_\delta,\vartheta_\delta)]^+ &\lesssim \left(1+\varrho_\delta|\log(\varrho_\delta)|+\varrho_\delta[\log(\vartheta_\delta)]^+ + \vartheta_\delta^3\right)\\ \varrho_\delta[s(\varrho_\delta,\vartheta_\delta)]^- &\gtrsim \varrho_\delta[\log(\vartheta_\delta)]^-.\end{aligned}$$

Consequently, $\varrho_\delta[s(\varrho_\delta,\vartheta_\delta)]^+[\mathbf{u}_B\cdot\mathbf{n}]^+$ is equi-integrable on $(0,T)\times\partial\Omega$, while the limit for $\varrho_\delta[s(\varrho_\delta,\vartheta_\delta)]^-[\mathbf{u}_B\cdot\mathbf{n}]^+$ can be performed via Fatou's lemma.

To establish the strong convergence claimed in (8.53), we introduce a cut-off function

$$T_k(r)=kT\left(\frac{r}{k}\right),\ r\ge 0,\ k\ge 1,$$

$$T\in C^\infty[0,\infty),\ T(z)=\begin{cases} z \text{ if } 0\le z\le 1,\\ \text{concave on } [0,\infty),.\\ 2 \text{ if } z\ge 3\end{cases}$$

Going back to Sect. 8.1.2, we consider $b(\varrho_\delta)=T_k(\varrho_\delta)$ in the integral identity (8.21) obtaining

$$\begin{aligned}&\int_0^T\int_\Omega p_\delta\left[\phi^2 T_k(\varrho_\delta)+\nabla_x\phi\cdot\nabla_x\Delta_x^{-1}[\phi T_k(\varrho_\delta)]\right]\,\mathrm{d}x\mathrm{d}t\\
&=\left[\int_\Omega \phi\varrho_\delta\mathbf{u}_\delta\cdot\nabla_x\Delta_x^{-1}[\phi T_k(\varrho_\delta)]\,\mathrm{d}x\right]_{t=0}^{t=T}\\
&+\int_0^T\int_\Omega\phi\varrho_\delta\mathbf{u}_\delta\cdot\nabla_x\Delta_x^{-1}\left[\phi\mathrm{div}_x(T_k(\varrho_\delta)\mathbf{u}_\delta)\right]\,\mathrm{d}x\mathrm{d}t\\
&-\int_0^T\int_\Omega\phi\varrho_\delta\mathbf{u}_\delta\cdot\nabla_x\Delta_x^{-1}\left[\phi(T_k(\varrho_\delta)-T_k'(\varrho_\delta)\varrho_\delta)\mathrm{div}_x\mathbf{u}_\delta\right]\,\mathrm{d}x\mathrm{d}t\\
&-\int_0^T\int_\Omega\phi\varrho_\delta\mathbf{u}_\delta\otimes\mathbf{u}_\delta:\nabla_x^2\Delta_x^{-1}[\phi T_k(\varrho_\delta)]\,\mathrm{d}x\mathrm{d}t\\
&-\int_0^T\int_\Omega\varrho_\delta\mathbf{u}_\delta\otimes\mathbf{u}_\delta\cdot\nabla_x\phi\cdot\nabla_x\Delta_x^{-1}[\phi T_k(\varrho_\delta)]\,\mathrm{d}x\mathrm{d}t\\
&+\int_0^T\int_\Omega\phi\mathbb{S}_\delta:\nabla_x^2\Delta_x^{-1}[\phi T_k(\varrho_\delta)]\,\mathrm{d}x\mathrm{d}t\\
&+\int_0^T\int_\Omega\mathbb{S}_\delta\cdot\nabla_x\phi\cdot\nabla_x\Delta_x^{-1}[\phi T_k(\varrho_\delta)]\,\mathrm{d}x\mathrm{d}t\\
&-\int_0^T\int_\Omega\phi\varrho_\delta\mathbf{g}\cdot\nabla_x\Delta_x^{-1}[\phi T_k(\varrho_\delta)]\,\mathrm{d}x\mathrm{d}t.\end{aligned}\tag{8.55}$$

In view of the regularizing properties of the operator $\nabla_x \Delta_x^{-1}$, we may extract a subsequence such that

$$\begin{aligned} T_k(\varrho_\delta) &\to \overline{T_k(\varrho)} \text{ in } C_{\text{weak}}([0,T]; L^q(\Omega)), \\ \nabla_x \Delta_x^{-1}[\phi T_k(\varrho_\delta)] &\to \nabla_x \Delta_x^{-1}[\phi \overline{T_k(\varrho)}] \text{ in } L^q((0,T) \times \Omega; R^d) \end{aligned} \tag{8.56}$$

for any $1 \le q < \infty$.

Now, we consider

$$\varphi(t,x) = \phi \nabla_x \Delta_x^{-1}[\phi \overline{T_k(\varrho)}], \quad \phi \in C_c^\infty(\Omega)$$

as a test function in the limit momentum balance (8.36) obtaining

$$\begin{aligned} &\int_0^T \int_\Omega \overline{p(\varrho,\vartheta)} \left[\phi^2 \overline{T_k(\varrho)} + \nabla_x \phi \cdot \nabla_x \Delta_x^{-1}[\phi \overline{T_k(\varrho)}] \right] \, \mathrm{d}x\mathrm{d}t \\ &= \left[\int_\Omega \phi \varrho \mathbf{u} \cdot \nabla_x \Delta_x^{-1}[\phi \overline{T_k(\varrho)}] \, \mathrm{d}x \right]_{t=0}^{t=T} \\ &+ \int_0^T \int_\Omega \phi \varrho \mathbf{u} \cdot \nabla_x \Delta_x^{-1} \left[\phi \mathrm{div}_x(\overline{T_k(\varrho)} \mathbf{u}) \right] \, \mathrm{d}x\mathrm{d}t \\ &- \int_0^T \int_\Omega \phi \varrho \mathbf{u} \cdot \nabla_x \Delta_x^{-1} \left[\phi \overline{(T_k(\varrho) - T_k'(\varrho)\varrho) \mathrm{div}_x \mathbf{u}} \right] \, \mathrm{d}x\mathrm{d}t \\ &- \int_0^T \int_\Omega \phi \varrho \mathbf{u} \otimes \mathbf{u} : \nabla_x^2 \Delta_x^{-1}[\phi \overline{T_k(\varrho)}] \, \mathrm{d}x\mathrm{d}t \\ &- \int_0^T \int_\Omega \varrho \mathbf{u} \otimes \mathbf{u} \cdot \nabla_x \phi \cdot \nabla_x \Delta_x^{-1}[\phi \overline{T_k(\varrho)}] \, \mathrm{d}x\mathrm{d}t \\ &+ \int_0^T \int_\Omega \phi \mathbb{S}(\vartheta, \nabla_x \mathbf{u}) : \nabla_x^2 \Delta_x^{-1}[\phi \overline{T_k(\varrho)}] \, \mathrm{d}x\mathrm{d}t \\ &+ \int_0^T \int_\Omega \mathbb{S}(\vartheta, \nabla_x \mathbf{u}) \cdot \nabla_x \phi \cdot \nabla_x \Delta_x^{-1}[\phi \overline{T_k(\varrho)}] \, \mathrm{d}x\mathrm{d}t \\ &- \int_0^T \int_\Omega \phi \varrho \mathbf{g} \cdot \nabla_x \Delta_x^{-1}[\phi \overline{T_k(\varrho)}] \, \mathrm{d}x\mathrm{d}t, \end{aligned} \tag{8.57}$$

where we have used the limit renormalized equation (8.34). Recall that, by virtue of the strong convergence of the approximate temperatures established in (8.38),

$$\overline{\mathbb{S}(\vartheta, \nabla_x \mathbf{u})} = \mathbb{S}(\vartheta, \nabla_x \mathbf{u}) = \mu(\vartheta) \left(\nabla_x \mathbf{u} + \nabla_x \mathbf{u}^t - \frac{2}{d} \mathrm{div}_x \mathbf{u} \mathbb{I} \right) + \eta(\vartheta) \mathrm{div}_x \mathbf{u} \mathbb{I},$$

and

$$\overline{p(\varrho,\vartheta)} = \overline{p_{\mathrm{el}}(\varrho)} + \overline{p_m(\varrho,\vartheta)} + \frac{a}{3}\vartheta^4.$$

Finally, we let $\delta \to 0$ in (8.55), use the convergence established in (8.56), and compare the limit with (8.57):

$$\begin{aligned}
&\int_0^T \int_\Omega \phi^2 \left(\overline{(p_{\rm el} + p_{\rm m})(\varrho, \vartheta) T_k(\varrho)} - \overline{(p_{\rm el} + p_{\rm m})(\varrho, \vartheta)}\ \overline{T_k(\varrho)} \right) \ {\rm d}x{\rm d}t \\
&= \lim_{\delta \to 0} \int_0^T \int_\Omega \phi \Big(\varrho_\delta \mathbf{u}_\delta \cdot \nabla_x \Delta_x^{-1} \Big[\phi {\rm div}_x (T_k(\varrho_\delta)\mathbf{u}_\delta) \Big] \\
&- \varrho \mathbf{u} \cdot \nabla_x \Delta_x^{-1} \Big[\phi {\rm div}_x (\overline{T_k(\varrho)}\mathbf{u}) \Big] \Big) {\rm d}x{\rm d}t \\
&- \lim_{\delta \to 0} \int_0^T \int_\Omega \phi \Big(\varrho_\delta \mathbf{u}_\delta \otimes \mathbf{u}_\delta : \nabla_x^2 \Delta_x^{-1} [\phi T_k(\varrho_\delta)] \\
&- \varrho \mathbf{u} \otimes \mathbf{u} : \nabla_x^2 \Delta_x^{-1} [\phi \overline{T_k(\varrho)}] \Big) {\rm d}x{\rm d}t \\
&+ \lim_{\delta \to 0} \int_0^T \int_\Omega \phi \Big(\mathbb{S}(\vartheta_\delta, \nabla_x \mathbf{u}_\delta) : \nabla_x^2 \Delta_x^{-1} [\phi T_k(\varrho_\delta)] \\
&- \mathbb{S}(\vartheta, \nabla_x \mathbf{u}) : \nabla_x^2 \Delta_x^{-1} [\phi \overline{T_k(\varrho)}] \Big) {\rm d}x{\rm d}t.
\end{aligned} \tag{8.58}$$

Now, we may rewrite

$$\begin{aligned}
&\lim_{\delta \to 0} \int_0^T \int_\Omega \phi \Big(\varrho_\delta \mathbf{u}_\delta \cdot \nabla_x \Delta_x^{-1} \Big[\phi {\rm div}_x (T_k(\varrho_\delta)\mathbf{u}_\delta) \Big] \\
&- \varrho \mathbf{u} \cdot \nabla_x \Delta_x^{-1} \Big[\phi {\rm div}_x (\overline{T_k(\varrho)}\mathbf{u}) \Big] \Big) {\rm d}x{\rm d}t \\
&- \lim_{\delta \to 0} \int_0^T \int_\Omega \phi \Big(\varrho_\delta \mathbf{u}_\delta \otimes \mathbf{u}_\delta : \nabla_x^2 \Delta_x^{-1} [\phi T_k(\varrho_\delta)] \\
&- \varrho \mathbf{u} \otimes \mathbf{u} : \nabla_x^2 \Delta_x^{-1} [\phi \overline{T_k(\varrho)}] \Big) {\rm d}x{\rm d}t \\
&= \lim_{\delta \to 0} \int_0^T \int_\Omega \phi \Big(T_k(\varrho_\delta) \mathbf{u}_\delta \cdot \nabla_x \Delta_x^{-1} {\rm div}_x [\phi \varrho_\delta \mathbf{u}_\delta] \\
&- \varrho_\delta \mathbf{u}_\delta \otimes \mathbf{u}_\delta : \nabla_x^2 \Delta_x^{-1} [\phi T_k(\varrho_\delta)] \Big) {\rm d}x{\rm d}t \\
&- \int_0^T \int_\Omega \phi \Big(\overline{T_k(\varrho)} \mathbf{u} \cdot \nabla_x \Delta_x^{-1} {\rm div}_x [\phi \varrho \mathbf{u}] \\
&- \varrho \mathbf{u} \otimes \mathbf{u} : \nabla_x^2 \Delta_x^{-1} [\phi \overline{T_k(\varrho)}] \Big) {\rm d}x{\rm d}t.
\end{aligned}$$

In the above formula, we easily identify the "commutator"

$$\phi \varrho_\delta \mathbf{u}_\delta \cdot \nabla_x^2 \Delta_x^{-1} [\phi T_k(\varrho_\delta)] - \phi T_k(\varrho_\delta) \cdot \nabla_x \Delta_x^{-1} {\rm div}_x [\phi \varrho_\delta \mathbf{u}_\delta].$$

Now, exactly as in Sect. 7.2.3, we use Compactness Lemma (Lemma 2) to deduce

$$\phi\varrho_\delta\mathbf{u}_\delta\cdot\nabla_x^2\Delta_x^{-1}[\phi T_k(\varrho_\delta)]-\phi T_k(\varrho_\delta)\cdot\nabla_x\Delta_x^{-1}\mathrm{div}_x[\phi\varrho_\delta\mathbf{u}_\delta]$$
$$\to$$
$$\phi\varrho\mathbf{u}\cdot\nabla_x^2\Delta_x^{-1}[\phi T_k(\varrho)]-\phi T_k(\varrho)\cdot\nabla_x\Delta_x^{-1}\mathrm{div}_x[\phi\varrho\mathbf{u}]$$

weakly in $L^r(\Omega;R^d)$, $1\le r<\frac{5}{4}$ pointwise for any $t\in[0,T]$. This, together with the convergence of the velocities established in (8.30), yields

$$\lim_{\delta\to0}\int_0^T\int_\Omega\phi\Big(T_k(\varrho_\delta)\mathbf{u}_\delta\cdot\nabla_x\Delta_x^{-1}\mathrm{div}_x[\phi\varrho_\delta\mathbf{u}_\delta]$$
$$-\varrho_\delta\mathbf{u}_\delta\otimes\mathbf{u}_\delta:\nabla_x^2\Delta_x^{-1}[\phi T_k(\varrho_\delta)]\Big)\mathrm{d}x\mathrm{d}t$$
$$=\int_0^T\int_\Omega\phi\Big(\overline{T_k(\varrho)}\mathbf{u}\cdot\nabla_x\Delta_x^{-1}\mathrm{div}_x[\phi\varrho\mathbf{u}]$$
$$-\varrho\mathbf{u}\otimes\mathbf{u}:\nabla_x^2\Delta_x^{-1}[\phi\overline{T_k(\varrho)}]\Big)\mathrm{d}x\mathrm{d}t.$$

Accordingly, relation (8.58) reduces to

$$\begin{aligned}&\int_0^T\int_\Omega\phi^2\left(\overline{(p_{\rm el}+p_{\rm m})(\varrho,\vartheta)T_k(\varrho)}-\overline{(p_{\rm el}+p_{\rm m})(\varrho,\vartheta)}\;\overline{T_k(\varrho)}\right)\,\mathrm{d}x\mathrm{d}t\\&=\lim_{\delta\to0}\int_0^T\int_\Omega\phi\Big(\mathbb{S}(\vartheta_\delta,\nabla_x\mathbf{u}_\delta):\nabla_x^2\Delta_x^{-1}[\phi T_k(\varrho_\delta)]\\&-\mathbb{S}(\vartheta,\nabla_x\mathbf{u}):\nabla_x^2\Delta_x^{-1}[\phi\overline{T_k(\varrho)}]\Big)\mathrm{d}x\mathrm{d}t.\end{aligned}\tag{8.59}$$

The integral on the right-hand side of (8.59) can be rewritten as

$$\int_\Omega\phi\Big(\mathbb{S}(\vartheta_\delta,\nabla_x\mathbf{u}_\delta):\nabla_x^2\Delta_x^{-1}[\phi T_k(\varrho_\delta)]$$
$$-\mathbb{S}(\vartheta,\nabla_x\mathbf{u}):\nabla_x^2\Delta_x^{-1}[\phi\overline{T_k(\varrho)}]\Big)\mathrm{d}x$$
$$=\int_\Omega\phi\Big(\nabla_x^2\Delta_x^{-1}:\Big[\phi\mathbb{S}(\vartheta_\delta,\nabla_x\mathbf{u}_\delta)\Big]T_k(\varrho_\delta)$$
$$-\nabla_x^2\Delta_x^{-1}:\Big[\phi\mathbb{S}(\vartheta,\nabla_x\mathbf{u})\Big]\overline{T_k(\varrho)}\Big)\mathrm{d}x.$$

Next,

$$\nabla_x^2\Delta_x^{-1}:\Big[\phi\mathbb{S}(\vartheta_\delta,\nabla_x\mathbf{u}_\delta)\Big]=\phi\left(\frac{4}{3}\mu(\vartheta_\delta)+\eta(\vartheta_\delta)\right)\mathrm{div}_x\mathbf{u}_\delta$$
$$+\nabla_x^2\Delta_x^{-1}:\Big[\phi\mathbb{S}(\vartheta_\delta,\nabla_x\mathbf{u}_\delta)\Big]-\phi\left(\frac{4}{3}\mu(\vartheta_\delta)+\eta(\vartheta_\delta)\right)\mathrm{div}_x\mathbf{u}_\delta,$$

and, similarly,

$$\nabla_x^2\Delta_x^{-1} : \Big[\phi\mathbb{S}(\vartheta,\nabla_x\mathbf{u})\Big] = \phi\left(\frac{4}{3}\mu(\vartheta)+\eta(\vartheta)\right)\mathrm{div}_x\mathbf{u}$$
$$+\nabla_x^2\Delta_x^{-1} : \Big[\phi\mathbb{S}(\vartheta,\nabla_x\mathbf{u})\Big] - \phi\left(\frac{4}{3}\mu(\vartheta)+\eta(\vartheta)\right)\mathrm{div}_x\mathbf{u}.$$

Thus, by virtue of Commutator Lemma (Lemma 3), we may infer that

$$\int_0^T\int_\Omega \phi T_k(\varrho_\delta)\Big(\nabla_x^2\Delta_x^{-1} : \Big[\phi\mathbb{S}(\vartheta_\delta,\nabla_x\mathbf{u}_\delta)\Big] - \phi\left(\frac{4}{3}\mu(\vartheta_\delta)+\eta(\vartheta_\delta)\right)\mathrm{div}_x\mathbf{u}_\delta\Big)\mathrm{d}x\mathrm{d}t$$
$$\to$$
$$\int_0^T\int_\Omega \phi\overline{T_k(\varrho)}\Big(\nabla_x^2\Delta_x^{-1} : \Big[\phi\mathbb{S}(\vartheta,\nabla_x\mathbf{u})\Big] - \phi\left(\frac{4}{3}\mu(\vartheta)+\eta(\vartheta)\right)\mathrm{div}_x\mathbf{u}\Big)\mathrm{d}x\mathrm{d}t.$$

Consequently, relation (8.59) takes the form

$$\int_0^T\int_\Omega \phi^2\left(\overline{(p_{\rm el}+p_{\rm m})(\varrho,\vartheta)T_k(\varrho)} - \overline{(p_{\rm el}+p_{\rm m})(\varrho,\vartheta)}\;\overline{T_k(\varrho)}\right)\mathrm{d}x\mathrm{d}t$$
$$= \int_0^T\int_\Omega \phi^2\left(\frac{4}{3}\mu(\vartheta)+\eta(\vartheta)\right)\left(\overline{T_k(\varrho)\mathrm{div}_x\mathbf{u}} - \overline{T_k(\varrho)}\mathrm{div}_x\mathbf{u}\right)\mathrm{d}x\mathrm{d}t \tag{8.60}$$

for any $\phi \in C_c^\infty(\Omega)$. This can be equivalently stated as (cf. (7.39))

the **effective viscous flux identity:**

$$\overline{(p_{\rm el}+p_{\rm m})(\varrho,\vartheta)T_k(\varrho)} - \overline{(p_{\rm el}+p_{\rm m})(\varrho,\vartheta)}\;\overline{T_k(\varrho)}$$
$$= \left(\frac{4}{3}\mu(\vartheta)+\eta(\vartheta)\right)\left(\overline{T_k(\varrho)\mathrm{div}_x\mathbf{u}} - \overline{T_k(\varrho)}\mathrm{div}_x\mathbf{u}\right). \tag{8.61}$$

At this stage, we introduce

the **oscillation defect measure**

$$\mathbf{osc}_q[\varrho_\delta\to\varrho](Q) = \sup_{k\geq 1}\left(\limsup_{\delta\to 0}\int_Q |T_k(\varrho_\delta)-T_k(\varrho)|^q\mathrm{d}x\mathrm{d}t\right).$$

The following result is crucial for studying density oscillations.

Lemma 9. *Let $\frac{1}{2} \le \Lambda \le 1$ be the constant introduced in the hypotheses (4.18) and (4.19). Let $\{\varrho_\delta\}_{\delta>0}$ be the sequence of approximate mass densities.*

Then

$$\mathbf{osc}_q[\varrho_\delta \to \varrho]((0,T)\times\Omega) \lesssim 1 \tag{8.62}$$

for some $q > \frac{8}{3+\Lambda}$.

Proof. Recall that

$$\frac{\partial p_{\rm m}(\varrho,\vartheta)}{\partial\varrho} \ge 0.$$

Consequently, as we have already established the pointwise convergence of the temperature in (8.38),

$$\overline{p_{\rm m}(\varrho,\vartheta)T_k(\varrho)} - \overline{p_{\rm m}(\varrho,\vartheta)}\;\overline{T_k(\varrho)} \ge 0$$

Next, as $p_{\rm el}(\varrho) = \overline{p}\varrho^{\frac{5}{3}}$ is convex and $\varrho \mapsto T_k(\varrho)$ concave, we get

$$\begin{aligned}
&\overline{p}\limsup_{\delta\to 0}\int_0^T\int_\Omega \varphi|T_k(\varrho_\delta) - T_k(\varrho)|^{\frac{8}{3}}\ {\rm d}x{\rm d}t\\
&\le a\int_0^T\int_\Omega \varphi\left(\overline{\varrho^{\frac{5}{3}}T_k(\varrho)} - \overline{\varrho^{\frac{5}{3}}}\;\overline{T_k(\varrho)}\right)\ {\rm d}x{\rm d}t\\
&+\int_0^T\int_\Omega \varphi\left(\varrho^{\frac{5}{3}} - \overline{\varrho^{\frac{5}{3}}}\right)\left(T_k(\varrho) - \overline{T_k(\varrho)}\right)\ {\rm d}x{\rm d}t\\
&\le \int_0^T\int_\Omega \varphi\left[\overline{(p_{\rm el}+p_{\rm m})(\varrho,\vartheta)T_k(\varrho)} - \overline{(p_{\rm el}+p_{\rm m})(\varrho,\vartheta)}\;\overline{T_k(\varrho)}\right]\ {\rm d}x{\rm d}t
\end{aligned} \tag{8.63}$$

for any $\varphi \in C_c^\infty((0,T)\times\Omega)$, $\varphi \ge 0$.

Now, we introduce the function

$$G_k(t,x,r) = |T_k(r) - T_k(\varrho(t,x))|^{\frac{8}{3}}.$$

On the one hand, it follows from (8.63) and the effective viscous flux identity (8.61) that

$$\overline{G_k(\cdot,\cdot,\varrho)} \lesssim \left(\frac{4}{3}\mu(\vartheta) + \eta(\vartheta)\right)\left(\overline{{\rm div}_x\mathbf{u}T_k(\varrho)} - {\rm div}_x\mathbf{u}\overline{T_k(\varrho)}\right) \tag{8.64}$$

uniformly for $k \to \infty$. On the other hand, in accordance with hypothesis (4.18) and by virtue of the uniform bound (8.17), we have

$$
\begin{aligned}
&\int_0^T \int_\Omega (1+\vartheta)^{-\Lambda} \overline{G_k(\cdot,\cdot,\varrho)} \, dxdt \\
&\quad \lesssim 1 + \sup_{\delta>0} \|\mathrm{div}_x \mathbf{u}_\delta\|_{L^{\frac{8}{5-\Lambda}}((0,T)\times\Omega)} \limsup_{\delta\to 0} \|T_k(\varrho_\delta) - T_k(\varrho)\|_{L^{\frac{8}{3+\Lambda}}((0,T)\times\Omega)} \\
&\quad \lesssim 1 + \limsup_{\delta\to 0} \|T_k(\varrho_\delta) - T_k(\varrho)\|_{L^{\frac{8}{3+\Lambda}}((0,T)\times\Omega)}.
\end{aligned}
\tag{8.65}
$$

Thus the choice

$$
\frac{8}{3+\Lambda} < q < \frac{8}{3}, \ \beta = \frac{3q\Lambda}{8}
$$

gives rise to

$$
\begin{aligned}
&\int_0^T \int_\Omega |T_k(\varrho_\delta) - T_k(\varrho)|^q \, dxdt \\
&\quad = \int_0^T \int_\Omega (1+\vartheta)^{-\beta}(1+\vartheta^\beta)|T_k(\varrho_\delta) - T_k(\varrho)|^q \, dxdt \\
&\quad \lesssim \int_0^T \int_\Omega (1+\vartheta)^{-\Lambda}|T_k(\varrho_\delta) - T_k(\varrho)|^{\frac{8}{3}} \, dxdt + \int_0^T \int_\Omega (1+\vartheta)^{\frac{3\Lambda q}{8-3q}} \, dxdt.
\end{aligned}
\tag{8.66}
$$

Finally, we fix

$$
\frac{8}{3+\Lambda} < q \leq \frac{32}{3\Lambda + 12}, \ \text{meaning,} \ \frac{3\Lambda q}{8-3q} \leq 4.
$$

Combining the estimates (8.65), (8.66) with (8.15), we obtain (8.62).

□

We have collected all necessary ingredients to show pointwise convergence of the approximate densities. By virtue of Theorem 19 in Appendix, boundedness of the oscillations defect measure stated in Lemma 9, together with the fact that the limit $(\varrho, \mathbf{u})$ satisfies the equation of continuity (8.32), implies that $(\varrho, \mathbf{u})$ satisfy

the **renormalized equation of continuity**

$$
\begin{aligned}
&\int_0^T \int_\Omega \Big[b(\varrho)\partial_t \varphi + b(\varrho)\mathbf{u}_\delta \cdot \nabla_x \varphi + \Big(b(\varrho) - b'(\varrho)\varrho_\delta \Big) \mathrm{div}_x \mathbf{u} \varphi \Big] \, dxdt \\
&\quad = - \int_\Omega b(\varrho_0)\varphi(0,\cdot) \, dx \\
&\quad + \int_0^T \int_{\partial\Omega} \varphi b(\varrho_B) \, [\mathbf{u}_B \cdot \mathbf{n}]^- d\sigma_x dt + \int_0^T \int_{\partial\Omega} \varphi b(\varrho) \, [\mathbf{u}_B \cdot \mathbf{n}]^+ d\sigma_x dt
\end{aligned}
\tag{8.67}
$$

for any $\varphi \in C^1_c([0,T) \times \overline{\Omega})$ and any $b \in C^1[0,\infty)$, $b' \in C_c[0,\infty)$.

The first observation is that (8.67) can be extended to

$$b(\varrho) = L_k(\varrho),\ L_k'(\varrho)\varrho - L_k(\varrho) = T_k(\varrho),\ T_k(\varrho) = \min\{\varrho, k\}.$$

Indeed, L_k is a compactly supported perturbation of an affine function, and the desired conclusion follows from (8.32), (8.67).

Now, subtracting (8.67) from (8.34), we get

$$\int_\Omega \left[\overline{L_k(\varrho)} - L_k(\varrho)\right](\tau,\cdot)\ \mathrm{d}x + \int_0^\tau \int_{\partial\Omega} \left[\overline{L_k(\varrho)} - L_k(\varrho)\right]\ [\mathbf{u}_B\cdot\mathbf{n}]^+ \mathrm{d}\sigma_x \mathrm{d}t$$
$$+ \int_0^\tau \int_\Omega \left[\overline{T_k(\varrho)\mathrm{div}_x\mathbf{u}} - T_k(\varrho)\mathrm{div}_x\mathbf{u}\right]\ \mathrm{d}x\mathrm{d}t = 0 \tag{8.68}$$

cf. (7.45). Note carefully that validity of (8.68) requires strong convergence of the initial densities (8.33).

Next, write

$$\overline{T_k(\varrho)\mathrm{div}_x\mathbf{u}} - T_k(\varrho)\mathrm{div}_x\mathbf{u}$$
$$= \overline{T_k(\varrho)\mathrm{div}_x\mathbf{u}} - \overline{T_k(\varrho)}\mathrm{div}_x\mathbf{u} + (\overline{T_k(\varrho)} - T_k(\varrho))\mathrm{div}_x\mathbf{u},$$

where, by virtue of the effective viscous flux identity (8.61),

$$\overline{T_k(\varrho)\mathrm{div}_x\mathbf{u}} - \overline{T_k(\varrho)}\mathrm{div}_x\mathbf{u} \geq 0.$$

Finally, as

$$\{\mathrm{div}_x\mathbf{u}_\varepsilon\}_{\delta>0} \text{ is bounded in } L^2(0,T;L^r(\Omega)),\ r = \frac{8}{5-\Lambda},$$

and (8.62) holds with $q > \frac{8}{3+\Lambda}$, we have, by interpolation,

$$\left\|(\overline{T_k(\varrho)} - T_k(\varrho))\mathrm{div}_x\mathbf{u}\right\|_{L^s((0,T)\times\Omega)} \lesssim 1 \text{ for some } s > 1.$$

Since

$$\overline{T_k(\varrho)} \to T_k(\varrho) \text{ in } L^1((0,T)\times\Omega) \text{ as } k \to \infty,$$

we may let $k \to \infty$ in (8.68) to obtain the desired conclusion

$$\int_\Omega \left[\overline{\varrho\log(\varrho)} - \varrho\log(\varrho)\right](\tau,\cdot)\ \mathrm{d}x$$
$$+ \int_0^\tau \int_{\partial\Omega} \left[\overline{\varrho\log(\varrho)} - \varrho\log(\varrho)\right]\ [\mathbf{u}_B\cdot\mathbf{n}]^+ \mathrm{d}\sigma_x \mathrm{d}t \leq 0. \tag{8.69}$$

This yields, up to a suitable subsequence, the strong convergence of the approximate densities claimed in (8.53).

8.2 Vanishing Artificial Pressure—Hard Sphere EOS

Our ultimate goal in this chapter is to adapt the arguments used in the preceding part to the hard sphere pressure EOS. Our starting point is therefore a sequence of approximate solutions $\{\varrho_\delta, \vartheta_\delta, \mathbf{u}_\delta\}_{\delta>0}$ of the approximate problem (8.1)–(8.4), where the pressure p_δ satisfies

$$p_\delta(\varrho, \vartheta) = p_{\delta,HS}(\varrho) + \delta\varrho^2 + ([\varrho - \overline{\varrho} - 1]^+)^\Gamma + \varrho\vartheta + \frac{a}{3}\vartheta^4,$$

with

$$p_{\delta,HS}(\varrho) = \begin{cases} \frac{A\varrho}{(\overline{\varrho}-\varrho)^r} \text{ if } 0 \le \varrho \le \overline{\varrho} - \delta, \\ \\ a_{1,\delta}\varrho + a_{2,\delta} \text{ otherwise} \end{cases}, \ p_{\delta,HS}(\varrho) \in C^1[0,\infty), \tag{8.70}$$

cf. (5.2).

We recall the crucial hypothesis concerning the boundary density and the heat flux,

$$0 < \varrho_B < \overline{\varrho}, \ f_{i,B} - \varrho_B e_{\rm el}(\varrho_B) > 0 \text{ in } [0,T] \times \partial\Omega,$$

where

$$e_{\rm el}(\varrho) = \int_{\overline{\varrho}/2}^{\varrho} \frac{A}{z(\overline{\varrho} - z)^r} {\rm d}z.$$

Many steps, in particular those concerning the limit passage in the entropy and energy equations, will be identical or even easier than in the preceding section as $\{\varrho_\delta\}_{\delta>0}$ enjoys better integrability determined by Γ. Thus the only part of the proof that calls for attention are the pressure estimates and the asymptotic limit of the singular components of the pressure p and the internal energy e. Note carefully that the interior pressure estimates (8.24) are not sufficient here as the pressure is singular at $\overline{\varrho}$.

8.2.1 Energy Estimates

The internal energy associated to the hard sphere pressure approximation p_δ reads

$$e_\delta(\varrho, \vartheta) = \int_{\overline{\varrho}/2}^{\varrho} \frac{p_{\delta,HS}(z)}{z^2} {\rm d}r + \frac{3}{2}\vartheta + \frac{a}{\varrho}\vartheta^4,$$

augmented by the added "elastic" perturbation

$$\delta\varrho + \int_{\overline{\varrho}}^{\varrho} \frac{([z - \overline{\varrho} - 1]^+)^\Gamma}{z^2} {\rm d}z.$$

The entropy is

$$s(\varrho, \vartheta) = \frac{3}{2}\log(\vartheta) - \log(\varrho) + \frac{4a}{3}\frac{\vartheta^3}{\varrho}.$$

Note that, in contrast with the real gas EOS considered in the preceding part, it is not necessary to modify e by the term $\delta\vartheta$.

Multiplying the entropy inequality by a positive constant $\overline{\vartheta}$ and subtracting the resulting expression from the energy balance (8.4), we get an analogue of (8.6):

$$\begin{aligned}
&\int_\Omega \left[\frac{1}{2}\varrho_\delta|\mathbf{u}_\delta - \mathbf{u}_B|^2 + \varrho_\delta e_\delta(\varrho_\delta, \vartheta_\delta) - \overline{\vartheta}\varrho_\delta s(\varrho_\delta, \vartheta_\delta)\right](\tau, \cdot)\,\mathrm{d}x \\
&+ \int_\Omega \left(\delta\varrho_\delta^2 + \varrho_\delta \int_{\overline{\varrho}}^{\varrho_\delta} \frac{([z - \overline{\varrho} - 1]^+)^\Gamma}{z^2}\mathrm{d}z\right)(\tau, \cdot)\,\mathrm{d}x \\
&+ \int_0^\tau \int_{\partial\Omega} (\varrho_\delta e_\delta - \overline{\vartheta}\varrho_\delta s_\delta)\,[\mathbf{u}_B \cdot \mathbf{n}]^+ \mathrm{d}\sigma_x \mathrm{d}t \\
&+ \int_0^\tau \int_{\partial\Omega} \left(1 - \frac{\overline{\vartheta}}{\vartheta_\delta}\right) f_{i,B}[\mathbf{u}_B \cdot \mathbf{n}]^- \mathrm{d}\sigma_x \mathrm{d}t \\
&+ \int_0^\tau \int_\Omega \frac{\overline{\vartheta}}{\vartheta_\delta}\left(\mathbb{S}_\delta : \nabla_x \mathbf{u}_\delta - \frac{\mathbf{q}_\delta \cdot \nabla_x \vartheta_\delta}{\vartheta_\delta}\right)\,\mathrm{d}x\mathrm{d}t \\
&\leq \int_\Omega \left[\frac{1}{2}\frac{|\mathbf{m}_0|^2}{\varrho_{0,\delta}} - \mathbf{m}_0 \cdot \mathbf{u}_B + \frac{1}{2}\varrho_{0,\delta}|\mathbf{u}_B|^2\right]\,\mathrm{d}x \\
&+ \int_\Omega \left[\delta\varrho_{0,\delta}^2 + \varrho_{0,\delta} e_\delta(\varrho_{0,\delta}, \vartheta_{0,\delta}) - \overline{\vartheta}\varrho_{0,\delta} s_\delta(\varrho_{0,\delta}, \vartheta_{0,\delta})\right]\,\mathrm{d}x \\
&- \int_0^\tau \int_\Omega \left[\varrho_\delta(\mathbf{u}_\delta - \mathbf{u}_B) \otimes (\mathbf{u}_\delta - \mathbf{u}_B) + p_\delta \mathbb{I}\right] : \mathbb{D}_x \mathbf{u}_B\,\mathrm{d}x\mathrm{d}t \\
&+ \int_0^\tau \int_\Omega \mathbb{S}_\delta(\vartheta_\delta, \mathbb{D}_x \mathbf{u}_\delta) : \mathbb{D}_x \mathbf{u}_B\,\mathrm{d}x\mathrm{d}t \\
&+ \int_0^\tau \int_\Omega \varrho_\delta(\mathbf{g} - \partial_t \mathbf{u}_B - \mathbf{u}_B \cdot \nabla_x \mathbf{u}_B) \cdot (\mathbf{u}_\delta - \mathbf{u}_B)\,\mathrm{d}x\mathrm{d}t \\
&+ \delta \int_0^\tau \int_\Omega \left(\frac{1}{\vartheta_\delta^2} - \frac{\overline{\vartheta}}{\vartheta_\delta^3}\right)\,\mathrm{d}x\mathrm{d}t - \delta \int_0^\tau \int_{\partial\Omega} \varrho_B^2[\mathbf{u}_B \cdot \mathbf{n}]^-\,\mathrm{d}\sigma_x \mathrm{d}t \\
&- \int_0^\tau \int_{\partial\Omega} \left(\frac{e(\varrho_B, \vartheta_\delta)}{\vartheta_\delta} - \overline{\vartheta}s(\varrho_B, \vartheta_\delta)\right)\varrho_B[\mathbf{u}_B \cdot \mathbf{n}]^- \mathrm{d}\sigma_x \mathrm{d}t
\end{aligned} \tag{8.71}$$

for a.a. $0 < \tau < T$. The inequality (8.71) gives rise to similar estimates as its counterpart (8.6) as long as we control the pressure term

$$-\int_0^\tau \int_\Omega p_\delta(\varrho_\delta, \vartheta_\delta)\mathrm{div}_x \mathbf{u}_B\,\mathrm{d}x.$$

Unlike in the real gas model, the hard sphere pressure p_{HS} is not controlled by the corresponding component e_{HS} of the internal energy near the singularity point $\overline{\varrho}$. Consequently, to apply Gronwall's argument to (8.71), we suppose that $\mathbf{u}_B$ can be extended inside Ω in such a way that

$$\mathrm{div}_x \mathbf{u}_B \geq 0 \text{ a.a. in } (0,T) \times \Omega \tag{8.72}$$

yielding

$$-\int_0^T \int_\Omega p_\delta(\varrho_\delta, \vartheta_\delta) \mathrm{div}_x \mathbf{u}_B \, \mathrm{d}x \leq 0$$

in (8.71). Hypothesis (8.72) entails further restrictions on the boundary velocity, namely

$$\int_{\partial\Omega} \mathbf{u}_B(\tau, \cdot) \cdot \mathbf{n} \, \mathrm{d}\sigma_x \geq 0 \text{ for a.a. } \tau \in (0,T). \tag{8.73}$$

In particular,

$$\Gamma_{\text{out}} = \emptyset \Rightarrow \Gamma_{\text{in}} = \emptyset. \tag{8.74}$$

Note that condition (8.74) is quite natural for the hard sphere gas as, on the one hand, the total mass satisfies the bound

$$\int_\Omega \varrho \, \mathrm{d}x \leq \overline{\varrho}|\Omega|,$$

while, on the other hand,

$$\frac{\mathrm{d}}{\mathrm{d}t} \int_\Omega \varrho \, \mathrm{d}x = -\int_{\partial\Omega} \varrho_B \, [\mathbf{u}_B \cdot \mathbf{n}]^- \mathrm{d}\sigma_x \text{ if } \Gamma_{\text{out}} = \emptyset.$$

These two conditions may not be compatible in the long run.

If (8.72) holds, we recover the uniform bounds (8.12)–(8.20), specifically,

$$\begin{aligned}
\operatorname*{ess\,sup}_{\tau\in(0,T)} \|\sqrt{\varrho_\delta}\mathbf{u}_\delta(\tau,\cdot)\|_{L^2(\Omega;R^d)} &\lesssim 1, \\
\operatorname*{ess\,sup}_{\tau\in(0,T)} \int_\Omega \varrho_\delta e_\delta(\varrho_\delta, \vartheta_\delta) \, \mathrm{d}x &\lesssim 1, \\
\operatorname*{ess\,sup}_{\tau\in(0,T)} \|\varrho_\delta(\tau,\cdot)\|_{L^\Gamma(\Omega)} &\lesssim 1, \\
\operatorname*{ess\,sup}_{\tau\in(0,T)} \|\vartheta_\delta(\tau,\cdot)\|_{L^4(\Omega)} &\lesssim 1,
\end{aligned} \tag{8.75}$$

$$\int_0^T \int_\Omega \left(|\nabla_x \log(\vartheta_\delta)|^2 + |\nabla_x \vartheta_\delta^{\frac{3}{2}}|^2 \right) \mathrm{d}x\mathrm{d}t \lesssim 1, \tag{8.76}$$

and

$$\int_0^T \int_\Omega \|\mathbf{u}_\delta\|^2_{W^{1,q}(\Omega;R^d)} \,\mathrm{d}x\mathrm{d}t \lesssim 1,\ q = \frac{8}{5-\Lambda},$$
$$\int_0^T \int_\Omega \|\mathbf{u}_\delta\|^r_{W^{1,s}(\Omega;R^d)} \,\mathrm{d}x\mathrm{d}t \lesssim 1,\ r = \frac{6}{4-\Lambda},\ s = \frac{18}{10-\Lambda},$$
$$\int_0^T \int_\Omega |\mathbb{S}_\delta(\vartheta_\delta, \nabla_x \mathbf{u}_\delta)|^q \,\mathrm{d}x\mathrm{d}t \lesssim 1 \text{ for some } q > 1. \tag{8.77}$$

Note that (8.76) is conditioned by another hypothesis on the boundary data, namely (8.27) and (8.28) discussed in Sect. 8.1.3.

With the bounds (8.75)–(8.77) at hand, we can repeat all steps of the proof of convergence from Sect. 8.1.4 as soon as we are able to establish uniform bounds on the pressure $p_\delta(\varrho, \vartheta)$. As already pointed out, boundedness of the internal energy is not enough to render the pressure bounded even in the space $L^1(\Omega)$.

8.2.2 Pressure Estimates

Using the arguments of Sect. 8.1.2 we are able to recover the pressure estimate (8.24). To this end, we use

$$\varphi(t,x) = \mathcal{B}\left[\phi\varrho_\delta - \frac{1}{|\Omega|}\int_\Omega \phi\varrho_\delta \,\mathrm{d}x\right],\ \phi \in C_c^\infty(\Omega),$$

where $\mathcal{B}$ is the Bogovskii operator, as a test function in the approximate momentum equation (8.3). After a straightforward manipulation similar to (8.21), we obtain

$$\begin{aligned}
&\int_0^T \int_\Omega p_\delta \left[\phi\varrho_\delta - \frac{1}{|\Omega|}\int_\Omega \phi\varrho_\delta \,\mathrm{d}x\right] \mathrm{d}x\mathrm{d}t \\
&= \left[\int_\Omega \varrho_\delta \mathbf{u}_\delta \cdot \mathcal{B}\left[\phi\varrho_\delta - \frac{1}{|\Omega|}\int_\Omega \phi\varrho_\delta \,\mathrm{d}x\right] \mathrm{d}x\right]_{t=0}^{t=T} \\
&- \int_0^T \int_\Omega \varrho_\delta \mathbf{u}_\delta \cdot \partial_t \mathcal{B}\left[\phi\varrho_\delta - \frac{1}{|\Omega|}\int_\Omega \phi\varrho_\delta \,\mathrm{d}x\right] \mathrm{d}x\mathrm{d}t \\
&- \int_0^T \int_\Omega \varrho_\delta \mathbf{u}_\delta \otimes \mathbf{u}_\delta : \nabla_x \mathcal{B}\left[\phi\varrho_\delta - \frac{1}{|\Omega|}\int_\Omega \phi\varrho_\delta \,\mathrm{d}x\right] \mathrm{d}x\mathrm{d}t \\
&+ \int_0^T \int_\Omega \mathbb{S}_\delta : \nabla_x \mathcal{B}\left[\phi\varrho_\delta - \frac{1}{|\Omega|}\int_\Omega \phi\varrho_\delta \,\mathrm{d}x\right] \mathrm{d}x\mathrm{d}t \\
&- \int_0^T \int_\Omega \varrho_\delta \mathbf{g} \cdot \mathcal{B}\left[\phi\varrho_\delta - \frac{1}{|\Omega|}\int_\Omega \phi\varrho_\delta \,\mathrm{d}x\right] \mathrm{d}x\mathrm{d}t,
\end{aligned} \tag{8.78}$$

where, as ϱ_δ, $\mathbf{u}_\delta$ satisfy the equation of continuity,

$$\partial_t \mathcal{B}\left[\phi\varrho_\delta - \frac{1}{|\Omega|}\int_\Omega \phi\varrho_\delta \, \mathrm{d}x\right]$$
$$= -\mathcal{B}\Big[\mathrm{div}_x(\phi\varrho_\delta\mathbf{u}_\delta)\Big] + \mathcal{B}\left[\varrho_\delta\mathbf{u}_\delta\cdot\nabla_x\phi - \frac{1}{|\Omega|}\int_\Omega \varrho_\delta\mathbf{u}_\delta\cdot\nabla_x\phi \, \mathrm{d}x\right]. \quad (8.79)$$

Seeing that the Bogovski operator enjoys the same regularization properties as $\nabla_x\Delta_x^{-1}$, we may repeat the arguments of Sect. 8.1.2 to conclude

$$\left|\int_0^T\int_\Omega p_\delta\left[\phi\varrho_\delta - \frac{1}{|\Omega|}\int_\Omega \phi\varrho_\delta \, \mathrm{d}x\right] \, \mathrm{d}x\mathrm{d}t\right| \le c(\phi),\ \phi\in C_c^\infty(\Omega),$$

in particular

$$\int_0^T\int_\Omega p_{\delta,HS}(\varrho_\delta)\left[\phi\varrho_\delta - \frac{1}{|\Omega|}\int_\Omega \phi\varrho_\delta \, \mathrm{d}x\right] \, \mathrm{d}x\mathrm{d}t \le c(\phi),\ \phi\in C_c^\infty(\Omega). \quad (8.80)$$

Our next goal is to deduce from (8.80) that

$$\int_0^T\int_K p_{\delta,HS}(\varrho_\delta) \, \mathrm{d}x\mathrm{d}t \le c(K) \text{ for any compact } K\subset\Omega. \quad (8.81)$$

Unlike in Sect. 8.1.2, this is not obvious as $p_{\delta,HS}$ becomes singular at $\overline{\varrho}$ in the asymptotic limit and the expression

$$\left[\phi\varrho_\delta - \frac{1}{|\Omega|}\int_\Omega \phi\varrho_\delta \, \mathrm{d}x\right]$$

in (8.80) may vanish. To prevent this to happen, we have to impose certain growth restrictions on the hard sphere component p_{HS}. For the moment, we suppose

$$r > 1$$

in (8.70), which renders

$$\varrho e_{HS}(\varrho) \approx (\overline{\varrho} - \varrho)^{1-r} \quad (8.82)$$

singular at $\overline{\varrho}$. Consequently, we may use the uniform energy estimates (8.75) to obtain

$$\operatorname*{ess\,sup}_{\tau\in(0,T)} \int_\Omega \varrho e_{\delta,HS}(\varrho_\delta)(\varrho) \, \mathrm{d}x \lesssim 1 \text{ uniformly for any fixed } \delta > 0,$$

where we have used the weak lower semi-continuity of the convex function $\varrho \mapsto \varrho e_{\delta,HS}(\varrho)$. Thus letting $\delta\to 0$, we conclude

$$\operatorname*{ess\,sup}_{\tau\in(0,T)} \int_\Omega \varrho e_{HS}(\varrho) \, \mathrm{d}x < \infty. \quad (8.83)$$

Combining (8.82) with (8.83), we obtain the desired conclusion

$$\sup_{\tau\in(0,T)} \frac{1}{|\Omega|}\int_\Omega \varrho(\tau,\cdot)\ \mathrm{d}x < \overline{\varrho}.$$

As

$$\left[\tau \mapsto \frac{1}{|\Omega|}\int_\Omega \varrho_\delta(\tau,\cdot)\ \mathrm{d}x\right] \to \left[\tau \mapsto \frac{1}{|\Omega|}\int_\Omega \varrho(\tau,\cdot)\ \mathrm{d}x\right] \text{ uniformly for } \tau \in [0,T],$$

the above relation yields

$$\limsup_{\delta\to 0}\left[\sup_{\tau\in(0,T)} \frac{1}{|\Omega|}\int_\Omega \varrho_\delta(\tau,\cdot)\ \mathrm{d}x\right] < \overline{\varrho}. \tag{8.84}$$

Now (8.84) together with (8.80) yields the bound (8.81).

Next, exactly as in Sect. 8.1.2, we have

$$\int_0^T \int_\Omega p_\delta(\varrho_\delta,\vartheta_\delta)\mathrm{dist}^{-\omega}[x,\partial\Omega]\ \mathrm{d}x\mathrm{d}t \stackrel{<}{\sim} 1 \text{ for some } \omega > 0. \tag{8.85}$$

The local estimates (8.81) together with (8.85) yield integrability of the pressure,

$$\int_0^T\int_\Omega p_\delta(\varrho_\delta,\vartheta_\delta)\ \mathrm{d}x\ \mathrm{d}t \stackrel{<}{\sim} 1 \ \Rightarrow\ \int_0^T\int_\Omega p_{\delta,HS}(\varrho_\delta)\ \mathrm{d}x\mathrm{d}t \stackrel{<}{\sim} 1. \tag{8.86}$$

To deduce equi-integrability of the pressure inside Ω, we evoke the estimate (8.21) with

$$b(\varrho) = p^{\omega}_{\delta,HS}(\varrho_\delta),\ \omega > 0,$$

obtaining

$$\begin{aligned}
&\int_0^T\int_\Omega p_\delta\left[\phi^2 p^{\omega}_{\delta,HS}(\varrho_\delta) + \nabla_x\phi\cdot\nabla_x\Delta_x^{-1}[\phi p^{\omega}_{\delta,HS}(\varrho_\delta)]\right]\ \mathrm{d}x\mathrm{d}t \\
&= \left[\int_\Omega \phi\varrho_\delta\mathbf{u}_\delta\cdot\nabla_x\Delta_x^{-1}[\phi p^{\omega}_{\delta,HS}(\varrho_\delta)]\ \mathrm{d}x\right]_{t=0}^{t=T} \\
&- \int_0^T\int_\Omega \phi\varrho_\delta\mathbf{u}_\delta\otimes\mathbf{u}_\delta : \nabla_x^2\Delta_x^{-1}[\phi p^{\omega}_{\delta,HS}(\varrho_\delta)]\ \mathrm{d}x\mathrm{d}t \\
&- \int_0^T\int_\Omega \varrho_\delta\mathbf{u}_\delta\otimes\mathbf{u}_\delta\cdot\nabla_x\phi\cdot\nabla_x\Delta_x^{-1}[\phi p^{\omega}_{\delta,HS}(\varrho_\delta)]\ \mathrm{d}x\mathrm{d}t \\
&+ \int_0^T\int_\Omega \phi\mathbb{S}_\delta : \nabla_x^2\Delta_x^{-1}[\phi p^{\omega}_{\delta,HS}(\varrho_\delta)]\ \mathrm{d}x\mathrm{d}t
\end{aligned}$$

$$+\int_0^T\int_\Omega \mathbb{S}_\delta\cdot\nabla_x\phi\cdot\nabla_x\Delta_x^{-1}[\phi p^\omega_{\delta,HS}(\varrho_\delta)]\ \mathrm{d}x\mathrm{d}t$$
$$-\int_0^T\int_\Omega \phi\varrho_\delta\mathbf{g}\cdot\nabla_x\Delta_x^{-1}[\phi p^\omega_{\delta,HS}(\varrho_\delta)]\ \mathrm{d}x\mathrm{d}t$$
$$-\int_0^T\int_\Omega \phi\varrho_\delta\mathbf{u}_\delta\cdot\partial_t\Big(\nabla_x\Delta_x^{-1}[\phi p^\omega_{\delta,HS}(\varrho_\delta)]\Big)\ \mathrm{d}x\mathrm{d}t, \tag{8.87}$$

where $\phi\in C^1_c(\Omega)$, and

$$\partial_t\Big(\nabla_x\Delta_x^{-1}[\phi p^\omega_{\delta,HS}(\varrho_\delta)]\Big)$$
$$=-\nabla_x\Delta_x^{-1}\Big[\phi\mathrm{div}_x(p^\omega_{\delta,HS}(\varrho_\delta)\mathbf{u}_\delta)\Big]$$
$$+\nabla_x\Delta_x^{-1}\Big[\phi\Big(p^\omega_{\delta,HS}(\varrho_\delta)-(p^\omega_{\delta,HS})'(\varrho_\delta)\varrho_\delta\Big)\mathrm{div}_x\mathbf{u}_\delta\Big]. \tag{8.88}$$

In view of the energy bounds (8.75)–(8.77) and (8.86), we can choose $\omega>0$ small so that all integrals on the right-hand side of (8.87) are bounded except the last one, namely

$$\int_0^T\int_\Omega \phi\varrho_\delta\mathbf{u}_\delta\cdot\partial_t\Big(\nabla_x\Delta_x^{-1}[\phi p^\omega_{\delta,HS}(\varrho_\delta)]\Big)\ \mathrm{d}x\mathrm{d}t.$$

Moreover, a short inspection of (8.88) reveals the only problematic term

$$I=\int_0^T\int_\Omega \phi\varrho_\delta\mathbf{u}_\delta\cdot\nabla_x\Delta_x^{-1}\Big[\phi(p^\omega_{\delta,HS})'(\varrho_\delta)\varrho_\delta\mathrm{div}_x\mathbf{u}_\delta\Big]\ \mathrm{d}x\mathrm{d}t.$$

To control I, first observe that

$$(p^\omega_{\delta,HS})'(\varrho)\approx\begin{cases}(\overline{\varrho}-\varrho)^{-(\omega r+1)} \text{ if } \varrho\le\overline{\varrho}-\delta\\ \delta^{-(\omega r+1)} \text{ if } \varrho>\overline{\varrho}-\delta.\end{cases}$$

Next, in view of the uniform bound on the internal energy, we get

$$\varrho_\delta e_\delta(\varrho_\delta)\gtrsim\varrho_\delta e_{\delta,HS}(\varrho_\delta)\approx\begin{cases}(\overline{\varrho}-\varrho_\delta)^{1-r} \text{ if } \varrho_\delta\le\overline{\varrho}-\delta\\ \text{convex, } C^1[\overline{\varrho}-\delta,\infty)\end{cases}$$

with the left-hand side bounded in $L^\infty(0,T;L^1(\Omega))$. Thus we conclude

$$(p^\omega_{\delta,HS})'(\varrho_\delta)\varrho_\delta \text{ bounded in } L^\infty(0,T;L^s(\Omega)) \text{ whenever } r\ge s\omega r+s+1. \tag{8.89}$$

Going back to I, we suppose, for the sake of simplicity, that the shear viscosity $\mu = \mu(\vartheta)$ satisfies hypothesis (4.18) with $\Lambda = 1$. Accordingly, in view of the estimates (8.77),

$$\{\mathbf{u}_\delta\}_{\delta>0} \text{ is bounded in } L^2(0,T;W^{1,2}(\Omega;R^d)).$$

Choosing $r > 3$, we can find, by virtue of (8.89), an $\omega > 0$ so that

$$(p^\omega_{\delta,HS})'(\varrho_\delta)\varrho_\delta \mathrm{div}_x \mathbf{u}_\delta \in L^2(0,T;L^\beta(\Omega)) \text{ for some } \beta > 1.$$

Consequently, by Hölder's inequality and the standard Sobolev embedding theorem (for $d = 3$), we get

$$\begin{aligned}
|I| &\leq \|\varrho_\delta\|_{L^\infty(0,T;L^\Gamma(\Omega))}\|\mathbf{u}_\delta\|_{L^2(0,T;L^6(\Omega;R^d))}\times \\
&\quad \times \left\|\nabla_x \Delta_x^{-1}\left[\phi(p^\omega_{\delta,HS})'(\varrho_\delta)\varrho_\delta \mathrm{div}_x \mathbf{u}_\delta\right]\right\|_{L^2(0,T;L^{\frac{3}{2}}(\Omega;R^d))} \\
&\leq \|\varrho_\delta\|_{L^\infty(0,T;L^\Gamma(\Omega))}\|\mathbf{u}_\delta\|_{L^2(0,T;W^{1,2}(\Omega;R^d))} \\
&\quad \times \|(p^\omega_{\delta,HS})'(\varrho_\delta)\varrho_\delta \mathrm{div}_x \mathbf{u}_\delta\|_{L^2(0,T;L^\beta(\Omega))}.
\end{aligned}$$

Thus we have deduced the desired pressure estimate

$$\int_0^T \int_K p_\delta(\varrho_\delta,\vartheta_\delta)p^\omega_{\delta,HS}(\varrho_\delta)\ \mathrm{d}x\mathrm{d}t \leq c(K) \text{ for some } \omega > 0$$
$$\text{and any compact } K \subset \Omega \text{ as long as } r > 3. \tag{8.90}$$

A more elaborate treatment presented in [38] would possibly improve the value of the critical exponent in (8.90) to

$$r > \frac{5}{2}.$$

8.2.3 Convergence $\delta \to 0$

With the pressure estimates (8.85), (8.86), and (8.90) at hand, the convergence proof can be done in the same way as in Sect. 8.1.4. Some steps are even easier thanks to better integrability (boundedness) of the approximate densities.

Chapter 9
Existence Theory: Main Results

The multilevel approximation scheme introduced in Chap. 5 produces, through the triple of successive limits $n \to \infty$, $\varepsilon \to 0$, and $\delta \to 0$, a weak solution to the Navier–Stokes–Fourier system for general large initial/boundary data defined globally in time. Let us summarize the principal achievements of the preceding four chapters and formulate the main results concerning the existence of weak solutions. Although we have systematically considered a compact time interval $[0, T]$, $T > 0$, the extension to $[0, \infty)$ is straightforward.

It turns out that the instantaneous state as well as the time evolution of the system can be conveniently described in terms of the *conservative-entropy variables*: the mass density $\varrho = \varrho(t, x)$, the momentum $\mathbf{m} = \varrho\mathbf{u}(t, x)$, and the total entropy $S = \varrho s(\varrho, \vartheta)$. Indeed, ϱ and $\mathbf{m}$ are weakly continuous, whereas S can be interpreted as a càglàd function of the time t. More precisely, the limits

$$
\begin{aligned}
\int_\Omega \varrho(\tau+, \cdot)\phi \ \mathrm{d}x &= \lim_{\delta\to 0+} \frac{1}{\delta}\int_\tau^{\tau+\delta}\int_\Omega \varrho(t,x)\phi(x) \ \mathrm{d}x\mathrm{d}t, \ \tau \in [0,T),\\
\int_\Omega \varrho(\tau-, \cdot)\phi \ \mathrm{d}x &= \lim_{\delta\to 0+} \frac{1}{\delta}\int_{\tau-\delta}^{\tau}\int_\Omega \varrho(t,x)\phi(x) \ \mathrm{d}x\mathrm{d}t, \ \tau \in (0,T],\\
&\phi \in C^1_c(\Omega)
\end{aligned}
\tag{9.1}
$$

$$
\begin{aligned}
\int_\Omega \varrho\mathbf{u}(\tau+, \cdot)\cdot\boldsymbol{\varphi} \ \mathrm{d}x &= \lim_{\delta\to 0+} \frac{1}{\delta}\int_\tau^{\tau+\delta}\int_\Omega \varrho\mathbf{u}(t,x)\cdot\boldsymbol{\varphi}(x) \ \mathrm{d}x\mathrm{d}t, \ \tau \in [0,T),\\
\int_\Omega \varrho\mathbf{u}(\tau-, \cdot)\cdot\boldsymbol{\varphi} \ \mathrm{d}x &= \lim_{\delta\to 0+} \frac{1}{\delta}\int_{\tau-\delta}^{\tau}\int_\Omega \varrho\mathbf{u}(t,x)\cdot\boldsymbol{\varphi}(x) \ \mathrm{d}x\mathrm{d}t, \tau \in (0,T],\\
&\boldsymbol{\varphi} \in C^1_c(\Omega; R^d),
\end{aligned}
\tag{9.2}
$$

$$
\int_\Omega \varrho s(\varrho,\vartheta)(\tau+, \cdot)\phi \ \mathrm{d}x = \lim_{\delta\to 0+} \frac{1}{\delta}\int_\tau^{\tau+\delta}\int_\Omega \varrho s(\varrho,\vartheta)(t,x)\phi(x) \ \mathrm{d}x\mathrm{d}t, \ \tau \in [0,T),
$$

E. Feireisl, A. Novotný, *Mathematics of Open Fluid Systems*, Nečas Center Series, https://doi.org/10.1007/978-3-030-94793-4_9

$$\int_{\Omega} \varrho s(\varrho, \vartheta)(\tau-, \cdot)\phi \, dx = \lim_{\delta \to 0+} \frac{1}{\delta} \int_{\tau-\delta}^{\tau} \int_{\Omega} \varrho s(t,x)\phi(x) \, dxdt, \ \tau \in (0,T],$$
$$\phi \in C^1_c(\Omega), \tag{9.3}$$

exist, and

$$\lim_{\delta \to 0+} \frac{1}{\delta} \int_{\tau}^{\tau+\delta} \int_{\Omega} \varrho(t,x)\phi(x) \, dxdt = \lim_{\delta \to 0+} \frac{1}{\delta} \int_{\tau-\delta}^{\tau} \int_{\Omega} \varrho(t,x)\phi(x) \, dxdt,$$
$$\lim_{\delta \to 0+} \frac{1}{\delta} \int_{\tau}^{\tau+\delta} \int_{\Omega} \varrho \mathbf{u}(t,x) \cdot \boldsymbol{\varphi}(x) \, dxdt = \lim_{\delta \to 0+} \frac{1}{\delta} \int_{\tau-\delta}^{\tau} \int_{\Omega} \varrho \mathbf{u}(t,x) \cdot \boldsymbol{\varphi}(x) \, dxdt$$

for any $\tau \in (0,T)$.

The equations of state $p = p(\varrho, \vartheta)$, $e = e(\varrho, \vartheta)$, the viscous stress $\mathbb{S} = \mathbb{S}(\vartheta, \mathbb{D}_x \mathbf{u})$, as well as the internal energy (heat) flux $\mathbf{q} = \mathbf{q}(\vartheta, \nabla_x \vartheta)$ are concisely formulated in terms of the standard variables ϑ and $\mathbf{u}$. However, in contrast with the conservative-entropy variables, the standard variables ϑ, $\mathbf{u}$ do not enjoy any time continuity and, accordingly, they are defined for a.a. $t \in (0,T)$ only.

9.1 Physical Domain and Boundary/Initial Data

We recall our main working hypothesis that the fluid is confined to a *bounded* domain $\Omega \subset R^d$, $d = 2, 3$. In addition, we suppose that Ω is Lipschitz and piecewise of class C^4- in the sense of Definition 5. It is an easy exercise to show that any Lipschitz domain in R^3 whose boundary consists of a finite number of smooth (C^4) surfaces separated by smooth curves fits in this category. We could definitely handle more general domains under the basic assumption that the singular part of their boundaries, where the outer normal vector is not defined, is compact and of $\mathcal{H}^{d-1}$-measure zero. Such a hypothesis is required by the theory of renormalization of the mass transport discussed in Sect. B.1 in Appendix.

9.1.1 Boundary Data

In order to perform renormalization of the equation of continuity up to the boundary, we have assumed certain compatibility between the boundary velocity $\mathbf{u}_B$ and the smoothness of the space–time cylinder $[0,T] \times \partial\Omega$, see (5.7). In particular, there exists a singular set $\Gamma_{\rm sing} \subset [0,T] \times \partial\Omega$, which is compact and of $\mathcal{H}^d$-measure zero (in the space–time), such that $[0,T] \times \partial\Omega \setminus \Gamma_{\rm sing}$ admits a decomposition specified in (5.7). Clearly, the singular set $\Gamma_{\rm sing}$ contains all points $(t, x) \in [0, T] \times \partial\Omega$, for which the outer normal $\mathbf{n}(x)$ does not exist as well as the boundary of the set Γ_0, where $\mathbf{u}_B \cdot \mathbf{n} = 0$.

If Ω is piecewise C^4 in the sense of Definition 5 and if $\mathbf{u}_B$ is at least C^2 as required, then hypothesis (5.7) is satisfied if the boundary of the set Γ^0_{wall}, where $\mathbf{u}_B \cdot \mathbf{n}$ vanishes, is of $\mathcal{H}^d$-measure zero.

The concept of C^k-regularity is based solely on the geometric properties of $\partial\Omega$, while (B.1) takes into account the behaviour of the boundary data. A close inspection of the convergence proof in Chap. 6 reveals that the C^4-regularity is in fact needed only for the in/out flow components of the boundary. In particular, this hypothesis is irrelevant wherever $\mathbf{u}_B \cdot \mathbf{n} = 0$. Similar relaxation is possible also at the level of hypothesis (B.1). In a similar fashion, the required C^2-regularity for the velocity $\mathbf{u}_B$ and the C^1-regularity for ϱ_B could be relaxed.

9.1.2 Initial Data

The initial data considered are quite general including basically any triple $(\varrho_0, \mathbf{m}_0, \vartheta_0)$ having finite energy and entropy, meaning

$$\int_\Omega \left[\frac{1}{2} \varrho_0 |\mathbf{u}_0|^2 + \varrho_0 e(\varrho_0, \vartheta_0) - \overline{\vartheta} \varrho_0 s(\varrho_0, \vartheta_0) \right] \, \mathrm{d}x < \infty, \ \overline{\vartheta} > 0.$$

As pointed out several times, it is more convenient to consider the initial data expressed in the conservative-entropy variables $(\varrho_0, \mathbf{m}_0, S_0)$ for which the initial (ballistic) energy satisfies

$$\int_\Omega \left[\frac{1}{2} \frac{|\mathbf{m}_0|^2}{\varrho_0} + \varrho_0 e(\varrho_0, S_0) - \overline{\vartheta} S_0 \right] \, \mathrm{d}x < \infty.$$

The latter definition somehow anticipates the initial entropy S_0 is an integrable function, although we know that the instantaneous values of the entropy $S(\tau, \cdot)$ are interpreted in general as finite measures on $\overline{\Omega}$. As the internal energy is a convex function of (ϱ, S), we could use the abstract approach of Demengel and Temam [27] to extend the functional $\int_\Omega \varrho e(\varrho, S) \, \mathrm{d}x$ to the space of measures. However, we do not need this generality here.

In accordance with (9.1), (9.2),

$$\varrho(0+,\cdot)=\varrho_0,\ \ \varrho\mathbf{u}(0+,\cdot)=\mathbf{m}_0,$$

while we may *set*

$$S(0-,\cdot)=\varrho s(\varrho,\vartheta)(0-,\cdot)=\varrho_0 s(\varrho_0,\vartheta_0). \tag{9.4}$$

Although it may seem that (9.4) imposes no restriction on the weak solution, we should keep in mind that both the entropy inequality (3.41) and the energy inequality (3.42) must be satisfied for the same trio $(\varrho_0,\mathbf{m}_0,\vartheta_0)$.

To conclude, we are well aware of the inconsistency in our hypotheses concerning the initial and boundary data. On the one hand, the initial data are quite general Lebesgue functions with finite energy/entropy. On the other hand, the boundary data $\mathbf{u}_B$, ϱ_B as well as $f_{i,B}$ are required to be smooth with ϱ_B separated from the vacuum. The hypothesis is smoothness of the boundary data can be definitely relaxed at the expense of several technical difficulties. Certain compatibility, however, corresponding to the fact that $\mathbf{u}_B$ is a trace of a Sobolev function must be retained. In Chap. 12, similar restrictions will be imposed on the boundary velocity ϑ_B.

9.1.3 Global Existence: Real Gas EOS

We first state our main result for the real gas described by the general equation of state (4.2),

$$p(\varrho,\vartheta)=p_{\rm el}(\varrho)+p_{\rm m}(\varrho,\vartheta)+p_{\rm rad}(\vartheta). \tag{9.5}$$

Theorem 10 (Global Existence—Real Gas).

- *Let $\Omega \subset R^d$, $d = 2, 3$, be a bounded piecewise C^4 domain in the sense of Definition 5.*
- *Suppose that the pressure $p = p(\varrho, \vartheta)$ is given by the equation of state (9.5), and, together with the associated internal energy e and the entropy s satisfy the hypotheses, (4.8)–(4.14).*
- *Let the transport coefficients $\mu = \mu(\vartheta)$, $\eta = \eta(\vartheta)$, $\kappa = \kappa(\vartheta)$ satisfy (4.18)–(4.20), with $\frac{1}{2} \leq \Lambda \leq 1$.*
- *Let the boundary data be given by the functions*

$$\mathbf{u}_B = \mathbf{u}_B(t,x), \mathbf{u}_B \in C^2(R^{d+1}; R^d), \ \varrho_B = \varrho_B(t,x), \ \varrho_B \in C^1(R^{d+1}),$$
$$f_{i,B} = f_{i,B}(t,x), \ f_{i,B} \in C(R^{d+1})$$

satisfying

$$\inf_{[0,T]\times\partial\Omega} \varrho_B > 0, \quad \inf_{[0,T]\times\partial\Omega} \Big(f_{i,B} - \varrho_B e_{\rm el}(\varrho_B) \Big) > 0. \tag{9.6}$$

In addition, suppose that the boundary $[0,T] \times \partial\Omega$ admits the decomposition (7.43), and either

$$\sup_{\tau\in(0,T)} \int_{\partial\Omega} [\mathbf{u}_B \cdot \mathbf{n}]^- {\rm d}\sigma_x < 0, \tag{9.7}$$

or

$$[\mathbf{u}_B \cdot \mathbf{n}]^+|_{\partial\Omega} = 0. \tag{9.8}$$

- *Let $\mathbf{g} \in L^\infty((0,T) \times \Omega; R^d)$.*

Then for any finite energy/entropy initial data $(\varrho_0, \vartheta_0, \mathbf{u}_0)$ and any $T > 0$, the Navier–Stokes–Fourier system admits a weak solution $(\varrho, \vartheta, \mathbf{u})$ in $(0,T) \times \Omega$ in the sense specified in Definition 4.

Remark 14. Although the proof was done for a finite $0 < T < \infty$, it is easy to adapt the same method to obtain global existence on the time interval $[0, \infty)$.

If the boundary velocity $\mathbf{u}_B = \mathbf{u}_B(x)$ is independent of t, then the hypotheses (9.7), (9.8) reduce to

$$\Gamma_{\rm in} = \emptyset \ \Rightarrow \ \Gamma_{\rm out} = \emptyset.$$

As pointed out in the course of the proof, this stipulation prevents the fluid to disappear entirely from the cavity Ω.

9.1.4 Global Existence—Hard Sphere Gas

The constitutive relation for the hard sphere pressure looks formally as (9.5),

$$p(\varrho, \vartheta) = p_{HS}(\varrho) + \varrho\vartheta + p_{\rm rad}(\vartheta), \tag{9.9}$$

where

$$p_{HS}(\varrho) = \frac{A\varrho}{(\overline{\varrho} - \varrho)^r}, A > 0,\ \overline{\varrho} > 0. \tag{9.10}$$

Theorem 11 (Global Existence—Hard Sphere Gas).

- *Let $\Omega \subset R^d$, $d = 2, 3$, be a bounded piecewise C^4 domain in the sense of Definition 5.*
- *Suppose that the pressure $p = p(\varrho, \vartheta)$ is given by the equation of state (9.9), (9.10), where $r > 3$, with the associated internal energy and entropy satisfying (4.54) and (4.55), respectively.*
- *Let the transport coefficients $\mu = \mu(\vartheta)$, $\eta = \eta(\vartheta)$, $\kappa = \kappa(\vartheta)$ satisfy (4.18)–(4.20), with $\Lambda = 1$.*
- *Let the boundary data be given by the functions*

$$\mathbf{u}_B = \mathbf{u}_B(t,x), \mathbf{u}_B \in C^2(R^{d+1}; R^d),\ \varrho_B = \varrho_B(t,x),\ \varrho_B \in C^1(R^{d+1}),$$
$$f_{i,B} = f_{i,B}(t,x),\ f_{i,B} \in C(R^{d+1})$$

satisfying

$$0 < \inf_{[0,T]\times\partial\Omega} \varrho_B \leq \sup_{[0,T]\times\partial\Omega} \varrho_B < \overline{\varrho},\ \inf_{[0,T]\times\partial\Omega} \Big(f_{i,B} - \varrho_B e_{\rm el}(\varrho_B) \Big) > 0.$$

In addition, suppose that the boundary $[0,T] \times \partial\Omega$ admits the decomposition (7.43), where

$$\int_{\partial\Omega} \mathbf{u}_B(\tau,\cdot)\cdot\mathbf{n}\ {\rm d}\sigma_x \geq 0 \text{ for any } 0 \leq \tau \leq T, \tag{9.11}$$

and either

$$\sup_{\tau\in(0,T)} \int_{\partial\Omega} [\mathbf{u}_B\cdot\mathbf{n}]^- {\rm d}\sigma_x < 0$$

or

$$[\mathbf{u}_B\cdot\mathbf{n}]^+|_{\partial\Omega} = 0. \tag{9.12}$$

- *Let $\mathbf{g} \in L^\infty((0,T)\times\Omega; R^d)$.*

Then for any finite energy/entropy initial data $(\varrho_0, \vartheta_0, \mathbf{u}_0)$ and any $T > 0$, the Navier–Stokes–Fourier system admits a weak solution $(\varrho, \vartheta, \mathbf{u})$ in $(0,T)\times\Omega$ in the sense specified in Definition 4.

Remark 15. Of course, hypothesis (9.11) in combination with (9.12) results in

$$\mathbf{u}_B \cdot \mathbf{n}|_{\partial\Omega} = 0.$$

9.2 Concluding Remarks

The origins of the mathematical theory of compressible fluid flows in the framework of weak solutions and in the multidimensional setting go back to the seminal work of P.-L. Lions [67, 68] concerning viscous barotropic fluids. These results have been improved by means of a more precise description of the density oscillations in [35, 45] up to the adiabatic exponents $\gamma > \frac{3}{2}$. More recently, Plotnikov and Weigant [81] succeeded in applying these ideas to the borderline *isothermal* case in two space dimensions.

Several new ideas related to the existence problem for the full Navier–Stokes–Fourier system with density dependent shear and bulk viscosities satisfying a particular differential relation have been developed recently in a series of papers by Bresch and Desjardins [12, 13]. Making a clever use of the structure of the equations, the authors discovered a new integral identity that allows to obtain uniform estimates on the density gradient and that may be used to prove existence of global in time solutions.

Very recently, Bresch and Jabin [14] developed a new approach relaxing certain hypotheses concerning the structure of the viscous stress tensor as well as the pressure required by the theory based on Lions' ideas.

The existence theory presented in this book can be viewed as a part of the program originated in the monograph [36] and continued in [43]. The essential new ingredient is the accommodation of the inhomogeneous boundary conditions. Note that related results for *barotropic* fluid flows can be found already in the original work of Lions [68], and, more recently, Girinon [58] and the monograph by Plotnikov and Sokolowski [80].

An existence theory to problems with boundary conditions, again in the context of barotropic fluids, better adapted to the problems of stability and weak–strong uniqueness has been proposed in [16, 19], and [64]. The origins of the results presented in this monograph can be found in [44].

The stabilizing effect of the boundary condition (9.6) was observed by Norman [77]. Of course, more complicated boundary conditions of *mixed* type combining the temperature and the heat flux can be handled by means of the approach described in Chap. 12 below.

Part III
Qualitative Properties

Chapter 10
Long Time Behaviour

The mathematical theory built up in the preceding part represents a rigorous basis to study the long time behaviour of thermodynamically open fluid systems. As we do not impose any restrictions on the size of the data, the model covers fluids in a turbulent regime. Unfortunately, however, the weak solutions are not (known to be) uniquely determined by the data; therefore, the standard dynamical systems approach based on the concept of solution semigroup is not directly applicable. Instead, we adopt the statistical description based on the Krylov–Bogolyubov theory generating a measure on the space of entire solutions to the Navier–Stokes–Fourier system. By *entire solution* we mean a solution defined at any time $t \in R$. The leading idea of replacing the phase space by the space of trajectories endowed with the natural operation of *time shift* goes back to the seminal work of Sell [83] and Málek, Nečas [70].

Our first task is to identify a suitable class of initial/boundary data that give rise to solutions with globally bounded energy. More precisely, we look for a universal constant $\mathcal{E}_\infty$ such that the total energy of the system satisfies

$$\limsup_{\tau \to \infty} \int_\Omega \left[\frac{1}{2} \frac{|\mathbf{m}|^2}{\varrho} + \varrho e(\varrho, S) \right] (\tau, \cdot) \, \mathrm{d}x \leq \mathcal{E}_\infty.$$

Here, it is convenient to describe the dynamics via the conservative-entropy variables $(\varrho, \mathbf{m}, S)$ that admit well defined instantaneous values, cf. Chap. 9.

As expected, the existence of $\mathcal{E}_\infty$ depends essentially on the choice of the boundary data and the mechanical force $\mathbf{g}$. There are examples of energetically closed systems ($\mathbf{u}_B \equiv 0$) driven by a non-conservative force $\mathbf{g}$ for which

$$\int_\Omega \left[\frac{1}{2} \frac{|\mathbf{m}|^2}{\varrho} + \varrho e(\varrho, S) \right] (\tau, \cdot) \, \mathrm{d}x \to \infty \text{ as } \tau \to \infty$$

E. Feireisl, A. Novotný, *Mathematics of Open Fluid Systems*, Nečas Center Series, https://doi.org/10.1007/978-3-030-94793-4_10

for any global in time solution, see [46]. If $\mathcal{E}_\infty < \infty$ exists, we say the system admits a *bounded absorbing set* or that it is dissipative in the sense of Levinson.

Given a globally bounded trajectory $(\varrho, \mathbf{m}, S)(t, \cdot)$, $t > 0$, we may consider the associated ω-limit set. The terminology is borrowed from the theory of dynamical systems; however, the meaning is slightly different in our setting. Supposing the solution $(\varrho, \mathbf{m}, S)$ ranges in a suitable (functional) phase space X for any $t \in [0, \infty)$, we define its ω-limit as

$$\omega[\varrho, \mathbf{m}, S] = \Big\{ (\tilde{\varrho}, \widetilde{\mathbf{m}}, \widetilde{S})(\tau, \cdot),\ \tau \in R \ \Big|\ \text{there exists } T_n \to \infty$$
$$\text{such that } (\varrho, \mathbf{m}, S)(t + T_n, \cdot) \to (\tilde{\varrho}, \widetilde{\mathbf{m}}, \widetilde{S})(t, \cdot) \text{ for any } t \in [-T, T] \Big\}.$$

The above definition is a bit vague as the notion of convergence is not specified. The exact formulation will be given later. In contrast with the standard definition, the ω-limit set is a subset of the space of entire trajectories that attracts the time shifts of a given solution locally in time. On the one hand, it is expected that "many" different global in time solutions will share the same ω-limit set that may capture the behaviour of the system in the long run. On the other hand, the existence of many different ω-limit sets cannot be excluded either.

We show that the ω-limit set generated by a bounded solution is non-empty and, what is more, consists of entire solutions to the Navier–Stokes–Fourier system. To see this, we have to show that global bounded solutions are asymptotically compact, meaning they approach a compact set in the phase space X as $t \to \infty$. This is a non-trivial issue because of possible density oscillations that may develop with growing time.

10.1 Asymptotic Compactness

Suppose that there is a weak solution $(\varrho, \mathbf{u}, \vartheta)$ of the Navier–Stokes–Fourier system defined on the time interval $[0, \infty)$ and such that

$$\mathcal{E}(\tau) \equiv \int_\Omega \left[\frac{1}{2} \varrho |\mathbf{u}|^2 + \varrho e(\varrho, \vartheta) \right] (\tau, \cdot)\ \mathrm{d}x \le \mathcal{E}_\infty,$$
$$\mathcal{S}(\tau) \equiv \int_\Omega \varrho s(\varrho, \vartheta)(\tau, \cdot)\ \mathrm{d}x \ge \mathcal{S}_\infty \tag{10.1}$$

for a.a. $\tau > 0$. Rather inconsistently with the introduction, we prefer to work with the standard variables $(\varrho, \vartheta, \mathbf{u})$ here.

Given a sequence of times $T_n \to \infty$, we consider the associated sequence of time shifts

$$\varrho_n(t, x) = \varrho(t + T_n, x),\ \mathbf{u}_n(t, x) = \mathbf{u}(t + T_n, x),\ \vartheta_n(t, x) = \vartheta(t + T_n, x).$$

Our goal is to establish (local) precompactness of the sequence $\{\varrho_n, \mathbf{u}_n, \vartheta_n\}_{n=1}^\infty$. Specifically, for any time interval $[-T, T]$, $T > 0$, we want to show

$$\varrho_n \to \tilde{\varrho},\ \mathbf{u}_n \to \tilde{\mathbf{u}},\ \vartheta_n \to \tilde{\vartheta} \text{ in a certain sense in } (-T,T),$$

where the limit $(\tilde{\varrho}, \tilde{\mathbf{u}}, \tilde{\vartheta})$ is a solution of the Navier–Stokes–Fourier system. To avoid technicalities, we focus on the *autonomous case*, where the boundary data $\mathbf{u}_B$, ϱ_B, and $f_{i,B}$ as well as the external driving force $\mathbf{g}$ are *independent of* the time t.

In view of the tools developed in the existence proof in Sect. 8, showing compactness of the time shifts may seem an easy task. Indeed the energy/entropy bounds (10.1) give rise to the same set of estimates on $(-T,T)$ including the pressure estimates as in Sect. 8.1.2. Accordingly, the same arguments as in Sect. 8 can be used to deduce the convergence of the velocity/momentum and the temperature exactly as in Sect. 8.1.4. The remaining issue is therefore the pointwise convergence of the densities. The proof presented in Sect. 8.1.8 is based on the control of the oscillation defect of the strictly convex function $\varrho \mapsto \varrho \log(\varrho)$, specifically

$$D(\tau) = \int_\Omega \left[\overline{\varrho \log(\varrho)} - \varrho \log(\varrho)\right](\tau, \cdot)\ \mathrm{d}x.$$

As a consequence of the effective viscous flux identity (8.61), the quantity D is a non-increasing function of time, more specifically,

$$\frac{\mathrm{d}}{\mathrm{d}t} D(\tau) + \int_{\partial\Omega} \left[\overline{\varrho \log(\varrho)} - \varrho \log(\varrho)\right] [\mathbf{u}_B \cdot \mathbf{n}]^+ \mathrm{d}\sigma_x \leq 0,$$

cf. (8.68). In particular, if $D(0) = 0$, then $D(t) = 0$ for any $t \geq 0$ yielding the strong convergence of the approximate densities. However, this piece of information is not available in the present situation as we do not control oscillations of the density shifts at the time $-T$. Instead we show that D satisfies a differential inequality

$$\frac{\mathrm{d}}{\mathrm{d}t} D(t) + \Psi(D) \leq 0,\ 0 \leq D(t) \leq \overline{D} \text{ for all } t \in R, \tag{10.2}$$

where $\Psi \in C[0,\infty)$, $\Psi(D) > 0$ whenever $D > 0$. It is straightforward to see that, necessarily, $D \equiv 0$. The function Ψ is not evaluated explicitly, but its existence follows from the intimate relation between D and the dissipation term identified in the inequality (8.68).

10.1.1 Statement of the Result

Here and hereafter, we tacitly assume that the thermodynamic functions p, e, and s, as well as the transport coefficients μ, η, and κ, comply with the hypotheses imposed either in Theorem 10 for the real gas EOS or in Theorem 11 for the hard sphere EOS.

Let us start with the exact formulation of the main result to be shown in the present section.

Theorem 12 (Asymptotic Compactness of Bounded Trajectories).

- *Let the boundary data* $\mathbf{u}_B$, ϱ_B, $f_{i,B}$ *as well as the driving force* $\mathbf{g}$ *be independent of* $t \in R$ *satisfying*

$$\varrho_B(x) > 0,\ f_{i,B}(x) > \varrho_B e_{\rm el}(x) \text{ in } \partial\Omega$$

for the real gas,

$$0 < \varrho_B(x) < \overline{\varrho},\ f_{i,B}(x) > \varrho_B e_{HS}(x) \text{ in } \partial\Omega$$

for the hard sphere gas.
- *Suppose the following implication*

$$[\mathbf{u}_B \cdot \mathbf{n}]^- |_{\partial\Omega} \equiv 0 \ \Rightarrow\ [\mathbf{u}_B \cdot \mathbf{n}]^+ |_{\partial\Omega} \equiv 0$$

holds. In addition, in the case of the hard sphere gas,

$$\int_{\partial\Omega} \mathbf{u}_B \cdot \mathbf{n}\ {\rm d}\sigma_x \geq 0.$$

Let $\{\varrho_n, \mathbf{u}_n, \vartheta_n\}_{n=1}^\infty$ *be a sequence of weak solutions of the Navier–Stokes–Fourier system in the sense of Definition 4 defined on the time interval* $[\tau_n, \infty)$,

$$\tau_n \geq -\infty, \tau_n \to -\infty \text{ as } n \to \infty,$$

satisfying

$$\int_\Omega \left[\frac{1}{2}\varrho|\mathbf{u}|^2 + \varrho e(\varrho, \vartheta) \right] (t, \cdot)\ {\rm d}x \leq \mathcal{E}_\infty < \infty,$$
$$\int_\Omega \varrho s(\varrho, \vartheta)(t, \cdot)\ {\rm d}x \geq \mathcal{S}_\infty > -\infty \tag{10.3}$$

for a.a. $t > \tau_n$.

Then there is a subsequence (not relabeled) such that

$$\begin{aligned}
&\varrho_n \to \varrho \text{ in } C_{\rm weak}([-T, T]; L^{\frac{5}{3}}(\Omega)) \text{ and in } C([-T, T]; L^1(\Omega)),\\
&\mathbf{u}_n \to \mathbf{u} \text{ weakly in } L^2(-T, T; W^{1,\alpha}(\Omega; R^d)),\\
&\vartheta_n \to \vartheta \text{ weakly-(*) in } L^\infty(-T, T; L^4(\Omega)),
\end{aligned} \tag{10.4}$$

for any $T > 0$, *where* $(\varrho, \mathbf{u}, \vartheta)$ *is an entire solution (defined for* $t \in R$*) of the Navier–Stokes–Fourier system, with the boundary data* ϱ_B, $\mathbf{u}_B$, $f_{i,B}$ *and the driving force* $\mathbf{g}$.

Remark 16. Recall that we have tacitly assumed all structural hypotheses on the nonlinearities imposed in Theorems 10, 11. In particular, in accordance with the bounds derived in Sect. 8,

$$\alpha = \frac{8}{5 - \Lambda},$$

where $\frac{1}{2} \leq \Lambda \leq 1$ for the real gas and $\Lambda = 1$ for the hard sphere gas.

The rest of this section is devoted to the proof of Theorem 12. Note that the result holds for both the real and the hard sphere equation of state. In the latter case, the density is bounded uniformly at any time. As $\tau_n \to -\infty$, the solutions $(\varrho_n, \mathbf{u}_n, \vartheta_n)$ are well defined on $(-T, T)$ as soon as $\tau_n \leq -T$ so (10.4) makes sense. As already pointed out, the compactness proof at the level of the temperature and the velocity/momentum essentially copies the ideas of Sect. 8. Accordingly, we mainly focus on the proof of the strong convergence of the density that requires non-standard modifications.

10.1.2 Time Evolution of the Density Oscillation Defect

Thanks to the boundedness of the total energy/entropy required by hypothesis (10.3), all uniform bounds established in Sect. 8 are available for any given time interval $(-T, T)$. Thus we may assume, extracting a suitable subsequence, that

$$\begin{aligned} &\varrho_n \to \varrho \text{ in } C_{\rm weak}([-T,T]; L^{\frac{5}{3}}(\Omega)) \\ &\mathbf{u}_n \to \mathbf{u} \text{ weakly in } L^2(-T,T; W^{1,\alpha}(\Omega; R^d)), \\ &\vartheta_n \to \vartheta \text{ weakly-(*) in } L^\infty(-T,T; L^4(\Omega)), \end{aligned}$$

for any $T > 0$. Besides, we have, similarly to Sect. 8.1.8, formula (8.68),

$$\begin{aligned} &\int_R \int_\Omega \Big(\left[\overline{L_k(\varrho)} - L_k(\varrho) \right] \partial_t \varphi + \left[\overline{L_k(\varrho)} - L_k(\varrho) \right] \mathbf{u} \cdot \nabla_x \varphi \\ &\quad + \varphi \left[T_k(\varrho) {\rm div}_x \mathbf{u} - \overline{T_k(\varrho) {\rm div}_x \mathbf{u}} \right] \Big) \, {\rm d}x {\rm d}t \\ &= \int_R \int_{\partial\Omega} \varphi \left[\overline{L_k(\varrho)} - L_k(\varrho) \right] \ [\mathbf{u}_B \cdot \mathbf{n}]^+ {\rm d}\sigma_x {\rm d}t \end{aligned} \tag{10.5}$$

for any $\varphi \in C^1_c(R \times \overline{\Omega})$. Recall that

$$T_k(\varrho) = \min\{\varrho, k\}, \ L_k'(\varrho)\varrho - L_k(\varrho) = T_k(\varrho),$$

and that the bars denote the associated weak limits. Consequently, for a spatially homogeneous $\phi = \psi(t)$, relation (10.5) gives rise to

$$\int_R \int_\Omega \left(\left[\overline{L_k(\varrho)} - L_k(\varrho) \right] \partial_t \psi + \psi \left[T_k(\varrho) \mathrm{div}_x \mathbf{u} - \overline{T_k(\varrho) \mathrm{div}_x \mathbf{u}} \right] \right) \mathrm{d}x \mathrm{d}t$$
$$= \int_R \psi \int_{\partial\Omega} \left[\overline{L_k(\varrho)} - L_k(\varrho) \right] \ [\mathbf{u}_B \cdot \mathbf{n}]^+ \mathrm{d}\sigma_x \mathrm{d}t \tag{10.6}$$

for any $\psi \in C_c^1(R)$; in other words

$$\frac{\mathrm{d}}{\mathrm{d}t} \int_\Omega \left[\overline{L_k(\varrho)} - L_k(\varrho) \right] \ \mathrm{d}x + \int_\Omega \left[\overline{T_k(\varrho) \mathrm{div}_x \mathbf{u}} - T_k(\varrho) \mathrm{div}_x \mathbf{u} \right] \ \mathrm{d}x$$
$$= - \int_{\partial\Omega} \left[\overline{L_k(\varrho)} - L_k(\varrho) \right] \ [\mathbf{u}_B \cdot \mathbf{n}]^+ \mathrm{d}\sigma_x \leq 0 \text{ in } \mathcal{D}'(R). \tag{10.7}$$

At this stage, we return to Sect. 8 evoking the inequality (8.64)

$$\text{weak-} \lim_{n\to\infty} |T_k(\varrho_n) - T_k(\varrho)|^{\frac{8}{3}}$$
$$\leq \left(\frac{4}{3}\mu(\vartheta) + \eta(\vartheta) \right) \left(\overline{T_k(\varrho) \mathrm{div}_x \mathbf{u}} - \overline{T_k(\varrho)} \mathrm{div}_x \mathbf{u} \right)$$
$$\lesssim (1 + \vartheta) \left(\overline{T_k(\varrho) \mathrm{div}_x \mathbf{u}} - \overline{T_k(\varrho)} \mathrm{div}_x \mathbf{u} \right) \tag{10.8}$$

uniformly for $k \to \infty$. Accordingly, we rewrite (10.7) as

$$\frac{\mathrm{d}}{\mathrm{d}t} \int_\Omega \left[\overline{L_k(\varrho)} - L_k(\varrho) \right] \ \mathrm{d}x + \int_\Omega \left[\overline{T_k(\varrho) \mathrm{div}_x \mathbf{u}} - \overline{T_k(\varrho)} \mathrm{div}_x \mathbf{u} \right] \ \mathrm{d}x$$
$$+ \int_\Omega \left[\overline{T_k(\varrho)} - T_k(\varrho) \right] \mathrm{div}_x \mathbf{u} \ \mathrm{d}x \leq 0 \text{ in } \mathcal{D}'(R). \tag{10.9}$$

Next, use Hölder's inequality to deduce

$$\left(\int_\Omega |T_k(\varrho_n) - T_k(\varrho)|^q \ \mathrm{d}x \right)^p$$
$$\leq \left(\int_\Omega (1 + \vartheta)^{\frac{1}{p-1}} \ \mathrm{d}x \right)^{p-1} \int_\Omega \frac{1}{1+\vartheta} |T_k(\varrho_n) - T_k(\varrho)|^{\frac{8}{3}} \ \mathrm{d}x \tag{10.10}$$

where

$$p > 1, \ pq = \frac{8}{3}.$$

As the temperature is uniformly bounded in $L^4(\Omega)$, we may combine (10.8)–(10.10) to conclude

$$\int_\Omega \left[\overline{\varrho L_k(\varrho)} - \varrho L_k(\varrho) \right] (\tau_2, \cdot) \ \mathrm{d}x$$
$$+ \Theta \int_{\tau_1}^{\tau_2} \left(\int_\Omega \text{weak} - \lim_{n\to\infty} |T_k(\varrho_n) - T_k(\varrho)|^q \ \mathrm{d}x \right)^p \mathrm{d}t$$

$$\leq \int_\Omega \left[\overline{\varrho L_k(\varrho)} - \varrho L_k(\varrho) \right] (\tau_1, \cdot) \ \mathrm{d}x + \int_{\tau_1}^{\tau_2} \int_\Omega \left(T_k(\varrho) - \overline{T_k(\varrho)} \right) \mathrm{div}_x \mathbf{u} \ \mathrm{d}x\mathrm{d}t \tag{10.11}$$

for any $\tau_1 < \tau_2$, where we have set

$$p \geq \frac{5}{4}, \ pq = \frac{8}{3}, \ \text{with } \Theta = \Theta(\mathcal{E}_\infty, \mathcal{S}_\infty) > 0.$$

Seeing that

$$\int_\Omega \left[\overline{\varrho L_k(\varrho)} - \varrho L_k(\varrho) \right] (\tau, \cdot) \ \mathrm{d}x \to \int_\Omega \left[\overline{\varrho \log(\varrho)} - \varrho \log(\varrho) \right] (\tau, \cdot) \ \mathrm{d}x$$

as $k \to \infty$ and any $\tau \in R$,

$$\int_{\tau_1}^{\tau_2} \int_\Omega \left(T_k(\varrho) - \overline{T_k(\varrho)} \right) \mathrm{div}_x \mathbf{u} \ \mathrm{d}x\mathrm{d}t \to 0 \text{ as } k \to \infty \text{ for any } \tau_1 < \tau_2,$$

we recover an inequality reminiscent of (10.2).

It remains to examine coercivity of the expression

$$\int_{\tau_1}^{\tau_2} \left(\int_\Omega \text{weak} - \lim_{n \to \infty} |T_k(\varrho_n) - T_k(\varrho)|^q \ \mathrm{d}x \right)^p \mathrm{d}t.$$

To this end, write

$$\begin{aligned} \varrho L_k(\varrho) &= \varrho L_k(\varrho) - (\log(k) + 1)(\varrho - k) - k \log(k) \\ &\quad + (\log(k) + 1)(\varrho - k) + k \log(k) \\ &= \mathcal{L}_k(\varrho) + (\log(k) + 1)(\varrho - k) + k \log(k), \end{aligned}$$

$$\mathcal{L}_k(\varrho) = \varrho L_k(\varrho) - (\log(k) + 1)\varrho + k, \ \mathcal{L}_k \in C^1_c(0, k] \cap C[0, k].$$

As $\mathcal{L}_k$ is an affine perturbation of $\varrho L_k(\varrho)$, we get

$$\overline{\varrho L_k(\varrho)} - \varrho L_k(\varrho) = \overline{\mathcal{L}_k(\varrho)} - \mathcal{L}_k(\varrho).$$

Our goal is to show

$$|\mathcal{L}_k(\varrho_1) - \mathcal{L}_k(\varrho_2)| \leq \Xi \left(|T_k(\varrho_1) - T_k(\varrho_2)|^{\frac{1}{2}} + |T_k(\varrho_1) - T_k(\varrho_2)|^2 \right) \tag{10.12}$$

for some $\Xi > 0$, which yields

$$\overline{\mathcal{L}_k(\varrho)} - \mathcal{L}_k(\varrho) \leq \Xi (\text{weak-} \lim_{n \to \infty}) \left(|T_k(\varrho_n) - T_k(\varrho)|^{\frac{1}{2}} + |T_k(\varrho_n) - T_k(\varrho)|^2 \right). \tag{10.13}$$

As $\varrho \to \varrho \log(\varrho)$ is locally Hölder continuous (with any exponent $\alpha < 1$), the inequality (10.12) obviously holds if either $0 \leq \varrho_1, \varrho_2 \leq k$ or $\varrho_1, \varrho_2 \geq k$. If $\varrho_1 < k$, $\varrho_2 \geq k$, (10.12) reduces to

$$\varrho_1|\log(\varrho_1) - \log(k)| + |k - \varrho_1| \le \Xi\left(|\varrho_1 - k|^{\frac{1}{2}} + |\varrho_1 - k|^2\right).$$

Seeing that

$$\varrho_1|\log(\varrho_1) - \log(k)| \le |\varrho_1 - k|$$

we conclude. As a matter of fact, the exponents $\frac{1}{2}$, 2 can be replaced by β^{-1}, β with $\beta > 1$.

Next, use Hölder's inequality to observe

$$\int_\Omega |T_k(\varrho_n) - T_k(\varrho)|^{\frac{1}{2}}\, \mathrm{d}x \lesssim \left(\int_\Omega |T_k(\varrho_n) - T_k(\varrho)|^2\, \mathrm{d}x\right)^{\frac{1}{4}},$$

whence (10.13) yields

$$\begin{aligned}&\int_\Omega \left[\overline{L_k(\varrho)} - L_k(\varrho)\right]\, \mathrm{d}x = \int_\Omega \overline{\mathcal{L}_k(\varrho)} - \mathcal{L}_k(\varrho)\, \mathrm{d}x \\ &\lesssim \lim_{n\to\infty}\left[\left(\int_\Omega |T_k(\varrho_n) - T_k(\varrho)|^2\, \mathrm{d}x\right)^{\frac{1}{4}} + \int_\Omega |T_k(\varrho_n) - T_k(\varrho)|^2\, \mathrm{d}x\right].\end{aligned}$$

Going back to (10.11), we choose $q = 2$, $p = \frac{4}{3}$ to obtain

$$\begin{aligned}&\int_\Omega \left[\overline{\varrho L_k(\varrho)} - \varrho L_k(\varrho)\right](\tau_2, \cdot)\, \mathrm{d}x \\ &+ \int_{\tau_1}^{\tau_2} \Psi\left(\int_\Omega \left[\overline{\varrho L_k(\varrho)} - \varrho L_k(\varrho)\right]\, \mathrm{d}x\right) \mathrm{d}t \\ &\le \int_\Omega \left[\overline{\varrho L_k(\varrho)} - \varrho L_k(\varrho)\right](\tau_1, \cdot)\, \mathrm{d}x + \int_{\tau_1}^{\tau_2}\int_\Omega \left(T_k(\varrho) - \overline{T_k(\varrho)}\right) \mathrm{div}_x \mathbf{u}\, \mathrm{d}x\mathrm{d}t,\end{aligned} \tag{10.14}$$

where

$$\Psi(D) \lesssim \min\left\{D^{\frac{16}{3}}, D^{\frac{4}{3}}\right\}.$$

Letting $k \to \infty$ in (10.14), we obtain (10.2), which yields the desired conclusion

$$\int_\Omega \left[\overline{\varrho\log(\varrho)} - \varrho\log(\varrho)\right](\tau, \cdot)\, \mathrm{d}x = 0 \text{ for any } \tau \in R. \tag{10.15}$$

10.1.3 Strong Convergence of the Density

In order to complete the proof of Theorem 12, we have to show the strong convergence of $\{\varrho_n\}_{n=1}^\infty$ claimed in (10.3), namely

$$\varrho_n \to \varrho \text{ in } C([-T, T]; L^1(\Omega)) \text{ for any } T > 0.$$

Thanks to (10.15), we already know that

$$\varrho_n(\tau,\cdot) \to \varrho(\tau,\cdot) \text{ in } L^r(\Omega),\ 1 \le r < \frac{5}{3} \tag{10.16}$$

for any $\tau \in R$. As ϱ_n, ϱ are uniformly bounded in $L^{\frac{5}{3}}(\Omega)$, we have

$$\|(T_k(\varrho_n) - \varrho_n)(\tau,\cdot)\|_{L^1(\Omega)} + \|(T_k(\varrho) - \varrho)(\tau,\cdot)\|_{L^1(\Omega)} \lesssim k^{-\frac{2}{3}}$$

for any $\tau \in [-T, T]$. Therefore, it is enough to show

$$T_k(\varrho_n) \to T_k(\varrho) \text{ in } C([-T,T]; L^1(\Omega)) \text{ for any } T > 0, \text{ and any } k > 0. \tag{10.17}$$

The functions $T_k(\varrho_n)$, $T_k^2(\varrho_n)$ satisfy the renormalized equation of continuity (8.67). In particular, they are precompact in the space $C_{\rm weak}([-T,T]; L^r(\Omega))$ for any $1 \le r < \infty$, and we may infer that

$$T_k(\varrho_n) \to T_k(\varrho),\ T_k^2(\varrho_n) \to T_k^2(\varrho) \text{ in } C_{\rm weak}([-T,T]; L^r(\Omega)) \text{ as } n \to \infty. \tag{10.18}$$

Note carefully that the limit is uniquely identified by (10.16) at *any* time $\tau \in [-M, M]$. Finally, write

$$|T_k(\varrho_n) - T_k(\varrho)|^2 = T_k^2(\varrho_n) - T_k^2(\varrho) - 2T_k(\varrho)(T_k(\varrho_n) - T_k(\varrho)).$$

In view of (10.18), the desired relation (10.17) follows as soon as we show that

$$2T_k(\varrho)(T_k(\varrho_n) - T_k(\varrho)) \to 0 \in C_{\rm weak}([-T,T]; L^1(\Omega)). \tag{10.19}$$

As a consequence of (10.18), we observe the limit $T_k(\varrho)$ is strongly continuous, specifically,

$$T_k(\varrho) \in C([-M,M]; L^r(\Omega)) \text{ for any } 1 \le r < \infty.$$

In particular, the curve $t \mapsto T_k(\varrho)(t,\cdot)$ being a continuous image of a compact set $[-T;T]$ is compact in $L^r(\Omega)$ for any $1 \le r < \infty$. Consequently, for any $\varepsilon > 0$, there is a finite number of time intervals I_i, $\cup_i I_i = [-T,T]$, and $w_i \in L^r(\Omega)$ such that

$$\|T_k(\varrho)(t,\cdot) - \sum_i 1_{I_i}(t) w_i\|_{L^r(\Omega)} < \varepsilon \text{ uniformly for } t \in [-T,T].$$

By virtue of (10.18),

$$\sum_i 1_{I_i}(t) w_i (T_k(\varrho_n) - T_k(\varrho)) \to 0 \in C_{\rm weak}([-T,T]; L^1(\Omega)),$$

whence, as $\varepsilon > 0$ can be chosen arbitrary small, (10.19) and then (10.17) follow.

We have shown Theorem 12.

10.2 Dissipativity, Bounded Absorbing Sets

Dissipativity in the sense of Levinson or, equivalently, boundedness of the energy/entropy of global in time solutions to the Navier–Stokes–Fourier system is a delicate issue. The key tool is the total energy/entropy balance that for classical solutions takes the form (cf. (1.4), (1.5), and (1.7))

$$\partial_t \left(\frac{1}{2}\varrho|\mathbf{u}|^2 + \varrho e(\varrho,\vartheta) \right) + \mathrm{div}_x \left[\left(\frac{1}{2}\varrho|\mathbf{u}|^2 + \varrho e(\varrho,\vartheta) \right) \mathbf{u} \right]$$
$$+\mathrm{div}_x \mathbf{q}(\vartheta, \nabla_x \vartheta) + \mathrm{div}_x \left[\Big(p(\varrho,\vartheta)\mathbb{I} - \mathbb{S}(\vartheta, \nabla_x \mathbf{u}) \Big) \cdot \mathbf{u} \right] = \varrho \mathbf{g} \cdot \mathbf{u}, \tag{10.20}$$

$$\partial_t(\varrho s(\varrho,\vartheta)) + \mathrm{div}_x \left[\varrho s(\varrho,\vartheta)\mathbf{u} \right] + \mathrm{div}_x \left(\frac{\mathbf{q}(\vartheta,\nabla_x\vartheta)}{\vartheta} \right)$$
$$= \frac{1}{\vartheta} \left[\mathbb{S}(\vartheta, \nabla_x \mathbf{u}) : \nabla_x \mathbf{u} - \frac{\mathbf{q}(\vartheta,\nabla_x\vartheta)\cdot\nabla_x\vartheta}{\vartheta} \right]. \tag{10.21}$$

In addition, performing a routine manipulation detailed in Sect. 3.2.2, the energy balance (10.20) can be written in the form

$$\partial_t \left(\frac{1}{2}\varrho|\mathbf{u}-\mathbf{u}_B|^2 + \varrho e(\varrho,\vartheta) \right) + \mathrm{div}_x \left[\left(\frac{1}{2}\varrho|\mathbf{u}-\mathbf{u}_B|^2 + \varrho e(\varrho,\vartheta) \right) \mathbf{u} \right]$$
$$+ \mathrm{div}_x \mathbf{q}(\vartheta,\nabla_x\vartheta) + \mathrm{div}_x (p(\varrho,\vartheta)(\mathbf{u}-\mathbf{u}_B)) + \mathrm{div}_x \left(\mathbb{S}(\vartheta,\nabla_x\mathbf{u})\cdot(\mathbf{u}_B-\mathbf{u}) \right)$$
$$= \mathbb{S}(\vartheta,\nabla_x\mathbf{u}) : \nabla_x\mathbf{u}_B - \left[\varrho(\mathbf{u}-\mathbf{u}_B)\otimes(\mathbf{u}-\mathbf{u}_B) + p(\varrho,\vartheta)\mathbb{I} \right] : \nabla_x\mathbf{u}_B$$
$$+ \varrho\left[\mathbf{g} - \partial_t\mathbf{u}_B - (\mathbf{u}_B\cdot\nabla_x)\mathbf{u}_B \right]\cdot(\mathbf{u}-\mathbf{u}_B). \tag{10.22}$$

Integrating (10.22) over Ω and taking the boundary conditions into account, we obtain

$$\frac{\mathrm{d}}{\mathrm{d}t} \int_\Omega \left(\frac{1}{2}\varrho|\mathbf{u}-\mathbf{u}_B|^2 + \varrho e(\varrho,\vartheta) \right) \mathrm{d}x + \int_{\partial\Omega} \varrho e(\varrho,\vartheta) \; [\mathbf{u}_B\cdot\mathbf{n}]^+ \mathrm{d}\sigma_x$$
$$= - \int_\Omega \left[\varrho(\mathbf{u}-\mathbf{u}_B)\otimes(\mathbf{u}-\mathbf{u}_B) + p(\varrho,\vartheta)\mathbb{I} \right] : \nabla_x\mathbf{u}_B \; \mathrm{d}x$$
$$+ \int_\Omega \mathbb{S}(\vartheta,\nabla_x\mathbf{u}) : \nabla_x\mathbf{u}_B \; \mathrm{d}x$$

$$+\int_\Omega \varrho\Big[\mathbf{g}-\partial_t\mathbf{u}_B-(\mathbf{u}_B\cdot\nabla_x)\mathbf{u}_B\Big]\cdot(\mathbf{u}-\mathbf{u}_B)\ \mathrm{d}x-\int_{\partial\Omega} f_{i,B}\ [\mathbf{u}_B\cdot\mathbf{n}]^-\mathrm{d}\sigma_x. \tag{10.23}$$

Up to the inequality sign, this is basically the total energy balance (3.42) satisfied by any weak solution.

Repeating the same procedure with the entropy balance (10.21), we get

$$\begin{aligned}
&\frac{\mathrm{d}}{\mathrm{d}t}\int_\Omega \varrho s(\varrho,\vartheta)\ \mathrm{d}x+\int_{\partial\Omega}\varrho s(\varrho,\vartheta)\ [\mathbf{u}_B\cdot\mathbf{n}]^+\mathrm{d}\sigma_x\\
&+\int_{\partial\Omega}\left(\varrho_B s(\varrho_B,\vartheta)-\frac{1}{\vartheta}\varrho_B e(\varrho_B,\vartheta)\right)[\mathbf{u}_B\cdot\mathbf{n}]^-\mathrm{d}\sigma_x\\
&=\int_\Omega\frac{1}{\vartheta}\left[\mathbb{S}(\vartheta,\nabla_x\mathbf{u}):\nabla_x\mathbf{u}-\frac{\mathbf{q}(\vartheta,\nabla_x\vartheta)\cdot\nabla_x\vartheta}{\vartheta}\right]\mathrm{d}x\\
&-\int_{\partial\Omega}\frac{f_{i,B}}{\vartheta}\ [\mathbf{u}_B\cdot\mathbf{n}]^-\mathrm{d}\sigma_x
\end{aligned} \tag{10.24}$$

that should be compared with its weak counterpart (3.41).

Finally, we multiply (10.24) on an arbitrary positive constant $\overline{\vartheta}$ and subtract the resulting expression from (10.23) obtaining

the **total dissipation balance:**

$$\begin{aligned}
&\frac{\mathrm{d}}{\mathrm{d}t}\int_\Omega\left(\frac{1}{2}\varrho|\mathbf{u}-\mathbf{u}_B|^2+\varrho e(\varrho,\vartheta)-\overline{\vartheta}\varrho s(\varrho,\vartheta)\right)\mathrm{d}x\\
&+\int_{\partial\Omega}\Big(\varrho e(\varrho,\vartheta)-\overline{\vartheta}\varrho s(\varrho,\vartheta)\Big)\ [\mathbf{u}_B\cdot\mathbf{n}]^+\mathrm{d}\sigma_x\\
&-\int_{\partial\Omega}\overline{\vartheta}\left(\varrho_B s(\varrho_B,\vartheta)-\frac{1}{\vartheta}\varrho_B e(\varrho_B,\vartheta)\right)\ [\mathbf{u}_B\cdot\mathbf{n}]^-\mathrm{d}\sigma_x\\
&+\int_\Omega\frac{\overline{\vartheta}}{\vartheta}\left[\mathbb{S}(\vartheta,\nabla_x\mathbf{u}):\nabla_x\mathbf{u}-\frac{\mathbf{q}(\vartheta,\nabla_x\vartheta)\cdot\nabla_x\vartheta}{\vartheta}\right]\mathrm{d}x\\
&-\int_{\partial\Omega}\frac{\overline{\vartheta}}{\vartheta}f_{i,B}\ [\mathbf{u}_B\cdot\mathbf{n}]^-\mathrm{d}\sigma_x\\
&=-\int_\Omega\Big[\varrho(\mathbf{u}-\mathbf{u}_B)\otimes(\mathbf{u}-\mathbf{u}_B)+p(\varrho,\vartheta)\mathbb{I}\Big]:\nabla_x\mathbf{u}_B\ \mathrm{d}x\\
&+\int_\Omega\mathbb{S}(\vartheta,\nabla_x\mathbf{u}):\nabla_x\mathbf{u}_B\ \mathrm{d}x\\
&+\int_\Omega\varrho\Big[\mathbf{g}-\partial_t\mathbf{u}_B-(\mathbf{u}_B\cdot\nabla_x)\mathbf{u}_B\Big]\cdot(\mathbf{u}-\mathbf{u}_B)\ \mathrm{d}x\\
&-\int_{\partial\Omega}f_{i,B}\ [\mathbf{u}_B\cdot\mathbf{n}]^-\mathrm{d}\sigma_x.
\end{aligned} \tag{10.25}$$

Note that the weak solutions satisfy the same relation with an inequality sign. At this level, the difference between strong and weak does not seem essential.

The quantity

$$\mathcal{E}_{\overline{\vartheta}}(\tau) \equiv \int_{\Omega} \left(\frac{1}{2} \varrho |\mathbf{u} - \mathbf{u}_B|^2 + \varrho e(\varrho, \vartheta) - \overline{\vartheta} \varrho s(\varrho, \vartheta) \right) (\tau, \cdot) \, \mathrm{d}x, \ \overline{\vartheta} > 0$$

is the ballistic energy introduced in Sect. 3.6. Boundedness of $\mathcal{E}_{\overline{\vartheta}}(\tau)$ for $\tau \to \infty$ implies the uniform bounds (10.3) on the total energy/entropy required in Theorem 12. Our goal will be to establish a differential inequality of the form

$$\frac{\mathrm{d}}{\mathrm{d}t} \mathcal{E}_{\overline{\vartheta}} + \Psi(\mathcal{E}_{\overline{\vartheta}}) \leq H, \tag{10.26}$$

where $H = H(t)$ is uniformly bounded and Ψ has the following property:

$$\Psi(\mathcal{E}_{\overline{\vartheta}}) \leq M \ \Rightarrow \ |\mathcal{E}_{\overline{\vartheta}}| \leq \chi(M). \tag{10.27}$$

The relations (10.26), (10.27) imply that either the ballistic energy at a given time is bounded above by a constant depending solely on $\sup_t H$ or it is strictly decreasing. Obviously this yields the desired uniform bound on $\mathcal{E}_{\overline{\vartheta}}$. As we shall see below, deducing (10.26) from the dissipation balance (10.25) is a non-trivial task; in particular, the inequality (10.26) may not be true in general.

10.2.1 Preliminary Discussion

To illuminate the difficulties in obtaining (10.25), consider a very simple case when $\mathbf{u}_B \equiv 0$, $\mathbf{g} = \mathbf{g}(x)$. The dissipation balance (10.26) obviously reduces to

$$\frac{\mathrm{d}}{\mathrm{d}t} \int_{\Omega} \left(\frac{1}{2} \varrho |\mathbf{u}|^2 + \varrho e(\varrho, \vartheta) - \overline{\vartheta} \varrho s(\varrho, \vartheta) \right) \mathrm{d}x$$
$$+ \int_{\Omega} \frac{\overline{\vartheta}}{\vartheta} \left[\mathbb{S}(\vartheta, \nabla_x \mathbf{u}) : \nabla_x \mathbf{u} - \frac{\mathbf{q}(\vartheta, \nabla_x \vartheta) \cdot \nabla_x \vartheta}{\vartheta} \right] \mathrm{d}x \leq \int_{\Omega} \varrho \mathbf{g} \cdot \mathbf{u} \, \mathrm{d}x$$

where we have put the inequality sign relevant for the weak solutions. In addition, for the sake of simplicity, suppose that the viscosity coefficients have linear growth in ϑ ($\Lambda = 1$ in hypotheses (4.18), (4.19)), and that the internal energy satisfies the hard sphere equations of state (4.53), (4.54). Accordingly, the dissipation balance takes the form

$$\frac{\mathrm{d}}{\mathrm{d}t}\int_\Omega \left[\frac{1}{2}\varrho|\mathbf{u}|^2 + \frac{3}{2}\varrho\vartheta + a\vartheta^4 + \varrho e_{HS}(\varrho) - \overline{\vartheta}\left(\varrho\frac{3}{2}\log(\vartheta) - \varrho\log(\varrho) + \frac{4}{3}\vartheta^3\right)\right] \mathrm{d}x$$
$$+ \int_\Omega \frac{\overline{\vartheta}}{\vartheta}\left[\mathbb{S}(\vartheta,\nabla_x\mathbf{u}) : \nabla_x\mathbf{u} - \frac{\mathbf{q}(\vartheta,\nabla_x\vartheta)\cdot\nabla_x\vartheta}{\vartheta}\right] \mathrm{d}x \leq \int_\Omega \varrho\mathbf{g}\cdot\mathbf{u}\ \mathrm{d}x,$$
$$e_{HS}(\varrho) = \int_1^\varrho \frac{\beta}{r(\overline{\varrho}-r)^\alpha}\mathrm{d}r,$$

where, furthermore,

$$\left|\int_\Omega \varrho\mathbf{g}\cdot\mathbf{u}\ \mathrm{d}x\right| \leq \varepsilon\|\mathbf{u}\|^2_{L^2(\Omega;R^d)} + c(\varepsilon,\overline{\varrho})\|\mathbf{g}\|^2_{L^2(\Omega;R^d)}.$$

Thus we may use the Korn–Poincaré inequality along with the structural hypotheses imposed on the transport coefficients to deduce from the dissipation balance the inequality

$$\frac{\mathrm{d}}{\mathrm{d}t}\int_\Omega \left[\frac{1}{2}\varrho|\mathbf{u}|^2 + \frac{3}{2}\varrho\vartheta + a\vartheta^4 + \varrho e_{HS}(\varrho) - \overline{\vartheta}\left(\varrho\frac{3}{2}\log(\vartheta) - \varrho\log(\varrho) + \frac{4}{3}\vartheta^3\right)\right] \mathrm{d}x$$
$$+ \int_\Omega \left[|\nabla_x\mathbf{u}|^2 + |\nabla_x\vartheta|^2 + |\nabla_x\log(\vartheta)|^2\right] \lesssim \|\mathbf{g}\|^2_{L^2(\Omega,R^d)}. \tag{10.28}$$

Inequality (10.28) seems optimal as we have controlled all "source" terms on the right-hand side of the total dissipation balance (10.25).

Does (10.28) guarantee boundedness of the ballistic energy $\mathcal{E}_{\overline{\vartheta}}$? Here we content ourselves with a sketchy formal argument that we make rigorous in the following section. Suppose that

$$\limsup_{t\to\infty} \mathcal{E}_{\overline{\vartheta}}(t) < \mathcal{E}_\infty. \tag{10.29}$$

Boundedness of $\mathcal{E}_{\overline{\vartheta}}$ implies boundedness of the total entropy $\int_\Omega \varrho s(\varrho,\vartheta)\ \mathrm{d}x$ that happens to be, in the present setting, a non-decreasing function of t; in particular, it tends to a finite limit as $t\to\infty$. It follows from the entropy inequality (3.41) that

$$\int_0^\infty \int_\Omega \frac{1}{\vartheta}\left(\mathbb{S}(\vartheta,\nabla_x\mathbf{u}) : \nabla_x\mathbf{u} - \frac{\mathbf{q}(\vartheta,\nabla_x\vartheta)\cdot\nabla_x\vartheta}{\vartheta}\right) \mathrm{d}x\mathrm{d}t < \infty. \tag{10.30}$$

Evoking the situation treated in Theorem 12, we can find a sequence of times $T_n\to\infty$ and the time shifts

$$\varrho_n(t,x) = \varrho(t+T_n,x),\ \mathbf{u}_n(t,x) = \mathbf{u}(t+T_n,x),\ \vartheta_n(t,x) = \vartheta(t+T_n,x),$$

such that

$$\begin{aligned}
&\varrho_n \to \tilde{\varrho} \text{ in } C_{\rm weak}([0,T]; L^{\frac{5}{3}}(\Omega)) \text{ and in } C([0,T]; L^1(\Omega)),\\
&\mathbf{u}_n \to \tilde{\mathbf{u}} \text{ weakly in } L^2(0,T; W^{1,\alpha}(\Omega; R^d)),\\
&\vartheta_n \to \tilde{\vartheta} \text{ weakly-(*) in } L^\infty(0,T; L^4(\Omega))
\end{aligned}$$

for any fixed T. Now, on the one hand, Theorem 12 yields that $(\tilde{\varrho}, \tilde{\mathbf{u}}, \tilde{\vartheta})$ is a weak solution of the Navier–Stokes–Fourier system in $(0,T)\times\Omega$. On the other hand, by virtue of (10.30), $\tilde{\mathbf{u}} \equiv 0$, $\tilde{\vartheta} = \tilde{\vartheta}(t)$ is independent of x. Moreover, it follows from the equation of continuity $\tilde{\varrho} = \tilde{\varrho}(x)$ is independent of t. Plugging this ansatz in the momentum equation, we get

$$\tilde{\vartheta}(t)\nabla_x\tilde{\varrho} + \beta\nabla_x\left(\frac{\tilde{\varrho}}{\overline{\varrho} - \tilde{\varrho}}\right) = \tilde{\varrho}\mathbf{g}. \tag{10.31}$$

As the total mass of the fluid is constant for our choice of the boundary conditions,

$$\int_\Omega \tilde{\varrho}\, {\rm d}x = M_0 > 0,$$

Equation (10.31) gives rise to the following alternatives:

- $$\tilde{\varrho} > 0 \text{ a positive constant } \Rightarrow \mathbf{g} \equiv 0.$$
- $$\nabla_x\tilde{\varrho} \neq 0 \Rightarrow \tilde{\vartheta} > 0 \text{ a positive constant.}$$

Moreover, in the latter case, we may divide (10.31) on $\tilde{\varrho}$ at least on each component of the domain of positivity $\mathcal{O}$ of $\tilde{\varrho}$ obtaining

$$\mathbf{g} = \tilde{\vartheta}\nabla_x \log(\tilde{\varrho}) + \beta\nabla_x F(\tilde{\varrho}),$$

where F is the primitive of $\frac{1}{\tilde{\varrho}}\left(\frac{\tilde{\varrho}}{\overline{\varrho}-\tilde{\varrho}}\right)'$. Thus, necessarily, the force $\mathbf{g}$ must be potential,

$$\mathbf{g} = \nabla_x G \text{ at least in } \mathcal{O}.$$

Accordingly,

$$G(x) = \tilde{\vartheta}\log(\tilde{\varrho}) + \beta F(\tilde{\varrho}) + C_{\mathcal{O}},$$

where $C_{\mathcal{O}}$ is a suitable constant. Assuming $\mathbf{g} \in L^\infty$ yields G bounded in Ω, whence, necessarily, $\mathcal{O} = \Omega$ and

$$\mathbf{g} = \nabla_x G \text{ in } \Omega.$$

To conclude, we have shown that even a very strong form of dissipation balance (10.28) need not guarantee uniform boundedness of the energy/entropy unless $\mathbf{g}$ is a potential force. As this observation was conditioned by a rather special choice of the boundary conditions, it *should not* be interpreted in the way that the relation $\mathbf{g} = \nabla_x G$ is *necessary* for global boundedness of the energy.

10.2.2 Bounded Absorbing Sets: Hard Sphere Gas

A short inspection of the total dissipation balance (10.25) reveals immediately the principal difficulty in obtaining uniform energy bounds, namely the source term

$$-\int_\Omega \left[\varrho(\mathbf{u}-\mathbf{u}_B)\otimes(\mathbf{u}-\mathbf{u}_B)+p(\varrho,\vartheta)\mathbb{I}\right] : \nabla_x \mathbf{u}_B \,\mathrm{d}x$$

proportional to the total energy. To control its possible growth by dissipation, either uniform bounds on the density or a special form of $\mathbf{u}_B$ seems necessary. The density is uniformly bounded for the hard sphere model, where

$$p(\varrho,\vartheta) = \varrho\vartheta + p_{HS}(\varrho) + \frac{a}{3}\vartheta^4,\ p_{HS}(\varrho) = \frac{A\varrho}{(\overline{\varrho}-\varrho)^r},\ A, r > 0, \tag{10.32}$$

with the associated internal energy

$$e(\varrho,\vartheta) = \frac{3}{2}\vartheta + e_{HS}(\varrho) + \frac{a}{\varrho}\vartheta^4,\ e_{HS}(\varrho) = \int_1^\varrho \frac{A}{r(\overline{\varrho}-z)^r}\mathrm{d}z, \tag{10.33}$$

and the entropy

$$s(\varrho,\vartheta) = \frac{3}{2}\log(\vartheta) - \log(\varrho) + \frac{4}{3\varrho}\vartheta^3. \tag{10.34}$$

Moreover, we fix the transport coefficients

$$\begin{aligned} 0 < \underline{\mu}(1+\vartheta) \leq \mu(\vartheta) \leq \overline{\mu}(1+\vartheta),\ 0 \leq \eta(\vartheta) \leq \overline{\eta}(1+\vartheta),\\ 0 < \underline{\kappa}(1+\vartheta^3) \leq \kappa(\vartheta) \leq \overline{\kappa}(1+\vartheta^3). \end{aligned} \tag{10.35}$$

In addition, we suppose

$$\begin{aligned} \varrho_B \in BC([0,\infty)\times R^d),\ \mathbf{u}_B \in BC^2([0,\infty)\times R^d; R^d),\\ \mathbf{g} \in L^\infty((0,\infty)\times\Omega; R^d), \end{aligned} \tag{10.36}$$

and

$$f_{i,B} \in BC([0,\infty)\times\partial\Omega), 0 < \underline{f} \leq \inf_{t,x}\left(f_{i,B} - \varrho_B e_{HS}(\varrho_B)\right). \tag{10.37}$$

Finally, being aware of the problems discussed in the previous sections, we suppose, in addition to hypothesis (8.27), that the outflow dominates the inflow, namely

$$0 < \underline{U} \leq \inf_{\tau \geq 0} \int_{\Omega} \mathrm{div}_x \mathbf{u}_B(\tau, \cdot) \, dx, \sup_{\tau \geq 0} \int_{\partial\Omega} [\mathbf{u}_B \cdot \mathbf{n}]^- \, d\sigma_x dt \leq -\underline{U} < 0. \quad (10.38)$$

The hypothesis (10.38) is absolutely crucial for the existence of an absorbing set. Note that (10.38) requires *non-zero* outflow; in particular, it is not satisfied for the no-slip boundary conditions.

10.2.2.1 Extension of the Boundary Velocity

We start by finding a suitable extension $\mathbf{u}_B$ of the boundary velocity. Let

$$U(t) = \frac{1}{|\Omega|} \int_{\Omega} \mathrm{div}_x \mathbf{u}_B(t, \cdot)) \, dx.$$

It is easy to find a smooth vector field $\mathbf{v}_B : R^d \to R^d$ such that

$$\mathrm{div}_x \mathbf{v}_B(t, x) = U(t) \text{ for } t \geq 0.$$

Write

$$\mathbf{u}_B = \mathbf{u}_B - \mathbf{v}_B + \mathbf{v}_B.$$

As

$$\int_{\Omega} (\mathrm{div}_x \mathbf{u}_B - \mathrm{div}_x \mathbf{v}_B)(t, \cdot) \, dx = 0 \text{ for any } t \geq 0,$$

there exists $\mathbf{w}_B$ such that

$$\mathrm{div}_x \mathbf{w}_B(t, \cdot) = (\mathrm{div}_x \mathbf{v}_B - \mathrm{div}_x \mathbf{u}_B)(t, \cdot), \ \mathbf{w}_B(t, \cdot)|_{\partial\Omega} = 0, t \geq 0.$$

The function $\mathbf{w}_B$ can be constructed by means of the Bogovskii operator, cf. Sect. 0.8. Finally, set

$$\widetilde{\mathbf{u}}_B = \mathbf{w}_B + \mathbf{u}_B - \mathbf{v}_B + \mathbf{v}_B.$$

Obviously, $\widetilde{\mathbf{u}}_B = \mathbf{u}_B$ on $[0, \infty) \times \partial\Omega$, and

$$\mathrm{div}_x \widetilde{\mathbf{u}}_b = \mathrm{div}_x \mathbf{v}_B = U(t).$$

Going back to the original notation, we shall assume hereafter that

$$\mathrm{div}_x \mathbf{u}_B(t, x) = U(t) \geq \underline{U} > 0 \text{ for any } t \geq 0, \ x \in \Omega. \quad (10.39)$$

Remark 17. It is important to notice that the extension constructed via the Bogovskii operator is less regular. Specifically, in accordance with Theorem 2, we have

$$\mathbf{u}_B \in BC^1(R; W^{1,q}(\Omega; R^d)) \text{ for any } 1 \leq q < \infty.$$

In particular, the field $\mathbf{u}_B$ need not be Lipschitz in x still retaining the property that its divergence is bounded.

10.2.2.2 Uniform Bounds

Revisiting the total dissipation balance (10.25), we have

$$\begin{aligned}
&\frac{\mathrm{d}}{\mathrm{d}t}\int_\Omega \Big[\frac{1}{2}\varrho|\mathbf{u}-\mathbf{u}_B|^2 + \frac{3}{2}\varrho\vartheta + \varrho e_{HS}(\varrho) + a\vartheta^4 \\
&\qquad - \overline{\vartheta}\left(\frac{4}{3}a\vartheta^3 + \frac{3}{2}\varrho\log(\vartheta) - \varrho\log(\varrho)\right)\Big]\,\mathrm{d}x \\
&\quad + \int_{\partial\Omega}\Big(\varrho e(\varrho,\vartheta) - \overline{\vartheta}\varrho s(\varrho,\vartheta)\Big)\,[\mathbf{u}_B\cdot\mathbf{n}]^+\mathrm{d}\sigma_x \\
&\quad - \int_{\partial\Omega}\overline{\vartheta}\left(\frac{a}{3}\vartheta^3 + \frac{3}{2}\varrho_B\log(\vartheta) - \varrho_B\log(\varrho_B) - \frac{3}{2}\varrho_B\right)\,[\mathbf{u}_B\cdot\mathbf{n}]^-\mathrm{d}\sigma_x \\
&\quad + \int_\Omega \frac{\overline{\vartheta}}{\vartheta}\left[\mathbb{S}(\vartheta,\nabla_x\mathbf{u}):\nabla_x\mathbf{u} - \frac{\mathbf{q}(\vartheta,\nabla_x\vartheta)\cdot\nabla_x\vartheta}{\vartheta}\right]\,\mathrm{d}x \\
&\quad + \int_\Omega\left[\varrho\vartheta + p_{HS}(\varrho) + \frac{a}{3}\vartheta^4\right]\mathrm{div}_x\mathbf{u}_B\,\mathrm{d}x \\
&\quad - \int_{\partial\Omega}\frac{\overline{\vartheta}}{\vartheta}\Big[f_{i,B} - \varrho_B e_{HS}(\varrho_B)\Big]\,[\mathbf{u}_B\cdot\mathbf{n}]^-\mathrm{d}\sigma_x \\
&\leq -\int_\Omega \varrho(\mathbf{u}-\mathbf{u}_B)\otimes(\mathbf{u}-\mathbf{u}_B):\nabla_x\mathbf{u}_B\,\mathrm{d}x \\
&\quad + \int_\Omega \mathbb{S}(\vartheta,\nabla_x\mathbf{u}):\nabla_x\mathbf{u}_B\,\mathrm{d}x \\
&\quad + \int_\Omega \varrho\Big[\mathbf{g} - \partial_t\mathbf{u}_B - (\mathbf{u}_B\cdot\nabla_x)\mathbf{u}_B\Big]\cdot(\mathbf{u}-\mathbf{u}_B)\,\mathrm{d}x - \int_{\partial\Omega} f_{i,B}\,[\mathbf{u}_B\cdot\mathbf{n}]^-\mathrm{d}\sigma_x
\end{aligned}$$

in $\mathcal{D}'(0,\infty)$. In view of Remark 17, the above inequality must be justified for the present extension $\mathbf{u}_B$ via a density argument.

First, we regroup the integrals that are uniformly bounded in terms of the data obtaining

$$\begin{aligned}
&\frac{\mathrm{d}}{\mathrm{d}t}\int_\Omega \Big[\frac{1}{2}\varrho|\mathbf{u}-\mathbf{u}_B|^2 + \frac{3}{2}\varrho\vartheta + \varrho e_{HS}(\varrho) + a\vartheta^4 \\
&\qquad - \overline{\vartheta}\left(\frac{4}{3}a\vartheta^3 + \frac{3}{2}\varrho\log(\vartheta) - \varrho\log(\varrho)\right)\Big]\mathrm{d}x \\
&+ \int_{\partial\Omega}\left(\varrho e(\varrho,\vartheta) - \overline{\vartheta}\varrho s(\varrho,\vartheta)\right)[\mathbf{u}_B\cdot\mathbf{n}]^+\mathrm{d}\sigma_x \\
&- \int_{\partial\Omega}\overline{\vartheta}\left(\frac{a}{3}\vartheta^3 + \frac{3}{2}\varrho_B\log(\vartheta)\right)[\mathbf{u}_B\cdot\mathbf{n}]^-\mathrm{d}\sigma_x \\
&+ \int_\Omega \frac{\overline{\vartheta}}{\vartheta}\left[\mathbb{S}(\vartheta,\nabla_x\mathbf{u}):\nabla_x\mathbf{u} - \frac{\mathbf{q}(\vartheta,\nabla_x\vartheta)\cdot\nabla_x\vartheta}{\vartheta}\right]\,\mathrm{d}x \\
&+ \int_\Omega\left[\varrho\vartheta + p_{HS}(\varrho) + \frac{a}{3}\vartheta^4\right]\mathrm{div}_x\mathbf{u}_B\,\mathrm{d}x \\
&- \int_{\partial\Omega}\frac{\overline{\vartheta}}{\vartheta}\Big[f_{i,B} - \varrho_B e_{HS}(\varrho_B)\Big]\,[\mathbf{u}_B\cdot\mathbf{n}]^-\mathrm{d}\sigma_x \\
&\leq -\int_\Omega \varrho(\mathbf{u}-\mathbf{u}_B)\otimes(\mathbf{u}-\mathbf{u}_B):\nabla_x\mathbf{u}_B\,\mathrm{d}x \\
&+ \int_\Omega \mathbb{S}(\vartheta,\nabla_x\mathbf{u}-\nabla_x\mathbf{u}_B):\nabla_x\mathbf{u}_B\,\mathrm{d}x \\
&+ \int_\Omega \varrho\Big[\mathbf{g} - \partial_t\mathbf{u}_B - (\mathbf{u}_B\cdot\nabla_x)\mathbf{u}_B\Big]\cdot(\mathbf{u}-\mathbf{u}_B)\,\mathrm{d}x \\
&+ M\left(\overline{\vartheta}, \|\varrho_B\|_{L^\infty}, \|\mathbf{u}_B\|_{W^{1,q}}, \|f_{i,B}\|_{L^\infty}\right).
\end{aligned}\tag{10.40}$$

Next, we use hypothesis (10.35) to evaluate the dissipative term

$$\begin{aligned}
&\overline{\vartheta}\int_\Omega \frac{1}{\vartheta}\mathbb{S}(\vartheta,\nabla_x\mathbf{u}):\nabla_x\mathbf{u}\,\mathrm{d}x \gtrsim \overline{\vartheta}\int_\Omega |\nabla_x\mathbf{u} + \nabla_x^t\mathbf{u} - \frac{2}{d}\mathrm{div}_x\mathbf{u}\mathbb{I}|^2\,\mathrm{d}x \\
&\gtrsim \overline{\vartheta}\int_\Omega \left|\nabla_x(\mathbf{u}-\mathbf{u}_B) + \nabla_x^t(\mathbf{u}-\mathbf{u}_B) - \frac{2}{d}\mathrm{div}_x(\mathbf{u}-\mathbf{u}_B)\mathbb{I}\right|^2\,\mathrm{d}x \\
&\quad - \overline{\vartheta}\|\nabla_x\mathbf{u}_B\|^2_{L^2(\Omega)} \gtrsim \overline{\vartheta}\,\|\mathbf{u}-\mathbf{u}_B\|^2_{W^{1,2}(\Omega)} - \overline{\vartheta}\|\nabla_x\mathbf{u}_B\|^2_{L^2(\Omega)},
\end{aligned}\tag{10.41}$$

where the last step follows from the Korn–Poincaré inequality.

Similarly, using again (10.35), we have

$$-\overline{\vartheta}\int_\Omega \frac{q(\vartheta,\nabla_x\vartheta)\cdot\nabla_x\vartheta}{\vartheta^2}\,\mathrm{d}x \gtrsim \overline{\vartheta}\int_\Omega\left(\frac{1}{\vartheta^2}+\vartheta\right)|\nabla_x\vartheta|^2\,\mathrm{d}x. \tag{10.42}$$

Moreover, by virtue of (10.39),

$$\int_\Omega \frac{a}{3}\vartheta^4\mathrm{div}_x\mathbf{u}_B\,\mathrm{d}x - \overline{\vartheta}\int_\Omega \frac{q(\vartheta,\nabla_x\vartheta)\cdot\nabla_x\vartheta}{\vartheta^2}\,\mathrm{d}x \gtrsim \underline{U}\|\vartheta\|^2_{W^{1,2}(\Omega)} - 1. \tag{10.43}$$

On the other hand, as $\varrho \leq \overline{\varrho}$, the integrals on the right-hand side of (10.40) can be handled as follows:

- By virtue of Sobolev's embedding $W^{1,2} \hookrightarrow L^6$ $(d = 3)$,

$$\left| \int_\Omega \varrho(\mathbf{u} - \mathbf{u}_B) \otimes (\mathbf{u} - \mathbf{u}_B) : \nabla_x \mathbf{u}_B \ \mathrm{d}x \right|$$
$$\leq c\left(\overline{\varrho}, \|\mathbf{u}_B\|_{W^{1,\frac{3}{2}}}\right) \|\mathbf{u} - \mathbf{u}_B\|^2_{L^6(\Omega;R^d)}$$
$$\leq c\left(\overline{\varrho}, \|\mathbf{u}_B\|_{W^{1,\frac{3}{2}}}\right) \|\mathbf{u} - \mathbf{u}_B\|^2_{W^{1,2}(\Omega;R^d)}$$

where the rightmost integral can be absorbed via (10.41) provided the constant $\overline{\vartheta} > 0$ is chosen large enough.

-

$$\left| \int_\Omega \mathbb{S}(\vartheta, \nabla_x \mathbf{u} - \nabla_x \mathbf{u}_B) : \nabla_x \mathbf{u}_B \ \mathrm{d}x \right|$$
$$\leq \varepsilon \int_\Omega (1 + \vartheta^2)|\nabla_x \mathbf{u}_B|^2 \ \mathrm{d}x + c(\varepsilon)\|\nabla_x \mathbf{u} - \nabla_x \mathbf{u}_B\|^2_{L^2(\Omega;R^{d\times d})} \tag{10.44}$$

for any $\varepsilon > 0$. Furthermore, similarly to the above,

$$\int_\Omega \vartheta^2 |\nabla_x \mathbf{u}_B|^2 \ \mathrm{d}x \lesssim \|\vartheta\|^2_{L^6(\Omega)} \|\nabla_x \mathbf{u}_B\|^2_{L^3(\Omega;R^{d\times d})}$$
$$\lesssim \|\vartheta\|^2_{W^{1,2}(\Omega)} \|\nabla_x \mathbf{u}_B\|^2_{L^3(\Omega;R^{d\times d})}.$$

Thus we first fix $\varepsilon > 0$ to control the former integral in (10.44) by (10.43) and then choose $\overline{\vartheta}$ large enough to dominate the latter via (10.41).

-

$$\left| \int_\Omega \varrho\Big[\mathbf{g} - \partial_t \mathbf{u}_B - (\mathbf{u}_B \cdot \nabla_x)\mathbf{u}_B\Big] \cdot (\mathbf{u} - \mathbf{u}_B) \ \mathrm{d}x \right|$$
$$\leq c(\overline{\varrho}, \|\mathbf{u}_B\|_{W^{1,q}}, \|\mathbf{g}\|_{L^\infty}) \left(1 + \|\mathbf{u} - \mathbf{u}_B\|^2_{L^2(\Omega;R^d)}\right).$$

In view of the previous discussion, we deduce from (10.40)

$$\frac{\mathrm{d}}{\mathrm{d}t} \int_\Omega \Bigg[\frac{1}{2}\varrho|\mathbf{u} - \mathbf{u}_B|^2 + \frac{3}{2}\varrho\vartheta + \varrho e_{HS}(\varrho) + a\vartheta^4$$
$$- \overline{\vartheta}\left(\frac{4}{3}a\vartheta^3 + \frac{3}{2}\varrho \log(\vartheta) - \varrho\log(\varrho)\right)\Bigg] \mathrm{d}x$$
$$+ \int_{\partial\Omega} \Big(\varrho e(\varrho, \vartheta) - \overline{\vartheta}\varrho s(\varrho, \vartheta)\Big) \ [\mathbf{u}_B \cdot \mathbf{n}]^+ \mathrm{d}\sigma_x$$
$$- \int_{\partial\Omega} \overline{\vartheta}\left(\frac{a}{3}\vartheta^3 + \frac{3}{2}\varrho_B \log(\vartheta)\right) \ [\mathbf{u}_B \cdot \mathbf{n}]^- \mathrm{d}\sigma_x$$

$$
\begin{aligned}
&+\frac{\overline{\vartheta}}{2}\|\mathbf{u}-\mathbf{u}_B\|^2_{L^2(\Omega)}+\frac{\overline{\vartheta}}{2}\int_\Omega\left(\frac{1}{\vartheta^2}+\vartheta\right)|\nabla_x\vartheta|^2\,\mathrm{d}x\\
&+\underline{U}\int_\Omega\left[\varrho\vartheta+p_{HS}(\varrho)+\frac{a}{3}\vartheta^4\right]\,\mathrm{d}x\\
&-\int_{\partial\Omega}\frac{\overline{\vartheta}}{\vartheta}\Big[f_{i,B}-\varrho_B e_{HS}(\varrho_B)\Big]\,[\mathbf{u}_B\cdot\mathbf{n}]^-\mathrm{d}\sigma_x\\
&\leq M\left(\overline{\vartheta},\|\varrho_B\|_{L^\infty},\|\mathbf{u}_B\|_{W^{1,q}},\|f_{i,B}\|_{L^\infty},\|\mathbf{g}\|_{L^\infty}\right).
\end{aligned}
\tag{10.45}
$$

Next, by virtue of hypothesis (10.38),

$$
\begin{aligned}
\overline{\vartheta}\left|\int_{\partial\Omega}\frac{3}{2}\varrho_B\log(\vartheta)\,[\mathbf{u}_B\cdot\mathbf{n}]^-\mathrm{d}\sigma_x\right| &\leq -\frac{\overline{\vartheta}}{2}\int_{\partial\Omega}\frac{a}{3}\vartheta^3\,[\mathbf{u}_B\cdot\mathbf{n}]^-\mathrm{d}\sigma_x\\
&-\frac{\overline{\vartheta}}{2}\int_{\partial\Omega}\frac{1}{\vartheta}\Big[f_{i,B}-\varrho_B e_{HS}(\varrho_B)\Big]\,[\mathbf{u}_B\cdot\mathbf{n}]^-\mathrm{d}\sigma_x.
\end{aligned}
$$

Finally, we have to control the integral

$$
-\overline{\vartheta}\int_\Omega\frac{3}{2}\varrho\log(\vartheta)\,\mathrm{d}x
$$

appearing in $\mathcal{E}_{\overline{\vartheta}}$ by dissipation. To this end, we evoke hypothesis (10.39) and use the same arguments as in Sect. 8.1.3 to conclude

$$
\overline{\vartheta}\left|\int_\Omega\frac{3}{2}\varrho\log(\vartheta)\,\mathrm{d}x\right|\leq\overline{\vartheta}\,\overline{\varrho}\left|\int_\Omega\frac{3}{2}\log(\vartheta)\,\mathrm{d}x\right|
$$

$$
-\frac{\overline{\vartheta}}{2}\int_{\partial\Omega}\frac{1}{\vartheta}\Big[f_{i,B}-\varrho_B e_{HS}(\varrho_B)\Big]\,[\mathbf{u}_B\cdot\mathbf{n}]^-\mathrm{d}\sigma_x \tag{10.46}
$$

$$
+\frac{\overline{\vartheta}}{2}\|\nabla_x\log(\vartheta)\|^2_{L^2(\Omega;R^d)}+c. \tag{10.47}
$$

Introducing the ballistic energy

$$
\begin{aligned}
\mathcal{E}_{\overline{\vartheta}}=\int_\Omega\Big[&\frac{1}{2}\varrho|\mathbf{u}-\mathbf{u}_B|^2+\frac{3}{2}\varrho\vartheta+\varrho e_{HS}(\varrho)+a\vartheta^4\\
&-\overline{\vartheta}\left(\frac{4}{3}a\vartheta^3+\frac{3}{2}\varrho\log(\vartheta)-\varrho\log(\varrho)\right)\Big]\mathrm{d}x
\end{aligned}
\tag{10.48}
$$

we therefore deduce from (10.45) the desired relation

$$
\frac{\mathrm{d}}{\mathrm{d}t}\mathcal{E}_{\overline{\vartheta}}+\nu\mathcal{E}_{\overline{\vartheta}}\leq M\left(\overline{\vartheta},\|\varrho_B\|_{L^\infty},\|\mathbf{u}_B\|_{W^{1,q}},\|f_{i,B}\|_{L^\infty},\|\mathbf{g}\|_{L^\infty}\right) \tag{10.49}
$$

for some positive constant $\nu>0$, which yields ultimate boundedness of $\mathcal{E}_{\overline{\vartheta}}$ in terms of the data.

We have shown the following result.

Theorem 13 (Bounded Absorbing Set for Hard Sphere Gas).

- *Let the pressure* p, *the internal energy* e, *and the entropy* s *be given by the hard sphere EOS* (10.32)–(10.34).
- *Let the transport coefficients* μ, η, *and* κ *obey* (10.35).
- *Let the data* ϱ_B, $\mathbf{u}_B$, $\mathbf{g}$, *and* $f_{i,B}$ *satisfy* (10.36), (10.37). *In addition, suppose there is a constant* $\underline{U} > 0$ *such that*

$$\sup_{\tau \geq 0} \int_{\partial\Omega} [\mathbf{u}_B \cdot \mathbf{n}]^- \leq -\underline{U}$$
$$\inf_{\tau \geq 0} \int_{\partial\Omega} \mathbf{u}_B \cdot \mathbf{n} \, \mathrm{d}\sigma_x \geq \underline{U}. \tag{10.50}$$

Then there exists a constant $\mathcal{E}_\infty$ *depending only on*

$$\overline{\vartheta}, \|\varrho_B\|_{L^\infty}, \|\mathbf{u}_B\|_{W^{1,q}}, \|f_{i,B}\|_{L^\infty}, \|\mathbf{g}\|_{L^\infty}, \ q < \infty,$$

such that

$$\operatorname*{ess\,lim\,sup}_{\tau \to \infty} \int_\Omega \Big[\frac{1}{2} \varrho |\mathbf{u} - \mathbf{u}_B|^2 + \frac{3}{2} \varrho \vartheta + \varrho e_{HS}(\varrho) + a\vartheta^4$$
$$- \overline{\vartheta} \left(\frac{4}{3} a \vartheta^3 + \frac{3}{2} \varrho \log(\vartheta) - \varrho \log(\varrho) \right) \Big] (\tau, \cdot) \mathrm{d}x \leq \mathcal{E}_\infty \tag{10.51}$$

for any finite energy weak solutions to the Navier–Stokes–Fourier system defined on $[T, \infty)$, $T \in R$.

10.3 ω-Limit Sets Generated by Bounded Trajectories

Suppose now that the Navier–Stokes–Fourier system admits a global in time weak solution defined for $t \in [0, \infty)$ with ultimately bounded ballistic energy,

$$\operatorname*{ess\,lim\,sup}_{\tau \to \infty} \mathcal{E}_{\overline{\vartheta}}(\tau) \leq \mathcal{E}_\infty.$$

Some sufficient conditions for the existence of bounded solutions were obtained in Theorem 13. At this point, it is convenient to work with the *conservative-entropy state variables* $(\varrho, \mathbf{m} = \varrho \mathbf{u}, S = \varrho s(\varrho, \vartheta))$ with well defined *instantaneous* values at any time t. Recall that

$$\varrho \in C_{\mathrm{weak,loc}}((0,\infty); L^{\gamma}(\Omega)),\ \gamma = \begin{cases} \frac{5}{3} \text{ for real gas EOS,} \\ \text{arbitrary finite for hard sphere EOS,} \end{cases}$$

and similarly

$$\mathbf{m} \in C_{\mathrm{weak,loc}}((0,\infty); L^{q}(\Omega; R^d)),\ q = \begin{cases} \frac{5}{4} \text{ for real gas EOS,} \\ 2 \text{ for hard sphere EOS.} \end{cases}$$

This can be equivalently reformulated as:

- $$\varrho \in L^{\infty}([T_1, T_2]; L^{\gamma}(\Omega)) \text{ for any } 0 < T_1 < T_2.$$
- The function of time
$$\tau \mapsto \int_{\Omega} \varrho(\tau, \cdot)\phi(x)\ \mathrm{d}x \equiv \lim_{\delta \to 0} \frac{1}{2\delta} \int_{\tau-\delta}^{\tau+\delta} \int_{\Omega} \varrho(t,x)\phi(x)\ \mathrm{d}x\mathrm{d}t$$
belongs to the space $C[T_1, T_2]$ for any $\phi \in C_c^{\infty}(\Omega)$.
- $$\mathbf{m} \in L^{\infty}([T_1, T_2]; L^{q}(\Omega; R^d)) \text{ for any } 0 < T_1 < T_2.$$
- The function of time
$$\tau \mapsto \int_{\Omega} \mathbf{m}(\tau, x) \cdot \boldsymbol{\varphi}(x)\ \mathrm{d}x \equiv \lim_{\delta \to 0} \frac{1}{2\delta} \int_{\tau-\delta}^{\tau+\delta} \int_{\Omega} \mathbf{m}(t,x) \cdot \boldsymbol{\varphi}(x)\ \mathrm{d}x\mathrm{d}t$$
belongs to the space $C[T_1, T_2]$ for any $\boldsymbol{\varphi} \in C_c^{\infty}(\Omega; R^d)$.

Unfortunately, the total entropy $S = \varrho s(\varrho, \vartheta)$ satisfies only the entropy inequality (3.41), from which we can deduce

$$\frac{\mathrm{d}}{\mathrm{d}t} \int_{\Omega} S(t,x)\phi(x)\ \mathrm{d}x - \int_{\Omega} \left(\varrho s \mathbf{u} + \frac{\mathbf{q}}{\vartheta} \right) \cdot \nabla_x \phi\ \mathrm{d}x \geq 0$$

in $\mathcal{D}'(T_1, T_2)$ for any $\phi \in C_c^{\infty}(\Omega)$, $\phi \geq 0$. Thus the function

$$\tau \mapsto \left[\int_{\Omega} S(\tau, x)\phi(x)\ \mathrm{d}x + \int_{T_1}^{\tau} \int_{\Omega} \left(\varrho s \mathbf{u} + \frac{\mathbf{q}}{\vartheta} \right) \cdot \nabla_x \phi\ \mathrm{d}x\mathrm{d}t \right] \tag{10.52}$$

is non-decreasing. In particular, we may correctly define the one-sided limits

$$\int_{\Omega} S(\tau-, x)\phi(x)\ \mathrm{d}x = \lim_{\delta \searrow 0} \frac{1}{\delta} \int_{\tau-\delta}^{\tau} \int_{\Omega} S(t,x)\phi(x)\ \mathrm{d}x,$$
$$\int_{\Omega} S(\tau+, x)\phi(x)\ \mathrm{d}x = \lim_{\delta \searrow 0} \frac{1}{\delta} \int_{\tau}^{\tau+\delta} \int_{\Omega} S(t,x)\phi(x)\ \mathrm{d}x$$

for any $\phi \in C_c^\infty(\Omega)$, $\phi \geq 0$, whence for any $\phi \in C_c^\infty(\Omega)$. Accordingly, we may identify the action of S on ϕ with either $\int_\Omega S(\tau+, x)\phi(x)\ \mathrm{d}x$ or $\int_\Omega S(\tau-, x)\phi(x)\ \mathrm{d}x$ obtaining a càdlàg or càglàd function of time, respectively.

Remark 18. A *càdlàg* function is right-continuous with limits from the left. Similarly, a *càglàd* function is left-continuous with limits from the right.

As we mostly consider the initial/boundary value problems, it is more convenient to use the càglàd version, the value at the initial time $\tau = 0$ being determined by the initial data. With this convention in mind, we may suppose:

- $$S \in L^\infty([T_1, T_2]; L^1(\Omega; R^d)) \text{ for any } 0 < T_1 < T_2.$$
- The function of time
$$\tau \mapsto \int_\Omega S(\tau, x)\phi(x)\ \mathrm{d}x \equiv \lim_{\delta \searrow 0} \frac{1}{\delta} \int_{\tau-\delta}^{\tau} \int_\Omega S(t, x)\phi(x)\ \mathrm{d}x\mathrm{d}t \tag{10.53}$$
belongs to the Skorokhod space $D_{\mathrm{loc}}(R)$ of càglàd functions for any $\phi \in C_c^\infty(\Omega)$, cf. Sect. A.1 in Appendix.

Strictly speaking, the function $S(\tau, x)$ is integrable only for a.a. τ; otherwise the one-sided limit (10.53) defines a signed measure on Ω.

10.3.1 Trajectory Space

The ultimate goal of this part of the monograph is to discuss *statistical solutions* of the Navier–Stokes–Fourier system interpreted as *measures* on the trajectory space. It is therefore convenient for the trajectory space to admit a topology of a Polish space (separable, metrizable, complete). Instead of working with the "natural" weak L^p-topologies that are not (globally) metrizable, we consider the Sobolev spaces $W_0^{k,2}(\Omega)$, $2k > d$ (compactly) embedded in $C_c(\Omega)$. In particular, the dual space $W^{-k,2}(\Omega)$ contains the space of Radon measures; therefore, it is large enough to accommodate the instantaneous values $(\varrho(\tau, \cdot), \mathbf{m}(\tau, \cdot), S(\tau, \cdot))$. Accordingly, we introduce the **trajectory space**

$$\mathcal{T} = \cup_{L=1}^{\infty} \mathcal{T}_L,$$

where

$$\begin{aligned}\mathcal{T}_L = \Big\{ (\varrho, S, \mathbf{m}) \ \Big| \ & \varrho \in L^\infty(R; W^{-k,2}(\Omega)),\ \langle \varrho; \phi_n \rangle \in C(R),\ n = 1,2,\dots, \\ & \sup_{t \in R} \|\varrho(t,\cdot)\|_{W^{-k,2}(\Omega)} \leq L, \\ & \mathbf{m} \in L^\infty(R; W^{-k,2}(\Omega; R^3)),\ \langle \mathbf{m}; \boldsymbol{\varphi}_n \rangle \in C(R),\ n = 1,2,\dots, \\ & \sup_{t \in R} \|\mathbf{m}(t,\cdot)\|_{W^{-k,2}(\Omega;R^3)} \leq L, \\ & S \in L^\infty(R; W^{-k,2}(\Omega)),\ \langle S; \phi_n \rangle \text{ càglàd in } R,\ n = 1,2,\dots, \\ & \sup_{t \in R} \|S(t,\cdot)\|_{W^{-k,2}(\Omega)} \leq L \Big\}.\end{aligned}$$

The Sobolev space $W_0^{k,2}(\Omega)$ is naturally embedded into the Hilbert space $L^2(\Omega)$ via the Gelfand identity

$$W_0^{k,2}(\Omega) \hookrightarrow L^2(\Omega) \approx [L^2(\Omega)]^* \hookrightarrow W^{-k,2}(\Omega),$$

where the Hilbert space L^2 is identified with its dual via the Riesz isometry. As $W_0^{k,2}(\Omega)$ can be identified with a domain of an elliptic operator of order $\frac{k}{2}$, there exists an L^2-orthonormal basis $\{\phi_j\}_{j=1}^{\infty}$ such that

$$\langle r, s \rangle_{W^{-k,2}(\Omega)} = \sum_{j=1}^{\infty} \lambda_j^{-\frac{k}{2}} \langle r, \phi_j \rangle \langle s, \phi_j \rangle,$$

for a suitable sequence of eigenvalues $\lambda_j \to \infty$.

Finally, we introduce metrics

$$\begin{aligned} & d_{\mathcal{T}} \left[(\varrho^1, S^1, \mathbf{m}^1); (\varrho^2, S^2, \mathbf{m}^2) \right] \\ & = \sum_{n=1}^{\infty} \frac{1}{2^n} \int_{-\infty}^{\infty} \exp\left(-t^2\right) G\left(\| \left\langle \varrho^1 - \varrho^2; \phi_n \right\rangle \|_{C[-t,t]} \right) \mathrm{d}t \\ & + \sum_{n=1}^{\infty} \frac{1}{2^n} \int_{-\infty}^{\infty} \exp\left(-t^2\right) G\left(\| \left\langle \mathbf{m}^1 - \mathbf{m}^2; \boldsymbol{\varphi}_n \right\rangle \|_{C([-t,t];R^3)} \right) \mathrm{d}t \\ & = \sum_{n=1}^{\infty} \frac{1}{2^n} \int_{-\infty}^{\infty} \exp\left(-t^2\right) G\left(\left[\left\langle S^1; \phi_n \right\rangle ; \left\langle S^2; \phi_n \right\rangle \right]_{D[-t,t]} \right) \mathrm{d}t, \end{aligned} \tag{10.54}$$

where

$$G(Z) = \frac{Z}{1+Z}$$

and $D[-t,t]$ denotes the Skorokhod space of càglàd functions defined on $[-t,t]$ with the associated complete metrics $[\cdot;\cdot]_{D[-t,t]}$, see Sect. A.1 in Appendix.

The bounded sets $\mathcal{T}_L$ endowed with the metrics $d_{\mathcal{T}}$ are complete separable metric spaces—Polish spaces.

10.3.2 ω-Limit Sets

Given a global in time solution $(\varrho, \mathbf{m}, S)$ of the Navier–Stokes–Fourier system on $[0,\infty)$, we may extend it for negative times by its initial value,

$$\varrho(-\tau,\cdot) = \varrho_0, \ \mathbf{m}(-\tau,\cdot) = \mathbf{m}_0, \ S(-\tau,\cdot) = S_0 \text{ for } \tau \geq 0.$$

Accordingly, we may assume $(\varrho, \mathbf{m}, S) \in \mathcal{T}$. We define the *$\omega$-limit set* associated to $(\varrho, \mathbf{m}, S)$ as

$$\omega(\varrho, \mathbf{m}, S) = \Big\{ (\tilde{\varrho}, \widetilde{\mathbf{m}}, \widetilde{S}) \in \mathcal{T} \ \Big| \ \text{there exists } T_n \to \infty, \\ (\varrho, \mathbf{m}, S)(\cdot + T_n, \cdot) \to (\tilde{\varrho}, \widetilde{\mathbf{m}}, \widetilde{S}) \text{ in } (\mathcal{T}, d_{\mathcal{T}}) \Big\}.$$

Remark 19. In contrast with the standard notion of the ω-limit set used in the dynamical systems theory, the ω-limit set as defined above is a subset of the trajectory spaces and not of the phase space.

The ω-limit set describes the behaviour of the system in the long run when its initial state becomes irrelevant. As we have partially observed in Sect. 10.2.1, the ω-limit sets generated by solutions of closed systems are typically singletons consisting of a stationary solution. A richer structure is expected to emerge for energetically open systems. We address this problem in Chap. 11. To conclude this part, we state a result that follows immediately from Theorem 12.

Theorem 14 (ω-Limit Set).

Under the hypotheses of Theorem 12, let the boundary data ϱ_B, $\mathbf{u}_B$, $f_{i,B}$ as well as the driving force $\mathbf{g}$ be independent of $t \in R$. Let $(\varrho, \mathbf{m} = \varrho\mathbf{u}, S = \varrho s(\varrho, \vartheta))$ be a weak solution of the Navier–Stokes–Fourier system on $[0, \infty) \times \Omega$ in the sense of Definition 4. Suppose, in addition, that

$$\operatorname{ess}\limsup_{t\to\infty} \int_\Omega \left[\frac{1}{2}\varrho|\mathbf{u} - \mathbf{u}_B|^2 + \varrho e - \overline{\vartheta}\varrho s(\varrho, \vartheta)\right](t, \cdot)\ \mathrm{d}x \leq \mathcal{E}_{b,\infty}$$

for some $\overline{\vartheta} > 0$.

Then:

- *The ω-limit set $\omega(\varrho, \mathbf{m}, S)$ is non-empty.*
- *$\omega(\varrho, \mathbf{m}, S)$ is a compact subset of $(\mathcal{T}, d_{\mathcal{T}})$.*
- *$\omega(\varrho, \mathbf{m}, S)$ is time shift invariant, meaning*

$$(\tilde{\varrho}, \widetilde{\mathbf{m}}, \widetilde{S}) \in \omega(\varrho, \mathbf{m}, S) \Rightarrow (\tilde{\varrho}, \widetilde{\mathbf{m}}, \widetilde{S})(\cdot + T, \cdot) \in \omega(\varrho, \mathbf{m}, S)$$

for any $T \in R$.

- *$\omega(\varrho, \mathbf{m}, S)$ consists of entire (defined for $t \in R$) solutions of the Navier–Stokes–Fourier system.*
- *There are constants $\mathcal{E}_\infty$, $\mathcal{S}_\infty$ such that*

$$\int_\Omega \left[\frac{1}{2}\tilde{\varrho}\left|\frac{\widetilde{\mathbf{m}}}{\tilde{\varrho}} - \mathbf{u}_B\right|^2 + \tilde{\varrho}e(\tilde{\varrho}, \widetilde{S})\right](\tau, \cdot)\ \mathrm{d}x \leq \mathcal{E}_\infty, \quad \int_\Omega \widetilde{S}(\tau, \cdot)\ \mathrm{d}x \geq \mathcal{S}_\infty$$

for a.a. $\tau \in R$ whenever $(\tilde{\varrho}, \widetilde{\mathbf{m}}, \widetilde{S}) \in \omega(\varrho, \mathbf{m}, S)$.

Proof. In view of Theorem 12, the only thing to observe is the convergence of the time shifts

$$S_n(t, \cdot) = S(t + T_n, \cdot) = \varrho s(\varrho, \vartheta)(t + T_n, \cdot)$$

in the Skorokhod space $D([-M; M]; W^{-k,2}(\Omega))$, meaning convergence of the integral averages

$$\tau \mapsto \int_\Omega S_n(\tau, \cdot)\phi\ \mathrm{d}x \text{ in } D[-M, M] \text{ for any } \phi \in C_c(\Omega).$$

In accordance with (10.52),

$$G_n(\tau) \equiv \int_\Omega S_n(\tau, \cdot)\phi\ \mathrm{d}x + \int_{-M}^{\tau} \int_\Omega \left(\varrho_n s_n \mathbf{u}_n + \frac{\mathbf{q}_n}{\vartheta_n}\right) \cdot \nabla_x \phi\ \mathrm{d}x$$

is a non-decreasing function of τ as soon as $\phi \in C^1_c(\Omega)$, $\phi \geq 0$, where the subscript n denotes the time shifts of the corresponding quantities. Moreover, by virtue of the uniform energy bounds, we have

$$\left[\tau \mapsto \int_{-M}^{\tau} \int_{\Omega} \left(\varrho_n s_n \mathbf{u}_n + \frac{\mathbf{q}_n}{\vartheta_n} \right) \cdot \nabla_x \phi \, \mathrm{d}x \right]$$
$$\to \left[\tau \mapsto \int_{-M}^{\tau} \int_{\Omega} \left(\widetilde{\varrho s \mathbf{u}} + \widetilde{\frac{\mathbf{q}}{\vartheta}} \right) \cdot \nabla_x \phi \, \mathrm{d}x \right] \text{ in } C[-M; M].$$

Consequently, it is enough to show convergence of non-decreasing functions G_n in the Skorokhod space $D[-M; M]$. We have

$$\langle G_n(\tau, \cdot); \phi \rangle \to \langle G(\tau, \cdot); \phi \rangle \text{ for a dense set of times for any compact interval } [-M, M],$$

whence, Lemma 12 in Sect. A.1 of Appendix,

$$\langle G_n(\tau, \cdot); \phi \rangle \to \langle G(\tau, \cdot); \phi \rangle \text{ in } D[-M, M] \text{ for a.a. } M > 0$$

yielding the desired conclusion

$$\langle S_n(\tau, \cdot); \phi \rangle \to \langle S(\tau, \cdot); \phi \rangle \text{ in } D[-M, M] \text{ for a.a. } M > 0.$$

Thus we have shown the convergence of the averages

$$\tau \mapsto \int_{\Omega} S_n(\tau, \cdot) \phi \, \mathrm{d}x$$

for any $\phi \geq 0$, whence the desired result follows by decomposition of a general function ϕ as $\phi = \phi^+ + \phi^-$.

□

Chapter 11
Statistical Solutions, Ergodic Hypothesis, and Turbulence

We conclude our discussion concerning qualitative properties of weak solutions to the Navier–Stokes–Fourier system by developing the theory of *statistical solutions*. More precisely, we show the existence of a *stationary* statistical solution on any shift invariant compact subset of the trajectory space $\mathcal{T}$ that consists of entire solutions. In particular, each ω-limit set generated by a global bounded solution $(\varrho, \mathbf{m}, S)$ will support a stationary statistical solution. Our principal working hypothesis throughout the whole chapter is that the boundary data ϱ_B, $\mathbf{u}_B$, $f_{i,B}$ as well as the driving force $\mathbf{g}$ are independent of t.

Let $\mathcal{T}$ be the space of all entire trajectories—time-dependent processes $\tau \in R \mapsto (\tilde{\varrho}, \widetilde{\mathbf{m}}, \widetilde{S})(t, \cdot) \to X$—ranging in the space

$$X = [W^{-k,2}(\Omega)]^{d+2}.$$

Very roughly indeed, a *stationary statistical solution* is a measure $\mathcal{V}$ on the space of entire trajectories $\mathcal{T}$ enjoying the following properties:

- $\mathcal{V} \in \mathfrak{P}[\mathcal{T}]$ is a probability measure on the space $\mathcal{T}$ of entire trajectories.
- For any trajectory $(\tilde{\varrho}, \widetilde{\mathbf{m}}, \widetilde{S})$, there holds that $(\tilde{\varrho}, \widetilde{\mathbf{m}}, \widetilde{S})$ is an entire solution of the Navier–Stokes–Fourier system $\mathcal{V}$-a.s.; in other words, the support of the measure $\mathcal{V}$ is a subset of the space of entire solutions.
- $\mathcal{V}$ is shift invariant, meaning

$$\mathcal{V}[\mathcal{B}] = \mathcal{V}[\mathcal{B}(\cdot + T)]$$

for any Borel set $\mathcal{B} \subset \mathcal{T}$ and any $T \in R$.

In particular, we show that any ω-limit set $\omega(\varrho, \mathbf{m}, S)$ generated by a globally bounded solution $(\varrho, \mathbf{m}, S)$ supports a stationary statistical solution

E. Feireisl, A. Novotný, *Mathematics of Open Fluid Systems*, Nečas Center Series, https://doi.org/10.1007/978-3-030-94793-4_11

$\mathcal{V}$. Such a measure is called *ergodic* if for any shift invariant Borel subset $\mathcal{B} \subset \mathcal{T}$ either $\mathcal{V}[\mathcal{B}] = 1$ or $\mathcal{V}[\mathcal{B}] = 0$.

11.1 Bounded Invariant Sets

Statistical solutions are identified with shift invariant probability measures on the trajectory space $\mathcal{T}$ supported by solutions of the Navier–Stokes–Fourier system. Next, we introduce the concept of *bounded invariant set.*

Definition 6 (Bounded Invariant Set).
$\mathcal{O} \subset \mathcal{T}$ is called *bounded invariant set* for the Navier–Stokes–Fourier system if:

- Any $(\varrho, \mathbf{m}, S) \in \mathcal{O}$ is an entire (weak) solution (defined for $t \in R$) of the Navier–Stokes–Fourier system.
- $\mathcal{O}$ is shift invariant, meaning

$$(\varrho, \mathbf{m}, S) \in \mathcal{O} \ \Rightarrow \ (\varrho, \mathbf{m}, S)(\cdot + T) \in \mathcal{O} \text{ for any } T \in R.$$

- The ballistic energy

$$\mathcal{E}_{\overline{\vartheta}}[\varrho, \mathbf{m}, S] = \int_{\Omega} \left[\frac{1}{2} \left| \frac{\mathbf{m}}{\varrho} - \mathbf{u}_B \right|^2 + \varrho e(\varrho, S) - \overline{\vartheta} S \right] \ \mathrm{d}x, \ \overline{\vartheta} > 0,$$

is uniformly bounded on $\mathcal{O}$, meaning there exists a constant $\mathcal{E}_{b,\infty}$ such that

$$\mathcal{E}_{\overline{\vartheta}}[\varrho, \mathbf{m}, S](\tau) \leq \mathcal{E}_{b,\infty} \text{ for a.a. } \tau \in R$$

for any $(\varrho, \mathbf{m}, S) \in \mathcal{O}$.

As a direct consequence of Theorems 12, 14, we get:

Lemma 10. *Any bounded invariant set is precompact in the trajectory space $(\mathcal{T}, d_{\mathcal{T}})$. The closure of a bounded invariant set is a compact invariant set, meaning it consists of entire solutions of the Navier–Stokes–Fourier system.*

Indeed Theorem 12 yields the desired compactness with respect to $(\varrho, \mathbf{m})$, while Theorem 14 yields compactness of the entropy S in the Skorokhod topology.

11.2 Krylov–Bogolyubov Approach

Given a bounded invariant set $\mathcal{O}$, we apply the classical Krylov–Bogolyubov approach to construct a statistical solution supported by $\mathcal{O}$. In view of Lemma 10, we may suppose $\mathcal{O}$ is closed meaning $\mathcal{O}$ endowed with the metrics $d_{\mathcal{T}}$ is a compact metric space. Given a trajectory $(\varrho, \mathbf{m}, S) \in \mathcal{O}$, we consider a family of probability measures $\mathcal{V}_T \in \mathfrak{P}[\mathcal{O}]$,

$$\mathcal{V}_T = \frac{1}{T} \int_0^T \delta_{(\varrho,\mathbf{m},S)(\cdot+t)} \mathrm{d}t, \ T > 0,$$

where δ_G denotes the Dirac measure supported by a trajectory G. The integral is defined in the weak-(*) sense, meaning

$$\langle \mathcal{V}_T; \mathcal{F} \rangle = \frac{1}{T} \int_0^T \mathcal{F}[(\varrho, \mathbf{m}, S)(\cdot + t)] \mathrm{d}t \text{ for any } \mathcal{F} \in BC(\mathcal{T}).$$

As the set $\mathcal{O}$ is shift invariant and the operation of time shift continuous with respect to the metrics $d_{\mathcal{T}}$ on $\mathcal{O}$, the integral is well defined in the Riemann sense, and we may infer that $\mathcal{V}_T \in \mathfrak{P}[\mathcal{O}]$.

As the measures $\mathcal{V}_T$ are supported by a compact set $\mathcal{O}$, the family $\{\mathcal{V}_T\}_{T>0}$ is tight. By Prokhorov theorem (Theorem 3), there is a sequence $T_n \to \infty$ and a probability measure $\mathcal{V} \in \mathfrak{P}[\mathcal{O}]$ such that

$$\mathcal{V}_{T_n} \to \mathcal{V} \text{ narrowly in } \mathfrak{P}[\mathcal{O}] \text{ as } n \to \infty,$$

meaning

$$\frac{1}{T_n} \int_0^{T_n} \mathcal{F}[(\varrho, \mathbf{m}, S)(\cdot + t)] \mathrm{d}t \to \int_{\mathcal{T}} \mathcal{F}(\tilde{\varrho}, \widetilde{\mathbf{m}}, \widetilde{S}) \, \mathrm{d}\mathcal{V}(\tilde{\varrho}, \widetilde{\mathbf{m}}, \widetilde{S}) \quad \text{for any } \mathcal{F} \in BC(\mathcal{T}). \tag{11.1}$$

Finally, we observe that the measure $\mathcal{V}$ is shift invariant. Indeed, in accordance with (11.1),

$$\begin{aligned}
\int_{\mathcal{T}} \mathcal{F}[(\tilde{\varrho}, \widetilde{\mathbf{m}}, \widetilde{S})(\cdot + T)] \, \mathrm{d}\mathcal{V}(\tilde{\varrho}, \widetilde{\mathbf{m}}, \widetilde{S}) &= \lim_{T_n \to \infty} \frac{1}{T_n} \int_0^{T_n} \mathcal{F}[(\tilde{\varrho}, \widetilde{\mathbf{m}}, \widetilde{S})(\cdot + t + T)] \mathrm{d}t \\
&= \frac{1}{T_n} \int_T^{T+T_n} \mathcal{F}[(\tilde{\varrho}, \widetilde{\mathbf{m}}, \widetilde{S})(\cdot + t)] \mathrm{d}t \\
&= \frac{1}{T_n} \int_0^{T_n} \mathcal{F}[(\tilde{\varrho}, \widetilde{\mathbf{m}}, \widetilde{S})(\cdot + t)] \mathrm{d}t \\
&+ \frac{1}{T_n} \left[\int_{T_n}^{T_n+T} \mathcal{F}[(\tilde{\varrho}, \widetilde{\mathbf{m}}, \widetilde{S})(\cdot + t)] \mathrm{d}t - \int_0^T \mathcal{F}[(\tilde{\varrho}, \widetilde{\mathbf{m}}, \widetilde{S})(\cdot + t)] \mathrm{d}t \right].
\end{aligned}$$

As $\mathcal{F}$ is bounded, we get

$$\frac{1}{T_n}\left[\int_{T_n}^{T_n+T}\mathcal{F}[(\tilde{\varrho},\widetilde{\mathbf{m}},\widetilde{S})(\cdot+t)]\mathrm{d}t-\int_0^T\mathcal{F}[(\tilde{\varrho},\widetilde{\mathbf{m}},\widetilde{S})(\cdot+t)]\mathrm{d}t\right]\to 0,$$

and the desired conclusion follows.

We have shown the following result.

Theorem 15 (Stationary Statistical Solution).
Let $\mathcal{O}\subset\mathcal{T}$ be a non-empty bounded invariant set for the Navier–Stokes–Fourier system.

Then there exists a stationary statistical solution $\mathcal{V}\in\mathfrak{P}[\overline{\mathcal{O}}]$. Specifically, $\mathcal{V}$ is a Borel probability measure on the trajectory space $\mathcal{T}$, shift invariant, and supported by $\overline{\mathcal{O}}$.

11.3 Application of Birkhoff–Khinchin Theorem

We associate to any stationary statistical solution $\mathcal{V}$ a random process ranging in the space

$$X=\left[W^{-k,2}(\Omega)\right]^{d+2}.$$

This can be done by considering the probability space $O=[\mathcal{T},\mathfrak{B}(\mathcal{T}),\mathcal{V}]$, where $\mathcal{T}$ is the trajectory space endowed with the metric $d_{\mathcal{T}}$, $\mathfrak{B}(\mathcal{T})$ the family of all Borel subsets of $\mathcal{T}$, and $\mathcal{V}$ the stationary solution—a Borel probability measure.

The associated random process is defined as

$$(\tilde{\varrho},\widetilde{\mathbf{m}},\widetilde{S})\in O,\ \ \tau\in R\mapsto(\tilde{\varrho}(\tau,\cdot),\widetilde{\mathbf{m}}(\tau,\cdot),\widetilde{S}(\tau-,\cdot))\in X.$$

As $\mathcal{V}$ is shift invariant, this is a stationary random process with càglàd paths ranging in the separable Hilbert space $W^{-k,2}(\Omega)$. We call the process $(\tilde{\varrho},\widetilde{\mathbf{m}},\widetilde{S})$ *canonical representation* of the stationary statistical solution $\mathcal{V}$.

For a Borel function $\mathcal{G}:X\to R$, we define its *expected value* with respect to $\mathcal{V}$,

$$\mathbb{E}[\mathcal{G}]=\int_{\mathcal{T}}\mathcal{G}\left[\tilde{\varrho}(\tau,\cdot),\widetilde{\mathbf{m}}(\tau,\cdot),\widetilde{S}(\tau-,\cdot)\right]\mathrm{d}\mathcal{V}(\tilde{\varrho},\widetilde{\mathbf{m}},\widetilde{S}).$$

Note carefully that the above definition is independent of τ since $\mathcal{V}$ is shift invariant.

Our goal is to show the following result that can be seen as a direct consequence of the celebrated *Birkhoff–Khinchin ergodic theorem.*

Theorem 16 (Convergence of Ergodic Averages).
Let $(\varrho, \mathbf{m}, S)$ be a canonical representation of a stationary statistical solution $\mathcal{V}$ of the Navier–Stokes–Fourier system. Let $\mathcal{G} : X \to R$ be a Borel measurable function such that

$$\mathbb{E}\left[|\mathcal{G}[\varrho(0,\cdot), \mathbf{m}(0,\cdot), S(0-,\cdot)]|\right] < \infty.$$

Then there exists a $\mathcal{V}$-measurable function $\overline{\mathcal{G}} : O \to R$ such that

$$\frac{1}{T}\int_0^T \mathcal{G}[\varrho(t,\cdot), \mathbf{m}(t,\cdot), S(t,\cdot)]\mathrm{d}t \to \overline{\mathcal{G}} \text{ as } T \to \infty,$$

$\mathcal{V}$–a.s. and in $L^1(O)$.

The convergence of the time averages

$$\frac{1}{T}\int_0^T \mathcal{G}[\varrho(t,\cdot), \mathbf{m}(t,\cdot), S(t,\cdot)]\mathrm{d}t \to \overline{\mathcal{G}} \text{ as } T \to \infty$$

for *any* trajectory is closely related to the validity of *ergodic hypothesis* mentioned in the introductory part of this book. Note that Theorem 16 provides only a partial answer for trajectories emanating from the support of statistical stationary solutions. Note also that the exceptional set of $\mathcal{V}$-measure zero may depend on the function $\mathcal{G}$.

The proof of Theorem 16 leans on an auxiliary result borrowing the main idea from Kolmogorov [62, Chapter 39].

Lemma 11. *Let $U : O \times [0,\infty) \to R$ be a measurable stationary stochastic process such that*

$$\mathbb{E}[|U(0)|] < \infty \text{ and } U \in L^1_{\mathrm{loc}}[0,\infty) \text{ a.s.}$$

Then

$$\lim_{T\to\infty} \frac{1}{T}\int_0^T U(t)\mathrm{d}t \to \overline{U} \text{ a.s. and in } L^1(O).$$

Proof. Splitting U into positive and negative parts, we observe that it is enough to show the result for non-negative U. The integral averages

$$U_n = \int_n^{n+1} U(t)\ \mathrm{d}t,\ n = 0, 1, \dots$$

are well defined and represent a stationary discrete process in the sense of Krylov [63, Chapter 4, Section 6, Definition 1]. Thus applying the discrete version of Birkhoff–Khinchin's Theorem [63, Chapter 4, Section 6, Theorem 11], we obtain the existence of $\overline{U} : O \to R$ such that for $N \to \infty$ where N takes only discrete values $N = 1, 2, \dots,$

$$\frac{1}{N}\sum_{n=1}^{N-1} U_n = \frac{1}{N}\int_0^N U(t)\ \mathrm{d}t \to \overline{U} \text{ a.s. and in } L^1(O).$$

Finally, denoting by $[T]$ the largest integer less than or equal to T and using non-negativity of U, we get

$$\begin{aligned} \overline{U} \leftarrow \frac{[T]}{T}\frac{1}{[T]}\int_0^{[T]} U(t)\ \mathrm{d}t &\le \frac{1}{T}\int_0^T U(t)\ \mathrm{d}t \\ &\le \frac{[T]+1}{T}\frac{1}{[T]+1}\int_0^{[T]+1} U(t)\ \mathrm{d}t \to \overline{U}, \end{aligned}$$

in the limit for $T \to \infty$.

□

Now, the conclusion of Theorem 16 follows from Lemma 11 as long as we observe that $t \mapsto \mathcal{G}[(\varrho(t,\cdot), \mathbf{m}(t,\cdot), S(t-,\cdot)]$ is locally integrable a.s. To see this, write

$$\begin{aligned} &\mathbb{E}\left[\int_{-M}^{M} |\mathcal{G}[\varrho(t,\cdot), \mathbf{m}(t,\cdot), S(t-,\cdot)]|\ \mathrm{d}t\right] \\ &= \int_{-M}^{M} \mathbb{E}\big[|\mathcal{G}[\varrho(t,\cdot), \mathbf{m}(t,\cdot), S(t-,\cdot)]|\big]\mathrm{d}t \\ &\le 2M\mathbb{E}\big[|\mathcal{G}[\varrho(0,\cdot), \mathbf{m}(0,\cdot), S(0-,\cdot)]|\big] < \infty. \end{aligned}$$

Hence, we deduce that

$$\mathcal{V}\left\{\int_{-M}^{M} |\mathcal{G}[\varrho(t,\cdot), \mathbf{m}(t,\cdot), S(t-,\cdot)]|\ \mathrm{d}t < \infty\right\} = 1 \text{ for any } M = 1, 2, \dots,$$

and consequently,

$$\begin{aligned} &\mathcal{V}\Big\{\mathcal{G}[\varrho(t,\cdot), \mathbf{m}(t,\cdot), S(t-,\cdot)] \in L^1_{\mathrm{loc}}(R)\Big\} \\ &= \mathcal{V}\left\{\bigcap_{M=1,2,\dots}\left\{\int_{-M}^{M} |\mathcal{G}[\varrho(t,\cdot), \mathbf{m}(t,\cdot), S(t-,\cdot)]|\ \mathrm{d}t < \infty\right\}\right\} = 1, \end{aligned}$$

which completes the proof of Theorem 16. Here, the symbol $\mathcal{V}\{M\}$ denotes the $\mathcal{V}$-measure of a set M,

$$\mathcal{V}\{M\} = \int_M 1 \mathrm{d}\mathcal{V}.$$

11.4 Concluding Remarks

Besides the work of Sell [83] and Málek, Nečas [70], the approach to the long time dynamics via the space of trajectories was used by Itô, Nisio [59] in the framework of stochastic differential equation. A similar strategy has been implemented by Foiaş, Rosa and Temam [50, 51] in studying convergence of time averages for the incompressible Navier–Stokes system driven by a non-conservative volume force.

Statistical solutions to problems in fluid mechanics have been introduced in the pioneering works of Foiaş [48] and Vishik, Fursikov [90] again in the context of incompressible fluid models. Similarly to this monograph, they are measures supported by the solution trajectories of the underlying system of equations. See also Constantin and Wu [22] and Foiaş, Rosa, and Temam [49] for related studies and additional details. When the measure is invariant with respect to (positive) time shifts, the statistical solution is said stationary.

More recently, a new class of (dynamical) statistical solutions for the *barotropic* Navier–Stokes system was introduced in [33]. The statistical solutions are identified with the pushforward measure associated to a semiflow selection among all possible (weak) solutions of the problem emanating from fixed initial data. Stationary statistical solutions, again in the context of barotropic Navier–Stokes system, were also constructed in [34].

Finally, relevant result including convergence to equilibria for *closed* systems is collected in the monograph [47].

Chapter 12
Systems with Prescribed Boundary Temperature

In this monograph we focused on *open* fluid systems interacting with the outer world through the boundary of the physical domain. The crucial factor was the possibility to control the time evolution of the total energy/entropy by means of the data. The resulting *a priori* bounds along with the pressure estimates are essential for the existence theory. We have considered general Dirichlet boundary conditions for the velocity field, with the associated mass flux on the inflow part of the boundary, and Robin type boundary conditions for the heat flux. Other types of boundary conditions could be handled in the same manner as long as the total energy flux through the boundary is controlled. Apparently this is not the case if the temperature instead of the heat flux is prescribed on the boundary. Such a situation, however, is physically relevant in many real world phenomena, in particular the Rayleigh–Bénard problem in the theory of turbulence, see Davidson [26, Chapter I]. To fill the gap in the theory, we shortly discuss the Dirichlet problem for the temperature in this concluding chapter.

12.1 Dirichlet Boundary Conditions for the Temperature

We consider

E. Feireisl, A. Novotný, *Mathematics of Open Fluid Systems*, Nečas Center Series, https://doi.org/10.1007/978-3-030-94793-4_12

the **Navier–Stokes–Fourier system** in the "entropy form"

$$\partial_t \varrho + \operatorname{div}_x(\varrho \mathbf{u}) = 0,$$
$$\partial_t(\varrho \mathbf{u}) + \operatorname{div}_x(\varrho \mathbf{u} \otimes \mathbf{u}) + \nabla_x p(\varrho, \vartheta) = \operatorname{div}_x \mathbb{S}(\vartheta, \mathbb{D}_x \mathbf{u}) + \varrho \mathbf{g},$$
$$\partial_t(\varrho s(\varrho, \vartheta)) + \operatorname{div}_x(\varrho s(\varrho, \vartheta)\mathbf{u}) + \operatorname{div}_x \left(\frac{\mathbf{q}(\vartheta, \nabla_x \vartheta)}{\vartheta} \right) = \frac{1}{\vartheta} \left(\mathbb{S} : \mathbb{D}_x \mathbf{u} - \frac{\mathbf{q} \cdot \nabla_x \vartheta}{\vartheta} \right), \tag{12.1}$$

supplemented with

Newton's rheological law

$$\mathbb{S}(\vartheta, \mathbb{D}_x \mathbf{u}) = \mu(\vartheta) \left(\nabla_x \mathbf{u} + \nabla_x^t \mathbf{u} - \frac{2}{d} \operatorname{div}_x \mathbf{u} \mathbb{I} \right) + \eta(\vartheta) \operatorname{div}_x \mathbf{u} \mathbb{I}, \tag{12.2}$$

and

Fourier's law

$$\mathbf{q} = -\kappa(\vartheta) \nabla_x \vartheta. \tag{12.3}$$

Similarly to Sect. 2.2, we prescribe

the **Dirichlet boundary conditions for the velocity**

$$\mathbf{u}|_{\partial \Omega} = \mathbf{u}_B, \tag{12.4}$$

with the corresponding mass flux condition on the inflow part of the boundary,

$$\varrho|_{\Gamma_{\rm in}} = \varrho_B. \tag{12.5}$$

Here, in contrast with the bulk part of this monograph, we consider

the **Dirichlet boundary conditions for the temperature**

$$\vartheta|_{\partial \Omega} = \vartheta_B, \tag{12.6}$$

where $\vartheta_B = \vartheta_B(t, x)$ is a given strictly positive function. To avoid technicalities, we suppose that both the boundary data and the boundary $\partial \Omega$ are sufficiently smooth; in particular, the outer normal vector $\mathbf{n}(x)$ exists at any $x \in \partial\Omega$.

In the remaining part of this chapter we focus on three major issues:

- A suitable **weak formulation** of the problem (12.1)–(12.6)
- **Weak–strong uniqueness principle** in the class of finite energy weak solutions
- **A priori bounds** and global existence via an approximation scheme

12.2 Weak Formulation

The concept of weak solution introduced in Chap. 3, Definition 4 is based on a combination of the entropy inequality

$$\partial_t(\varrho s(\varrho,\vartheta)) + \mathrm{div}_x(\varrho s(\varrho,\vartheta)\mathbf{u}) + \mathrm{div}_x\left(\frac{\mathbf{q}}{\vartheta}\right) \geq \frac{1}{\vartheta}\left(\mathbb{S} : \mathbb{D}_x\mathbf{u} - \frac{\mathbf{q}\cdot\nabla_x\vartheta}{\vartheta}\right) \tag{12.7}$$

with the total energy balance, written in the form of another inequality

$$\begin{aligned}
&\partial_t\left(\frac{1}{2}\varrho|\mathbf{u}-\mathbf{u}_B|^2 + \varrho e\right) + \mathrm{div}_x\left[\left(\frac{1}{2}\varrho|\mathbf{u}-\mathbf{u}_B|^2 + \varrho e\right)\mathbf{u}\right] + \mathrm{div}_x\mathbf{q} \\
&+ \mathrm{div}_x(p(\mathbf{u}-\mathbf{u}_B)) + \mathrm{div}_x\left(\mathbb{S}\cdot(\mathbf{u}_B-\mathbf{u})\right) \\
&\leq \mathbb{S} : \mathbb{D}_x\mathbf{u}_B - \left[\varrho(\mathbf{u}-\mathbf{u}_B)\otimes(\mathbf{u}-\mathbf{u}_B) + p\mathbb{I}\right] : \mathbb{D}_x\mathbf{u}_B \\
&+ \varrho\Big[\mathbf{g} - \partial_t\mathbf{u}_B - (\mathbf{u}_B\cdot\nabla_x)\mathbf{u}_B\Big]\cdot(\mathbf{u}-\mathbf{u}_B),
\end{aligned} \tag{12.8}$$

cf. (3.16). The weak formulation was obtained as a combination of the variational formulation of the entropy inequality (12.7) combined with the total energy balance obtained by simple integration of (12.8) over the spatial domain Ω. Now, the integration of (12.8) yields

$$\begin{aligned}
&\frac{\mathrm{d}}{\mathrm{d}t}\int_\Omega\left(\frac{1}{2}\varrho|\mathbf{u}-\mathbf{u}_B|^2 + \varrho e\right)\,\mathrm{d}x + \int_{\partial\Omega}\varrho e(\varrho,\vartheta_B)\mathbf{u}_B\cdot\mathbf{n}\mathrm{d}\sigma_x + \int_{\partial\Omega}\mathbf{q}\cdot\mathbf{n}\,\mathrm{d}\sigma_x \\
&\leq -\int_\Omega\left[\varrho(\mathbf{u}-\mathbf{u}_B)\otimes(\mathbf{u}-\mathbf{u}_B) + p\mathbb{I} - \mathbb{S}\right] : \mathbb{D}_x\mathbf{u}_B\,\mathrm{d}x \\
&+ \int_\Omega\varrho\Big[\mathbf{g} - \partial_t\mathbf{u}_B - (\mathbf{u}_B\cdot\nabla_x)\mathbf{u}_B\Big]\cdot(\mathbf{u}-\mathbf{u}_B)\,\mathrm{d}x,
\end{aligned} \tag{12.9}$$

where, unlike in Chap. 3, the heat flux $\mathbf{q}\cdot\mathbf{n}$ is not known on $\partial\Omega$.

A simple idea on how to remove the problem is to replace the total energy balance by a variant of *ballistic energy* balance similar to (10.25). To this end, consider an (smooth) extension $\tilde\vartheta$ of the boundary temperature inside Ω,

$$\tilde\vartheta|_{\partial\Omega} = \vartheta_B,\ \tilde\vartheta > 0.$$

Now, multiplying the entropy inequality on $\tilde{\vartheta}$ and integrating by parts, we obtain

$$
\begin{aligned}
&-\frac{\mathrm{d}}{\mathrm{d}t}\int_\Omega \tilde{\vartheta}\varrho s(\varrho,\vartheta)\ \mathrm{d}x - \int_{\partial\Omega} \varrho s(\varrho,\vartheta_B)\vartheta_B[\mathbf{u}_B\cdot\mathbf{n}]\ \mathrm{d}\sigma_x \\
-\int_{\partial\Omega}\mathbf{q}\cdot\mathbf{n}\ \mathrm{d}\sigma_x &\le -\int_\Omega \left[\varrho s(\varrho,\vartheta)\partial_t\tilde{\vartheta} + \varrho s(\varrho,\vartheta)\mathbf{u}\cdot\nabla_x\tilde{\vartheta} + \frac{\mathbf{q}}{\vartheta}\cdot\nabla_x\tilde{\vartheta}\right]\mathrm{d}x \\
&\quad -\int_\Omega \frac{\tilde{\vartheta}}{\vartheta}\left(\mathbb{S}:\mathbb{D}_x\mathbf{u} - \frac{\mathbf{q}\cdot\nabla_x\vartheta}{\vartheta}\right)\mathrm{d}x. \qquad (12.10)
\end{aligned}
$$

Summing up the inequalities (12.9), (12.10), we obtain

the **ballistic energy balance:**

$$
\begin{aligned}
&\frac{\mathrm{d}}{\mathrm{d}t}\int_\Omega\left(\frac{1}{2}\varrho|\mathbf{u}-\mathbf{u}_B|^2 + \varrho e - \tilde{\vartheta}\varrho s(\varrho,\vartheta)\right)\mathrm{d}x \\
&+\int_{\partial\Omega}\Big(\varrho e(\varrho,\vartheta_B) - \vartheta_B\varrho s(\varrho,\vartheta_B)\Big)\ [\mathbf{u}_B\cdot\mathbf{n}]^+\mathrm{d}\sigma_x \\
&+\int_{\partial\Omega}\Big(\varrho_B e(\varrho_B,\vartheta_B) - \vartheta_B\varrho_B s(\varrho_B,\vartheta_B)\Big)\ [\mathbf{u}_B\cdot\mathbf{n}]^-\mathrm{d}\sigma_x \\
&+\int_\Omega\frac{\tilde{\vartheta}}{\vartheta}\left(\mathbb{S}:\mathbb{D}_x\mathbf{u} - \frac{\mathbf{q}\cdot\nabla_x\vartheta}{\vartheta}\right)\mathrm{d}x \\
&\le -\int_\Omega\Big[\varrho(\mathbf{u}-\mathbf{u}_B)\otimes(\mathbf{u}-\mathbf{u}_B) + p\mathbb{I} - \mathbb{S}\Big]:\mathbb{D}_x\mathbf{u}_B\ \mathrm{d}x \\
&+\int_\Omega\varrho\Big[\mathbf{g} - \partial_t\mathbf{u}_B - (\mathbf{u}_B\cdot\nabla_x)\mathbf{u}_B\Big]\cdot(\mathbf{u}-\mathbf{u}_B)\ \mathrm{d}x \\
&-\int_\Omega\left[\varrho s(\varrho,\vartheta)\partial_t\tilde{\vartheta} + \varrho s(\varrho,\vartheta)\mathbf{u}\cdot\nabla_x\tilde{\vartheta} + \frac{\mathbf{q}}{\vartheta}\cdot\nabla_x\tilde{\vartheta}\right]\mathrm{d}x \qquad (12.11)
\end{aligned}
$$

for any

$$
\tilde{\vartheta}\in C^1([0,T]\times\overline{\Omega}),\ \tilde{\vartheta}>0,\ \tilde{\vartheta}|_{\partial\Omega} = \vartheta_B.
$$

Unlike the total energy balance (12.9), the ballistic energy balance (12.11) does not contain the boundary flux $\mathbf{q}\cdot\mathbf{n}$ and therefore is convenient for the weak formulation. To avoid technicalities, we shall assume that $\Omega\subset R^d$, $d=2,3$, is a bounded domain with regular boundary; in particular, the outer normal vector $\mathbf{n}$ is well defined.

Definition 7 (Weak Solution, Dirichlet Problem).
We say that a trio of functions $(\varrho,\vartheta,\mathbf{u})$ is a weak solution of the Navier–Stokes–Fourier system (12.1)–(12.3), with the boundary conditions (12.4)–(12.6), if the following holds:

- **Equation of continuity.** $\varrho \geq 0$ and the integral identity

$$\int_0^\tau \int_\Omega \left[\varrho \partial_t \varphi + \varrho \mathbf{u} \cdot \nabla_x \varphi \right] \, \mathrm{d}x\mathrm{d}t$$
$$= \int_0^\tau \int_{\partial\Omega} \varphi \varrho_B \, [\mathbf{u}_B \cdot \mathbf{n}]^- \mathrm{d}\sigma_x \mathrm{d}t + \int_0^\tau \int_{\partial\Omega} \varphi \varrho \, [\mathbf{u}_B \cdot \mathbf{n}]^+ \mathrm{d}\sigma_x \mathrm{d}t$$
$$+ \left[\int_\Omega \varphi \varrho \, \mathrm{d}x \right]_{t=0}^{t=\tau} \tag{12.12}$$

holds for any $0 \leq \tau \leq T$, $\varphi \in C^1([0,T] \times \overline{\Omega})$.

- **Momentum equation.**

$$\mathbf{u} \in L^r(0,T; W^{1,r}(\Omega; R^d)) \text{ for some } r > 1,$$
$$(\mathbf{u} - \mathbf{u}_B) \in L^r(0,T; W_0^{1,r}(\Omega; R^d)), \tag{12.13}$$

and

$$\int_0^\tau \int_\Omega \left[\varrho \mathbf{u} \cdot \partial_t \boldsymbol{\varphi} + \varrho \mathbf{u} \otimes \mathbf{u} : \nabla_x \boldsymbol{\varphi} + p(\varrho, \vartheta) \mathrm{div}_x \boldsymbol{\varphi} \right] \, \mathrm{d}x\mathrm{d}t$$
$$= \int_0^\tau \int_\Omega \left[\mathbb{S}(\vartheta, \mathbb{D}_x \mathbf{u}) : \mathbb{D}_x \boldsymbol{\varphi} - \varrho \mathbf{g} \cdot \boldsymbol{\varphi} \right] \, \mathrm{d}x\mathrm{d}t + \left[\int_\Omega \varrho \mathbf{u} \cdot \boldsymbol{\varphi} \, \mathrm{d}x \right]_{t=0}^{t=\tau} \tag{12.14}$$

for any $0 \leq \tau \leq T$, $\boldsymbol{\varphi} \in C_c^1([0,T] \times \Omega; R^d)$.

- **Entropy inequality.**

$$\left[\int_\Omega \varrho s \varphi \, \mathrm{d}x \right]_{t=\tau_1}^{t=\tau_2} - \int_{\tau_1}^{\tau_2} \int_\Omega \left[\varrho s \partial_t \varphi + \varrho s \mathbf{u} \cdot \nabla_x \varphi + \frac{\mathbf{q}}{\vartheta} \cdot \nabla_x \varphi \right] \, \mathrm{d}x\mathrm{d}t$$
$$\geq \int_{\tau_1}^{\tau_2} \int_\Omega \frac{\varphi}{\vartheta} \left(\mathbb{S}(\vartheta, \mathbb{D}_x \mathbf{u}) : \mathbb{D}_x \mathbf{u} - \frac{\mathbf{q}(\vartheta, \nabla_x \vartheta) \cdot \nabla_x \vartheta}{\vartheta} \right) \, \mathrm{d}x\mathrm{d}t \tag{12.15}$$

for any $0 \leq \tau_1 < \tau_2 \leq T$, and any $\varphi \in C_c^1([0,T] \times \Omega)$, $\varphi \geq 0$.

- **Ballistic energy inequality.** For any

$$\tilde{\vartheta} \in C^1([0,T] \times \overline{\Omega}), \ \tilde{\vartheta} > 0, \ \tilde{\vartheta}|_{\partial\Omega} = \vartheta_B,$$

there holds

$$
\begin{aligned}
&- \int_0^T \partial_t \psi \int_\Omega \left(\frac{1}{2} \varrho |\mathbf{u} - \mathbf{u}_B|^2 + \varrho e - \tilde{\vartheta} \varrho s \right) \, dx dt \\
&+ \int_0^T \psi \int_{\partial\Omega} \Big[\varrho_B e(\varrho_B, \vartheta_B) - \vartheta_B \varrho_B s(\varrho_B, \vartheta_B) \Big] \, [\mathbf{u}_B \cdot \mathbf{n}]^- \mathrm{d}\sigma_x \mathrm{d}t \\
&+ \int_0^T \psi \int_{\partial\Omega} \Big[\varrho e(\varrho, \vartheta_B) - \vartheta_B \varrho s(\varrho, \vartheta_B) \Big] \, [\mathbf{u}_B \cdot \mathbf{n}]^+ \mathrm{d}\sigma_x \mathrm{d}t \\
&+ \int_0^T \psi \int_\Omega \frac{\tilde{\vartheta}}{\vartheta} \left(\mathbb{S} : \mathbb{D}_x \mathbf{u} - \frac{\mathbf{q} \cdot \nabla_x \vartheta}{\vartheta} \right) \, dx dt \\
&\leq \psi(0) \int_\Omega \left(\frac{1}{2} \varrho |\mathbf{u} - \mathbf{u}_B|^2 + \varrho e - \tilde{\vartheta} \varrho s \right) (0, \cdot) \, \mathrm{d}x \\
&- \int_0^T \psi \int_\Omega \Big[\varrho (\mathbf{u} - \mathbf{u}_B) \otimes (\mathbf{u} - \mathbf{u}_B) + p\mathbb{I} - \mathbb{S} \Big] : \mathbb{D}_x \mathbf{u}_B \, dx dt \\
&+ \int_0^T \psi \int_\Omega \varrho (\mathbf{u} - \mathbf{u}_B) \cdot (\mathbf{g} - \partial_t \mathbf{u}_B - \mathbf{u}_B \cdot \nabla_x \mathbf{u}_B) \, dx dt \\
&- \int_0^T \psi \int_\Omega \left[\varrho s \left(\partial_t \tilde{\vartheta} + \mathbf{u} \cdot \nabla_x \tilde{\vartheta} \right) + \frac{\mathbf{q}}{\vartheta} \cdot \nabla_x \tilde{\vartheta} \right] \, dx dt
\end{aligned} \tag{12.16}
$$

for any $\psi \in C_c^1[0, T)$, $\psi \geq 0$.

Remark 20. As usual, the initial values of the conservative-entropy variables should be identified with the (given) initial data,

$$
\varrho(0, \cdot) = \varrho_0, \ \varrho \mathbf{u}(0, \cdot) = \mathbf{m}_0 = \varrho_0 \mathbf{u}_0, \ \varrho s(0-, \cdot) = S_0 = \varrho_0 s(\varrho_0, \vartheta_0)
$$

for suitably chosen $\mathbf{u}_0$, ϑ_0.

Note carefully a tiny difference between our choice of the extensions $\mathbf{u}_B$ and $\tilde{\vartheta}$ of the boundary data in the ballistic energy balance (12.16). As observed in Chap. 3, Lemma 5, if (12.16) holds for *some* extension $\mathbf{u}_B$, then it holds for *any* other extension. The same conclusion is not clear for $\tilde{\vartheta}$ as the latter satisfies only the entropy *inequality* (12.15).

12.3 Relative Energy and Weak–Strong Uniqueness

Our goal is to justify the concept of weak solution introduced in Definition 7 by showing the weak–strong uniqueness property.

12.3.1 Relative Energy

The relative energy is defined in the same way as in Sect. 3.1:

$$
\begin{aligned}
& E\left(\varrho, \vartheta, \mathbf{u} \middle| \tilde{\varrho}, \tilde{\vartheta}, \tilde{\mathbf{u}}\right) \\
&= \frac{1}{2}\varrho|\mathbf{u}-\tilde{\mathbf{u}}|^2 + \varrho e - \tilde{\vartheta}\Big(\varrho s - \tilde{\varrho} s(\tilde{\varrho}, \tilde{\vartheta})\Big) \\
&- \left(e(\tilde{\varrho}, \tilde{\vartheta}) - \tilde{\vartheta} s(\tilde{\varrho}, \tilde{\vartheta}) + \frac{p(\tilde{\varrho}, \tilde{\vartheta})}{\tilde{\varrho}}\right)(\varrho - \tilde{\varrho}) - \tilde{\varrho} e(\tilde{\varrho}, \tilde{\vartheta}) \\
&= \frac{1}{2}\varrho|\mathbf{u}-\tilde{\mathbf{u}}|^2 + \varrho e - \tilde{\vartheta}\varrho s - \left(e(\tilde{\varrho}, \tilde{\vartheta}) - \tilde{\vartheta} s(\tilde{\varrho}, \tilde{\vartheta}) + \frac{p(\tilde{\varrho}, \tilde{\vartheta})}{\tilde{\varrho}}\right)\varrho + p(\tilde{\varrho}, \tilde{\vartheta}).
\end{aligned}
\tag{12.17}
$$

Recall that under the hypothesis of thermodynamic stability (1.19), the relative energy represents a Bregman distance between the quantities $(\varrho, \vartheta, \mathbf{u})$ and $(\tilde{\varrho}, \tilde{\vartheta}, \tilde{\mathbf{u}})$ if expressed in the conservative-entropy variables $(\varrho, S, \mathbf{m})$ and $(\tilde{\varrho}, \widetilde{S}, \widetilde{\mathbf{m}})$.

12.3.2 Relative Energy Inequality

Since

$$
\begin{aligned}
\int_\Omega E\left(\varrho, \vartheta, \mathbf{u} \middle| \tilde{\varrho}, \tilde{\vartheta}, \tilde{\mathbf{u}}\right) \mathrm{d}x &= \int_\Omega \left[\frac{1}{2}\varrho|\mathbf{u}-\tilde{\mathbf{u}}|^2 + \varrho e - \tilde{\vartheta}\varrho s\right] \mathrm{d}x \\
&- \int_\Omega \left(e(\tilde{\varrho}, \tilde{\vartheta}) - \tilde{\vartheta} s(\tilde{\varrho}, \tilde{\vartheta}) + \frac{p(\tilde{\varrho}, \tilde{\vartheta})}{\tilde{\varrho}}\right)\varrho \, \mathrm{d}x + \int_\Omega p(\tilde{\varrho}, \tilde{\vartheta}) \, \mathrm{d}x,
\end{aligned}
$$

the time evolution of $\int_\Omega E\left(\varrho, \vartheta, \mathbf{u} \middle| \tilde{\varrho}, \tilde{\vartheta}, \tilde{\mathbf{u}}\right) \mathrm{d}x$ for any weak solution $(\varrho, \vartheta, \mathbf{u})$ of the Navier–Stokes–Fourier system can be expressed by means of the weak formulation specified in Definition 7. Using the same arguments as in Sect. 3.4, we deduce the **relative energy inequality**

$$
\begin{aligned}
&\left[\int_\Omega E\left(\varrho, \vartheta, \mathbf{u} \Big| \tilde{\varrho}, \tilde{\vartheta}, \tilde{\mathbf{u}}\right) \, \mathrm{d}x \right]_{t=0}^{t=\tau} \\
&+ \int_0^\tau \int_{\partial\Omega} \Big(\varrho_B e(\varrho_B, \vartheta_B) - \vartheta_B \varrho_B s(\varrho_B, \vartheta_B) \Big) \, [\mathbf{u}_B \cdot \mathbf{n}]^- \mathrm{d}\sigma_x \mathrm{d}t \\
&- \int_0^\tau \int_{\partial\Omega} \left(e(\tilde{\varrho}, \vartheta_B) - \vartheta_B s(\tilde{\varrho}, \vartheta_B) + \frac{p(\tilde{\varrho}, \vartheta_B)}{\tilde{\varrho}} \right) \varrho_B \, [\mathbf{u}_B \cdot \mathbf{n}]^- \mathrm{d}\sigma_x \mathrm{d}t \\
&+ \int_0^\tau \int_{\partial\Omega} \Big(\varrho e(\varrho, \vartheta_B) - \vartheta_B \varrho s(\varrho, \vartheta_B) \Big) \, [\mathbf{u}_B \cdot \mathbf{n}]^+ \mathrm{d}\sigma_x \mathrm{d}t \\
&- \int_0^\tau \int_{\partial\Omega} \left(e(\tilde{\varrho}, \vartheta_B) - \vartheta_B s(\tilde{\varrho}, \vartheta_B) + \frac{p(\tilde{\varrho}, \vartheta_B)}{\tilde{\varrho}} \right) \varrho \, [\mathbf{u}_B \cdot \mathbf{n}]^+ \mathrm{d}\sigma_x \mathrm{d}t \\
&+ \int_0^\tau \int_\Omega \frac{\tilde{\vartheta}}{\vartheta} \left(\mathbb{S}(\vartheta, \mathbb{D}_x \mathbf{u}) : \mathbb{D}_x \mathbf{u} - \frac{\mathbf{q}(\vartheta, \nabla_x \vartheta)}{\vartheta} \right) \, dx dt \\
&\leq \int_0^\tau \int_\Omega \frac{\varrho}{\tilde{\varrho}} (\mathbf{u} - \tilde{\mathbf{u}}) \cdot \nabla_x p(\tilde{\varrho}, \tilde{\vartheta}) \, dx dt \\
&- \int_0^\tau \int_\Omega \Big(\varrho (s - s(\tilde{\varrho}, \tilde{\vartheta})) \partial_t \tilde{\vartheta} + \varrho (s - s(\tilde{\varrho}, \tilde{\vartheta})) \mathbf{u} \cdot \nabla_x \tilde{\vartheta} \Big) \, dx dt \\
&- \int_0^\tau \int_\Omega \left(\frac{\mathbf{q}(\vartheta, \nabla_x \vartheta)}{\vartheta} \right) \cdot \nabla_x \tilde{\vartheta} \, dx dt \\
&- \int_0^\tau \int_\Omega \Big[\varrho (\mathbf{u} - \tilde{\mathbf{u}}) \otimes (\mathbf{u} - \tilde{\mathbf{u}}) + p(\varrho, \vartheta) \mathbb{I} - \mathbb{S}(\vartheta, \mathbb{D}_x \mathbf{u}) \Big] : \mathbb{D}_x \tilde{\mathbf{u}} \, dx dt \\
&+ \int_0^\tau \int_\Omega \varrho \left[\mathbf{g} - \partial_t \tilde{\mathbf{u}} - (\tilde{\mathbf{u}} \cdot \nabla_x) \tilde{\mathbf{u}} - \frac{1}{\tilde{\varrho}} \nabla_x p(\tilde{\varrho}, \tilde{\vartheta}) \right] \cdot (\mathbf{u} - \tilde{\mathbf{u}}) \, dx dt \\
&+ \int_0^\tau \int_\Omega \left[\left(1 - \frac{\varrho}{\tilde{\varrho}} \right) \partial_t p(\tilde{\varrho}, \tilde{\vartheta}) - \frac{\varrho}{\tilde{\varrho}} \mathbf{u} \cdot \nabla_x p(\tilde{\varrho}, \tilde{\vartheta}) \right] \, dx dt \qquad (12.18)
\end{aligned}
$$

for a.a. $0 < \tau < T$, for any weak solution $(\varrho, \vartheta, \mathbf{u})$ of the Navier–Stokes–Fourier system in the sense of Definition 7 and any trio of test functions

$$
\begin{aligned}
&\tilde{\varrho} \in C^1([0,T] \times \overline{\Omega}), \ \inf \tilde{\varrho} > 0, \ \tilde{\vartheta} \in C^1([0,T] \times \overline{\Omega}), \ \inf \tilde{\vartheta} > 0, \ \tilde{\vartheta}|_{\partial\Omega} = \vartheta_B, \\
&\tilde{\mathbf{u}} \in C^1([0,T] \times \overline{\Omega}; R^d), \ \tilde{\mathbf{u}}|_{\partial\Omega} = \mathbf{u}_B. \qquad (12.19)
\end{aligned}
$$

12.3.3 Weak–Strong Uniqueness

We show the weak–strong uniqueness property assuming that the pressure p, the internal energy e, and the entropy s satisfy the hypotheses of *real gas* specified in Sect. 4.1.1 and that the transport coefficients μ, η, and κ are as in (4.18)–(4.20). This motivates the following hypotheses concerning regularity/integrability of weak solutions:

$$
\begin{aligned}
\varrho &\in C_{\rm weak}([0,T]; L^{\frac{5}{3}}(\Omega)) \cap L^{\frac{5}{3}}((0,T) \times \partial\Omega; [\mathbf{u}_B \cdot \mathbf{n}]^+ ({\rm d}t \otimes {\rm d}\sigma_x)),\\
\varrho\mathbf{u} &\in C_{\rm weak}([0,T]; L^{\frac{5}{4}}(\Omega; R^d)),\\
\mathbf{u} &\in L^r(0,T; W^{1,r}(\Omega; R^d)),\ r = \frac{8}{5-\Lambda},\\
(\mathbf{u} - \mathbf{u}_B) &\in L^r(0,T; W^{1,r}_0(\Omega; R^d))\\
\vartheta &\in L^\infty(0,T; L^4(\Omega)) \cap L^2(0,T; W^{1,2}(\Omega)),\\
(\vartheta - \vartheta_B) &\in L^2(0,T; W^{1,2}_0(\Omega)),\\
\log(\vartheta) &\in L^2(0,T; W^{1,2}(\Omega)),
\end{aligned}
\tag{12.20}
$$

cf. Chap. 9.

Suppose that $(\varrho, \vartheta, \mathbf{u})$ is a weak solution and $(\tilde\varrho, \tilde\vartheta, \tilde{\mathbf{u}})$ a strong solution of the Navier–Stokes–Fourier system sharing the same initial/boundary data as well as the driving force $\mathbf{g}$. The proof of weak–strong uniqueness property follows the lines of Chap. 4 with the necessary modifications to accommodate the Dirichlet boundary conditions ϑ_B. The obvious idea is to use $(\tilde\varrho, \tilde\vartheta, \tilde{\mathbf{u}})$ as test functions in the relative energy inequality (12.18). As the weak and strong solutions share the data, the first simplification of (12.18) reads

$$
\begin{aligned}
&\int_\Omega E\left(\varrho, \vartheta, \mathbf{u} \Big| \tilde\varrho, \tilde\vartheta, \tilde{\mathbf{u}}\right)(\tau, \cdot)\ {\rm d}x - \int_0^\tau \int_{\partial\Omega} p(\varrho_B, \vartheta_B)\ [\mathbf{u}_B \cdot \mathbf{n}]^- {\rm d}\sigma_x {\rm d}t\\
&+ \int_0^\tau \int_{\partial\Omega} [\varrho e(\varrho, \vartheta_B) - \vartheta_B \varrho s(\varrho, \vartheta_B)]\ [\mathbf{u}_B \cdot \mathbf{n}]^+ {\rm d}\sigma_x {\rm d}t\\
&- \int_0^\tau \int_{\partial\Omega} \varrho \left(e(\tilde\varrho, \vartheta_B) - \vartheta_B s(\tilde\varrho, \vartheta_B) + \frac{p(\tilde\varrho, \vartheta_B)}{\tilde\varrho} \right)\ [\mathbf{u}_B \cdot \mathbf{n}]^- {\rm d}\sigma_x {\rm d}t\\
&+ \int_0^\tau \int_\Omega \frac{\tilde\vartheta}{\vartheta} \left(\mathbb{S}(\vartheta, \mathbb{D}_x \mathbf{u}) : \mathbb{D}_x \mathbf{u} - \frac{\mathbf{q}(\vartheta, \nabla_x \vartheta) \cdot \nabla_x \vartheta}{\vartheta} \right)\ {\rm d}x{\rm d}t\\
&\leq \int_0^\tau \int_\Omega \frac{\varrho}{\tilde\varrho} (\mathbf{u} - \tilde{\mathbf{u}}) \cdot \nabla_x p(\tilde\varrho, \tilde\vartheta)\ {\rm d}x{\rm d}t\\
&- \int_0^\tau \int_\Omega \left(\varrho(s - s(\tilde\varrho, \tilde\vartheta)) \partial_t \tilde\vartheta + \varrho(s - s(\tilde\varrho, \tilde\vartheta)) \mathbf{u} \cdot \nabla_x \tilde\vartheta \right)\ {\rm d}x{\rm d}t\\
&- \int_0^\tau \int_\Omega \frac{\mathbf{q}(\vartheta, \nabla_x \vartheta)}{\vartheta} \cdot \nabla_x \tilde\vartheta\ {\rm d}x{\rm d}t\\
&- \int_0^\tau \int_\Omega \Big[\varrho(\mathbf{u} - \tilde{\mathbf{u}}) \otimes (\mathbf{u} - \tilde{\mathbf{u}}) + p(\varrho, \vartheta)\mathbb{I} - \mathbb{S}(\vartheta, \mathbb{D}_x \mathbf{u}) \Big] : \mathbb{D}_x \tilde{\mathbf{u}}\ {\rm d}x{\rm d}t
\end{aligned}
$$

$$+ \int_0^\tau \int_\Omega \frac{\varrho}{\tilde\varrho} \mathrm{div}_x \mathbb{S}(\tilde\vartheta, \mathbb{D}_x \tilde{\mathbf{u}}) \cdot (\tilde{\mathbf{u}} - \mathbf{u}) \, \mathrm{d}x\mathrm{d}t$$

$$+ \int_0^\tau \int_\Omega \left[\left(1 - \frac{\varrho}{\tilde\varrho} \right) \partial_t p(\tilde\varrho, \tilde\vartheta) - \frac{\varrho}{\tilde\varrho} \mathbf{u} \cdot \nabla_x p(\tilde\varrho, \tilde\vartheta) \right] \, \mathrm{d}x\mathrm{d}t.$$

Moreover, regrouping several terms, we obtain

$$\begin{aligned}
&\int_\Omega E\left(\varrho, \vartheta, \mathbf{u} \middle| \tilde\varrho, \tilde\vartheta, \tilde{\mathbf{u}}\right)(\tau, \cdot) \, \mathrm{d}x - \int_0^\tau \int_{\partial\Omega} p(\varrho_B, \vartheta_B) \, [\mathbf{u}_B \cdot \mathbf{n}]^- \mathrm{d}\sigma_x \mathrm{d}t \\
&+ \int_0^\tau \int_{\partial\Omega} [\varrho e(\varrho, \vartheta_B) - \vartheta_B \varrho s(\varrho, \vartheta_B)] \, [\mathbf{u}_B \cdot \mathbf{n}]^+ \mathrm{d}\sigma_x \mathrm{d}t \\
&- \int_0^\tau \int_{\partial\Omega} \varrho \left(e(\tilde\varrho, \vartheta_B) - \vartheta_B s(\tilde\varrho, \vartheta_B) + \frac{p(\tilde\varrho, \vartheta_B)}{\tilde\varrho} \right) [\mathbf{u}_B \cdot \mathbf{n}]^+ \mathrm{d}\sigma_x \mathrm{d}t \\
&+ \int_0^\tau \int_\Omega \left(\frac{\tilde\vartheta}{\vartheta} - 1 \right) \mathbb{S}(\vartheta, \mathbb{D}_x \mathbf{u}) : \mathbb{D}_x \mathbf{u} \, \mathrm{d}x\mathrm{d}t \\
&+ \int_0^\tau \int_\Omega \left(1 - \frac{\tilde\vartheta}{\vartheta} \right) \left(\frac{\mathbf{q}(\vartheta, \nabla_x \vartheta) \cdot \nabla_x \vartheta}{\vartheta} \right) \mathrm{d}x\mathrm{d}t \\
&+ \int_0^\tau \int_\Omega \frac{\mathbf{q}(\vartheta, \nabla_x \vartheta) \cdot \nabla_x (\tilde\vartheta - \vartheta)}{\vartheta} \, \mathrm{d}x\mathrm{d}t \\
&+ \int_0^\tau \int_\Omega \Big(\mathbb{S}(\tilde\vartheta, \mathbb{D}_x \tilde{\mathbf{u}}) - \mathbb{S}(\vartheta, \mathbb{D}_x \mathbf{u}) \Big) : \mathbb{D}_x (\tilde{\mathbf{u}} - \mathbf{u}) \, \mathrm{d}x\mathrm{d}t \\
&\leq \int_0^\tau \int_\Omega (\mathbf{u} - \tilde{\mathbf{u}}) \cdot \nabla_x p(\tilde\varrho, \tilde\vartheta) \, \mathrm{d}x\mathrm{d}t \\
&- \int_0^\tau \int_\Omega \Big(\tilde\varrho (s - s(\tilde\varrho, \tilde\vartheta)) \partial_t \tilde\vartheta + \tilde\varrho (s - s(\tilde\varrho, \tilde\vartheta)) \tilde{\mathbf{u}} \cdot \nabla_x \tilde\vartheta \Big) \, \mathrm{d}x\mathrm{d}t \\
&- \int_0^\tau \int_\Omega p(\varrho, \vartheta) \mathrm{div}_x \tilde{\mathbf{u}} \, \mathrm{d}x\mathrm{d}t \\
&+ \int_0^\tau \int_\Omega \left[\left(1 - \frac{\varrho}{\tilde\varrho} \right) \partial_t p(\tilde\varrho, \tilde\vartheta) - \frac{\varrho}{\tilde\varrho} \mathbf{u} \cdot \nabla_x p(\tilde\varrho, \tilde\vartheta) \right] \mathrm{d}x\mathrm{d}t + \int_0^\tau \int_\Omega F_1 \, \mathrm{d}x\mathrm{d}t
\end{aligned} \tag{12.21}$$

with the quadratic error

$$\begin{aligned}
F_1 =& \varrho (\tilde{\mathbf{u}} - \mathbf{u}) \otimes (\mathbf{u} - \tilde{\mathbf{u}}) : \mathbb{D}_x \tilde{\mathbf{u}} + \left(\frac{\varrho}{\tilde\varrho} - 1 \right) \mathrm{div}_x \mathbb{S}(\tilde\vartheta, \mathbb{D}_x \tilde{\mathbf{u}})(\tilde{\mathbf{u}} - \mathbf{u}) \\
&+ \left(\frac{\varrho}{\tilde\varrho} - 1 \right) (\mathbf{u} - \tilde{\mathbf{u}}) \cdot \nabla_x p(\tilde\varrho, \tilde\vartheta) + \varrho (s - s(\tilde\varrho, \tilde\vartheta))(\tilde{\mathbf{u}} - \mathbf{u}) \cdot \nabla_x \tilde\vartheta \\
&- (\varrho - \tilde\varrho)(s - s(\tilde\varrho, \tilde\vartheta))(\partial_t \tilde\vartheta + \tilde{\mathbf{u}} \cdot \nabla_x \tilde\vartheta).
\end{aligned} \tag{12.22}$$

Next, we use the identity

$$
\begin{aligned}
&-\int_\Omega \tilde{\mathbf{u}} \cdot \nabla_x p(\tilde\varrho, \tilde\vartheta)\ \mathrm{d}x \\
&= - \int_{\partial\Omega} p(\varrho_B, \vartheta_B)\ [\mathbf{u}_B \cdot \mathbf{n}]^- \mathrm{d}\sigma_x - \int_{\partial\Omega} p(\tilde\varrho, \vartheta_B)\ [\mathbf{u}_B \cdot \mathbf{n}]^+ \mathrm{d}\sigma_x \\
&+ \int_\Omega p(\tilde\varrho, \tilde\vartheta) \mathrm{div}_x \tilde{\mathbf{u}}\ \mathrm{d}x
\end{aligned}
$$

to rewrite (12.22) in the form

$$
\begin{aligned}
&\int_\Omega E\left(\varrho, \vartheta, \mathbf{u} \middle| \tilde\varrho, \tilde\vartheta, \tilde{\mathbf{u}}\right)(\tau, \cdot)\ \mathrm{d}x \\
&+ \int_0^\tau \int_{\partial\Omega} \left[\varrho e(\varrho, \vartheta_B) - \vartheta_B \varrho s(\varrho, \vartheta_B)\right]\ [\mathbf{u}_B \cdot \mathbf{n}]^+ \mathrm{d}\sigma_x \mathrm{d}t \\
&- \int_0^\tau \int_{\partial\Omega} \varrho \left(e(\tilde\varrho, \vartheta_B) - \vartheta_B s(\tilde\varrho, \vartheta_B) + \frac{p(\tilde\varrho, \vartheta_B)}{\tilde\varrho} \right)\ [\mathbf{u}_B \cdot \mathbf{n}]^+ \mathrm{d}\sigma_x \mathrm{d}t \\
&+ \int_0^\tau \int_{\partial\Omega} p(\tilde\varrho, \vartheta_B)\ [\mathbf{u}_B \cdot \mathbf{n}]^+ \mathrm{d}\sigma_x \mathrm{d}t \\
&+ \int_0^\tau \int_\Omega \left(\frac{\tilde\vartheta}{\vartheta} - 1 \right) \mathbb{S}(\vartheta, \mathbb{D}_x \mathbf{u}) : \mathbb{D}_x \mathbf{u}\ \mathrm{d}x\mathrm{d}t \\
&+ \int_0^\tau \int_\Omega \left(1 - \frac{\tilde\vartheta}{\vartheta} \right) \left(\frac{\mathbf{q}(\vartheta, \nabla_x \vartheta) \cdot \nabla_x \vartheta}{\vartheta} \right)\ \mathrm{d}x\mathrm{d}t \\
&+ \int_0^\tau \int_\Omega \frac{\mathbf{q}(\vartheta, \nabla_x \vartheta) \cdot \nabla_x (\tilde\vartheta - \vartheta)}{\vartheta}\ \mathrm{d}x\mathrm{d}t \\
&+ \int_0^\tau \int_\Omega \left(\mathbb{S}(\tilde\vartheta, \mathbb{D}_x \tilde{\mathbf{u}}) - \mathbb{S}(\vartheta, \mathbb{D}_x \mathbf{u}) \right) : \mathbb{D}_x (\tilde{\mathbf{u}} - \mathbf{u})\ \mathrm{d}x\mathrm{d}t \\
&\leq - \int_0^\tau \int_\Omega \left(\tilde\varrho (s - s(\tilde\varrho, \tilde\vartheta)) \partial_t \tilde\vartheta + \tilde\varrho (s - s(\tilde\varrho, \tilde\vartheta)) \tilde{\mathbf{u}} \cdot \nabla_x \tilde\vartheta \right)\ \mathrm{d}x\mathrm{d}t \\
&+ \int_0^\tau \int_\Omega (p(\tilde\varrho, \tilde\vartheta) - p(\varrho, \vartheta)) \mathrm{div}_x \tilde{\mathbf{u}}\ \mathrm{d}x\mathrm{d}t \\
&+ \int_0^\tau \int_\Omega \left[\left(1 - \frac{\varrho}{\tilde\varrho} \right) \left(\partial_t p(\tilde\varrho, \tilde\vartheta) + \tilde{\mathbf{u}} \cdot \nabla_x p(\tilde\varrho, \tilde\vartheta) \right) \right]\ \mathrm{d}x\mathrm{d}t + \int_0^\tau \int_\Omega F_2\ \mathrm{d}x\mathrm{d}t,
\end{aligned} \tag{12.23}
$$

with

$$
F_2 = F_1 + \left(1 - \frac{\varrho}{\tilde\varrho} \right) (\mathbf{u} - \tilde{\mathbf{u}}) \cdot \nabla_x p(\tilde\varrho, \tilde\vartheta). \tag{12.24}
$$

Now, using

$$
\frac{\partial s(\tilde\varrho, \tilde\vartheta)}{\partial \varrho} = - \frac{1}{\tilde\varrho^2} \frac{\partial p(\tilde\varrho, \tilde\vartheta)}{\partial \vartheta},
$$

we compute

$$\left(1-\frac{\varrho}{\tilde{\varrho}}\right)\left(\partial_t p(\tilde{\varrho},\tilde{\vartheta}) + \tilde{\mathbf{u}}\cdot\nabla_x p(\tilde{\varrho},\tilde{\vartheta})\right) + \mathrm{div}_x\tilde{\mathbf{u}}\Big(p(\tilde{\varrho},\tilde{\vartheta}) - p(\varrho,\vartheta)\Big)$$

$$=\mathrm{div}_x\tilde{\mathbf{u}}\Big(p(\tilde{\varrho},\tilde{\vartheta}) - \frac{\partial p(\tilde{\varrho},\tilde{\vartheta})}{\partial\varrho}(\tilde{\varrho}-\varrho) - \frac{\partial p(\tilde{\varrho},\tilde{\vartheta})}{\partial\vartheta}(\tilde{\vartheta}-\vartheta) - p(\varrho,\vartheta)\Big)$$

$$+\left(1-\frac{\varrho}{\tilde{\varrho}}\right)\frac{\partial p(\tilde{\varrho},\tilde{\vartheta})}{\partial\vartheta}\left(\partial_t\tilde{\vartheta} + \tilde{\mathbf{u}}\cdot\nabla_x\tilde{\vartheta}\right) - \frac{\tilde{\vartheta}-\vartheta}{\tilde{\varrho}}\frac{\partial p(\tilde{\varrho},\tilde{\vartheta})}{\partial\vartheta}\left(\partial_t\tilde{\varrho} + \tilde{\mathbf{u}}\cdot\nabla_x\tilde{\varrho}\right)$$

$$=\mathrm{div}_x\tilde{\mathbf{u}}\Big(p(\tilde{\varrho},\tilde{\vartheta}) - \frac{\partial p(\tilde{\varrho},\tilde{\vartheta})}{\partial\varrho}(\tilde{\varrho}-\varrho) - \frac{\partial p(\tilde{\varrho},\tilde{\vartheta})}{\partial\vartheta}(\tilde{\vartheta}-\vartheta) - p(\varrho,\vartheta)\Big)$$

$$-\tilde{\varrho}(\tilde{\varrho}-\varrho)\frac{\partial s(\tilde{\varrho},\tilde{\vartheta})}{\partial\varrho}\left(\partial_t\tilde{\vartheta} + \tilde{\mathbf{u}}\cdot\nabla_x\tilde{\vartheta}\right) + \tilde{\varrho}(\tilde{\vartheta}-\vartheta)\frac{\partial s(\tilde{\varrho},\tilde{\vartheta})}{\partial\varrho}\left(\partial_t\tilde{\varrho} + \tilde{\mathbf{u}}\cdot\nabla_x\tilde{\varrho}\right)$$

$$=\mathrm{div}_x\tilde{\mathbf{u}}\Big(p(\tilde{\varrho},\tilde{\vartheta}) - \frac{\partial p(\tilde{\varrho},\tilde{\vartheta})}{\partial\varrho}(\tilde{\varrho}-\varrho) - \frac{\partial p(\tilde{\varrho},\tilde{\vartheta})}{\partial\vartheta}(\tilde{\vartheta}-\vartheta) - p(\varrho,\vartheta)\Big)$$

$$-\tilde{\varrho}(\tilde{\varrho}-\varrho)\frac{\partial s(\tilde{\varrho},\tilde{\vartheta})}{\partial\varrho}\left(\partial_t\tilde{\vartheta} + \tilde{\mathbf{u}}\cdot\nabla_x\tilde{\vartheta}\right) - \tilde{\varrho}(\tilde{\vartheta}-\vartheta)\frac{\partial s(\tilde{\varrho},\tilde{\vartheta})}{\partial\vartheta}\left(\partial_t\tilde{\vartheta} + \tilde{\mathbf{u}}\cdot\nabla_x\tilde{\vartheta}\right)$$

$$-(\tilde{\vartheta}-\vartheta)\mathrm{div}_x\left(\frac{\tilde{\mathbf{q}}}{\tilde{\vartheta}}\right) + \left(1-\frac{\vartheta}{\tilde{\vartheta}}\right)\left(\tilde{\mathbb{S}}:\nabla_x\tilde{\mathbf{u}} - \frac{\tilde{\mathbf{q}}\cdot\nabla_x\tilde{\vartheta}}{\tilde{\vartheta}}\right).$$

Consequently, relation (12.23) gives rise to

$$\int_\Omega E\left(\varrho,\vartheta,\mathbf{u}\Big|\tilde{\varrho},\tilde{\vartheta},\tilde{\mathbf{u}}\right)(\tau,\cdot)\,\mathrm{d}x$$

$$+\int_0^\tau\int_{\partial\Omega}\left[\varrho e(\varrho,\vartheta_B) - \vartheta_B\varrho s(\varrho,\vartheta_B)\right]\,[\mathbf{u}_B\cdot\mathbf{n}]^+\mathrm{d}\sigma_x\mathrm{d}t$$

$$-\int_0^\tau\int_{\partial\Omega}\varrho\left(e(\tilde{\varrho},\vartheta_B) - \vartheta_B s(\tilde{\varrho},\vartheta_B) + \frac{p(\tilde{\varrho},\vartheta_B)}{\tilde{\varrho}}\right)\,[\mathbf{u}_B\cdot\mathbf{n}]^+\mathrm{d}\sigma_x\mathrm{d}t$$

$$+\int_0^\tau\int_{\partial\Omega}p(\tilde{\varrho},\vartheta_B)\,[\mathbf{u}_B\cdot\mathbf{n}]^+\mathrm{d}\sigma_x\mathrm{d}t$$

$$+\int_0^\tau\int_\Omega\left(\frac{\tilde{\vartheta}}{\vartheta}-1\right)\mathbb{S}(\vartheta,\mathbb{D}_x\mathbf{u}):\mathbb{D}_x\mathbf{u}\,\mathrm{d}x\mathrm{d}t$$

$$+\int_0^\tau\int_\Omega\left(1-\frac{\tilde{\vartheta}}{\vartheta}\right)\left(\frac{\mathbf{q}(\vartheta,\nabla_x\vartheta)\cdot\nabla_x\vartheta}{\vartheta}\right)\,\mathrm{d}x\mathrm{d}t$$

$$+\int_0^\tau\int_\Omega\frac{\mathbf{q}(\vartheta,\nabla_x\vartheta)\cdot\nabla_x(\tilde{\vartheta}-\vartheta)}{\vartheta}\,\mathrm{d}x\mathrm{d}t$$

$$+\int_0^\tau\int_\Omega\Big(\mathbb{S}(\tilde{\vartheta},\mathbb{D}_x\tilde{\mathbf{u}}) - \mathbb{S}(\vartheta,\mathbb{D}_x\mathbf{u})\Big):\mathbb{D}_x(\tilde{\mathbf{u}}-\mathbf{u})\,\mathrm{d}x\mathrm{d}t$$

$$\leq -\int_0^\tau\int_\Omega(\tilde{\vartheta}-\vartheta)\mathrm{div}_x\left(\frac{\mathbf{q}(\tilde{\vartheta},\nabla_x\tilde{\vartheta})}{\tilde{\vartheta}}\right)\,\mathrm{d}x\mathrm{d}t$$

$$+\int_0^\tau \int_\Omega \left(1 - \frac{\vartheta}{\tilde\vartheta}\right) \left(\mathbb{S}(\tilde\vartheta, \mathbb{D}_x \tilde{\mathbf{u}}) : \mathbb{D}_x \tilde{\mathbf{u}} - \frac{\mathbf{q}(\tilde\vartheta, \nabla_x \tilde\vartheta)}{\tilde\vartheta} \right) \mathrm{d}x\mathrm{d}t + \int_0^\tau \int_\Omega F_3 \ \mathrm{d}x\mathrm{d}t \tag{12.25}$$

with the quadratic error

$$F_3 = F_2 + \mathrm{div}_x \tilde{\mathbf{u}} \Big(p(\tilde\varrho, \tilde\vartheta) - \frac{\partial p(\tilde\varrho, \tilde\vartheta)}{\partial \varrho}(\tilde\varrho - \varrho) - \frac{\partial p(\tilde\varrho, \tilde\vartheta)}{\partial \vartheta}(\tilde\vartheta - \vartheta) - p(\varrho, \vartheta) \Big)$$
$$+ \tilde\varrho \left(s(\tilde\varrho, \tilde\vartheta) - \frac{\partial s(\tilde\varrho, \tilde\vartheta)}{\partial \varrho}(\tilde\varrho - \varrho) - \frac{\partial s(\tilde\varrho, \tilde\vartheta)}{\partial \vartheta}(\tilde\vartheta - \vartheta) - s \right) (\partial_t \tilde\vartheta + \tilde{\mathbf{u}} \cdot \nabla_x \tilde\vartheta). \tag{12.26}$$

Finally, as $\vartheta - \tilde\vartheta$ vanishes on the boundary, we obtain

$$-\int_\Omega (\tilde\vartheta - \vartheta) \mathrm{div}_x \left(\frac{\mathbf{q}(\tilde\vartheta, \nabla_x \tilde\vartheta)}{\tilde\vartheta} \right) \mathrm{d}x = \int_\Omega \left(\frac{\mathbf{q}(\tilde\vartheta, \nabla_x \tilde\vartheta)}{\tilde\vartheta} \right) \cdot \nabla_x (\tilde\vartheta - \vartheta) \ \mathrm{d}x$$

to conclude

$$\int_\Omega E\left(\varrho, \vartheta, \mathbf{u} \Big| \tilde\varrho, \tilde\vartheta, \tilde{\mathbf{u}}\right)(\tau, \cdot) \ \mathrm{d}x$$
$$+ \int_0^\tau \int_{\partial\Omega} [\varrho e(\varrho, \vartheta_B) - \vartheta_B \varrho s(\varrho, \vartheta_B)] \ [\mathbf{u}_B \cdot \mathbf{n}]^+ \mathrm{d}\sigma_x \mathrm{d}t$$
$$- \int_0^\tau \int_{\partial\Omega} \varrho \left(e(\tilde\varrho, \vartheta_B) - \vartheta_B s(\tilde\varrho, \vartheta_B) + \frac{p(\tilde\varrho, \vartheta_B)}{\tilde\varrho} \right) \ [\mathbf{u}_B \cdot \mathbf{n}]^+ \mathrm{d}\sigma_x \mathrm{d}t$$
$$+ \int_0^\tau \int_{\partial\Omega} p(\tilde\varrho, \vartheta_B) \ [\mathbf{u}_B \cdot \mathbf{n}]^+ \mathrm{d}\sigma_x$$
$$+ \int_0^\tau \int_\Omega \left(\frac{\tilde\vartheta}{\vartheta} - 1 \right) \mathbb{S}(\vartheta, \mathbb{D}_x \mathbf{u}) : \mathbb{D}_x \mathbf{u} \ \mathrm{d}x\mathrm{d}t$$
$$+ \int_0^\tau \int_\Omega \left(\frac{\vartheta}{\tilde\vartheta} - 1 \right) \mathbb{S}(\tilde\vartheta, \mathbb{D}_x \tilde{\mathbf{u}}) : \mathbb{D}_x \tilde{\mathbf{u}} \ \mathrm{d}x\mathrm{d}t$$
$$+ \int_0^\tau \int_\Omega \left(1 - \frac{\tilde\vartheta}{\vartheta} \right) \left(\frac{\mathbf{q}(\vartheta, \nabla_x \vartheta) \cdot \nabla_x \vartheta}{\vartheta} \right) \ \mathrm{d}x\mathrm{d}t$$
$$+ \int_0^\tau \int_\Omega \left(1 - \frac{\vartheta}{\tilde\vartheta} \right) \frac{\mathbf{q}(\tilde\vartheta, \nabla_x \tilde\vartheta)}{\tilde\vartheta} \ \mathrm{d}x\mathrm{d}t$$
$$+ \int_0^\tau \int_\Omega \left(\frac{\mathbf{q}(\vartheta, \nabla_x \vartheta)}{\vartheta} - \frac{\mathbf{q}(\tilde\vartheta, \nabla_x \tilde\vartheta)}{\tilde\vartheta} \right) \cdot \nabla_x (\tilde\vartheta - \vartheta) \ \mathrm{d}x\mathrm{d}t$$
$$+ \int_0^\tau \int_\Omega \Big(\mathbb{S}(\tilde\vartheta, \mathbb{D}_x \tilde{\mathbf{u}}) - \mathbb{S}(\vartheta, \mathbb{D}_x \mathbf{u}) \Big) : \mathbb{D}_x (\tilde{\mathbf{u}} - \mathbf{u}) \ \mathrm{d}x\mathrm{d}t \le \int_0^\tau \int_\Omega F_3 \ \mathrm{d}x\mathrm{d}t, \tag{12.27}$$

with the quadratic error F_3 determined by (12.22), (12.24), and (12.26).

Now, we realize that the left-hand side of the inequality (12.27) coincides with its counterpart in Chap. 4, formula (4.21), while the error terms are the same as in (4.22). Thus using exactly the same arguments as in Sect. 4.2, we conclude

$$\int_\Omega E\left(\varrho, \vartheta, \mathbf{u} \middle| \tilde{\varrho}, \tilde{\vartheta}, \tilde{\mathbf{u}}\right)(\tau, \cdot)\,\mathrm{d}x = 0 \text{ for a.a. } \tau \in (0,T).$$

We have shown the following result.

Theorem 17 (Weak–Strong Uniqueness, Dirichlet Boundary Conditions).
Let $\Omega \subset R^d$, $d = 2, 3$, be a bounded domain of class C^4. Let the pressure $p = p(\varrho, \vartheta)$ be given by (4.2), *with the associated e, s satisfying* (4.8)–(4.14). *Let the transport coefficients μ, η, κ satisfy the structural hypotheses* (4.18)–(4.20). *Let $(\varrho, \vartheta, \mathbf{u})$ be a weak solution of the Navier–Stokes–Fourier system in $(0,T) \times \Omega$ with the Dirichlet boundary conditions in the sense of Definition 7 belonging to the regularity class* (12.20). *Suppose that the same problem (with the same initial and boundary data) admits a strong solution $(\tilde{\varrho}, \tilde{\vartheta}, \tilde{\mathbf{u}})$ in the class*

$$\begin{aligned} &\tilde{\varrho},\ \tilde{\vartheta} \in C^1([0,T] \times \overline{\Omega}),\ \tilde{\mathbf{u}} \in C^1([0,T] \times \overline{\Omega}; R^d),\\ &D_x^2 \tilde{\mathbf{u}},\ D_x^2 \tilde{\vartheta} \in C([0,T] \times \overline{\Omega})\\ &\inf_{(0,T)\times\Omega} \tilde{\varrho} > 0, \quad \inf_{(0,T)\times\Omega} \tilde{\vartheta} > 0. \end{aligned}$$

Then

$$\varrho = \tilde{\varrho},\ \vartheta = \tilde{\vartheta},\ \mathbf{u} = \tilde{\mathbf{u}} \text{ in } [0,T] \times \overline{\Omega}.$$

12.4 Existence of Weak Solutions

We conclude this chapter by a brief discussion concerning the existence of global in time weak solutions to the Navier–Stokes–Fourier system with the Dirichlet boundary conditions. The main issue in comparison with the problem with the heat flux boundary conditions considered in the bulk of this monograph are *energy bounds*. Roughly speaking, the energy bounds are easy to obtain from the associated total energy balance equation as long as the heat/mass flux through the boundary is controlled, cf. (3.42). If the temperature is prescribed on the boundary, the relevant estimates must be derived from the ballistic energy balance (12.11). In comparison with (3.42), the latter contains the integral

$$I_{\tilde{\vartheta}} = -\int_{\Omega} \left[\varrho s(\varrho, \vartheta)\partial_t \tilde{\vartheta} + \varrho s(\varrho, \vartheta)\mathbf{u} \cdot \nabla_x \tilde{\vartheta} + \frac{\mathbf{q}}{\vartheta} \cdot \nabla_x \tilde{\vartheta} \right] \mathrm{d}x$$

that must be dominated by the ballistic energy

$$\mathcal{E}_{\tilde{\vartheta}}(\varrho, \vartheta, \mathbf{u}) = \int_{\Omega} \left(\frac{1}{2}\varrho |\mathbf{u} - \mathbf{u}_B|^2 + \varrho e - \tilde{\vartheta} \varrho s(\varrho, \vartheta) \right) \mathrm{d}x$$

for a *suitable* extension $\tilde{\vartheta}$ of the boundary data ϑ_B.

12.4.1 Spatially Homogeneous Boundary Temperature

The first simple but still important from the point of view of applications observation is that the integral $I_{\tilde{\vartheta}}$ vanishes if $\vartheta_B > 0$ is a positive *constant* as we can take $\tilde{\vartheta} = \vartheta_B$ inside Ω. If $\vartheta_B = \vartheta_B(t)$ is a positive spatially homogeneous smooth function, then I_{ϑ_B} reduces to

$$I_{\tilde{\vartheta}} = -\int_{\Omega} \varrho s(\varrho, \vartheta) \partial_t \vartheta_B \, \mathrm{d}x$$

that can be dominated by $\mathcal{E}_{\vartheta_B}(\varrho, \vartheta, \mathbf{u})$ as long as ϑ_B is bounded below away from zero.

In both cases, the approximation scheme introduced in Sect. 5.1 can be easily adapted to yield global in time existence results for the Dirichlet problem. Indeed, the only modification consists in changing the boundary condition (5.19) to

$$\vartheta|_{\partial\Omega} = \vartheta_B. \tag{12.28}$$

12.4.2 General Boundary Conditions

To simplify, we consider the viscosity coefficients satisfying the hypotheses (4.18), (4.19) with $\Lambda = 1$. Accordingly, by virtue of Korn–Poincaré inequality,

$$\int_{\Omega} \frac{\tilde{\vartheta}}{\vartheta} \mathbb{S}(\vartheta, \mathbb{D}_x \mathbf{u}) : \mathbb{D}_x \mathbf{u} \, \mathrm{d}x \gtrsim \inf_{\Omega} \{\tilde{\vartheta}\} \left(\|\mathbf{u}\|^2_{W^{1,2}(\Omega)} - c(\mathbf{u}_B) \right). \tag{12.29}$$

Similarly,

$$\left| \int_{\Omega} \mathbb{S}(\vartheta, \mathbb{D}_x \mathbf{u}) : \mathbb{D}_x \mathbf{u}_B \, \mathrm{d}x \right| \leq \omega \|\mathbb{D}_x \mathbf{u}\|^2_{L^2(\Omega; R^{d\times d})} + c(\mathbf{u}_B)(1 + \int_{\Omega} \vartheta^2 \, \mathrm{d}x).$$

Consequently, the ballistic energy balance (12.16) gives rise to

$$\int_\Omega \left(\frac{1}{2}\varrho|\mathbf{u}-\mathbf{u}_B|^2 + \varrho e - \tilde\vartheta\varrho s \right)(\tau,\cdot)\ \mathrm{d}x$$
$$+ \int_0^\tau \int_{\partial\Omega} \Big[\varrho e(\varrho,\vartheta_B) - \vartheta_B \varrho_B s(\varrho,\vartheta_B) \Big]\ [\mathbf{u}_B\cdot\mathbf{n}]^+ \mathrm{d}\sigma_x \mathrm{d}t$$
$$+ \int_0^\tau \inf_\Omega\{\tilde\vartheta\} \int_\Omega \left(\|\mathbf{u}\|^2_{W^{1,2}(\Omega;R^d)} + \frac{\kappa(\vartheta)|\nabla_x\vartheta|^2}{\vartheta^2} \right)\ \mathrm{d}x\mathrm{d}t$$
$$\leq \int_\Omega \left(\frac{1}{2}\varrho_0|\mathbf{u}_0-\mathbf{u}_B|^2 + \varrho_0 e(\varrho_0,\vartheta_0) - \tilde\vartheta\varrho_0 s(\varrho_0,\vartheta_0) \right)\ \mathrm{d}x$$
$$+ c(\varrho_B,\vartheta_B,\mathbf{u}_B,\mathbf{g}) \Bigg[1 + \int_0^\tau \int_\Omega \left(\frac{1}{2}\varrho|\mathbf{u}-\mathbf{u}_B|^2 + \varrho e - \tilde\vartheta\varrho s \right)\ \mathrm{d}x\mathrm{d}t$$
$$- \int_0^\tau \int_\Omega \left[\varrho s \left(\partial_t\tilde\vartheta + \mathbf{u}\cdot\nabla_x\tilde\vartheta \right) + \frac{\mathbf{q}}{\vartheta}\cdot\nabla_x\tilde\vartheta \right]\ \mathrm{d}x\mathrm{d}t \Bigg]. \tag{12.30}$$

To control the rightmost integral, we consider $\tilde\vartheta$ to be the unique solution of the Laplace equation,

$$\Delta_x\tilde\vartheta(\tau,\cdot) = 0 \text{ in } \Omega,\ \tilde\vartheta(\tau,\cdot)|_{\partial\Omega} = \vartheta_B \text{ for any } \tau\in[0,T].$$

It follows from the standard maximum principle for harmonic functions that

$$\min_{[0,T]\times\partial\Omega} \vartheta_B \leq \tilde\vartheta(t,x) \leq \max_{[0,T]\times\partial\Omega} \vartheta_B \text{ for any } (t,x)\in(0,T)\times\Omega. \tag{12.31}$$

Let us denote this particular extension as ϑ_B.

Next, a simple integration by parts yields

$$\begin{aligned} -\int_\Omega \frac{\mathbf{q}}{\vartheta}\cdot\nabla_x\vartheta_B\ \mathrm{d}x &= \int_\Omega \frac{\kappa(\vartheta)}{\vartheta}\nabla_x\vartheta\cdot\nabla_x\vartheta_B\ \mathrm{d}x \\ &= \int_\Omega \nabla_x K(\vartheta)\cdot\nabla_x\vartheta_B\ \mathrm{d}x = \int_{\partial\Omega} K(\vartheta_B)\nabla_x\vartheta_B\cdot\mathbf{n}, \end{aligned} \tag{12.32}$$

where $K'(\vartheta) = \frac{\kappa(\vartheta)}{\vartheta}$. Moreover, if ϑ_B is continuously differentiable in time, we get, by virtue of the maximum principle (12.31),

$$-\int_\Omega \varrho s\partial_t\vartheta_B\ \mathrm{d}x \lesssim \left[c(\vartheta_B) + \int_\Omega \left(\frac{1}{2}\varrho|\mathbf{u}-\mathbf{u}_B|^2 + \varrho e - \vartheta_B\varrho s \right)\ \mathrm{d}x \right]. \tag{12.33}$$

In view of (12.32), (12.33), the inequality (12.30) reduces to

$$\int_\Omega \left(\frac{1}{2}\varrho|\mathbf{u}-\mathbf{u}_B|^2 + \varrho e - \vartheta_B\varrho s \right)(\tau,\cdot)\ \mathrm{d}x$$
$$+ \int_0^\tau \int_{\partial\Omega} \Big[\varrho e(\varrho,\vartheta_B) - \vartheta_B \varrho_B s(\varrho,\vartheta_B) \Big]\ [\mathbf{u}_B\cdot\mathbf{n}]^+ \mathrm{d}\sigma_x \mathrm{d}t$$

$$
\begin{aligned}
&+ \inf_{[0,T]\times\partial\Omega}\{\vartheta_B\}\int_0^T\int_\Omega\left(\|\mathbf{u}\|^2_{W^{1,2}(\Omega;R^d)} + \frac{\kappa(\vartheta)|\nabla_x\vartheta|^2}{\vartheta^2}\right)\,\mathrm{d}x\mathrm{d}t \\
&\le \int_\Omega\left(\frac{1}{2}\varrho_0|\mathbf{u}_0-\mathbf{u}_B|^2 + \varrho_0 e(\varrho_0,\vartheta_0) - \vartheta_B\varrho_0 s(\varrho_0,\vartheta_0)\right)\,\mathrm{d}x \\
&+ c(\varrho_B,\vartheta_B,\mathbf{u}_B,\mathbf{g})\left[1 + \int_0^T\int_\Omega\left(\frac{1}{2}\varrho|\mathbf{u}-\mathbf{u}_B|^2 + \varrho e - \vartheta_B\varrho s\right)\,\mathrm{d}x\mathrm{d}t\right. \\
&\left. - \int_0^T\int_\Omega \varrho s\mathbf{u}\cdot\nabla_x\vartheta_B\,\mathrm{d}x\mathrm{d}t\right]. \qquad (12.34)
\end{aligned}
$$

Recall that the entropy can be written as

$$
s(\varrho,\vartheta) = s_{\mathrm{m}}(\varrho,\vartheta) + s_{\mathrm{rad}}(\varrho,\vartheta),\ s_{\mathrm{rad}}(\vartheta,\varrho) = \frac{4}{3}\frac{\vartheta^3}{\varrho}.
$$

In addition to the hypotheses introduced in Sect. 4.1.1, we suppose validity of the third law of thermodynamics, namely

$$
\lim_{\vartheta\to 0} s_{\mathrm{m}}(\varrho,\vartheta) = 0 \text{ for any } \varrho > 0,
$$

see also Sect. 1.3, formula (1.21). This means, in particular, that the estimate (4.15) can be improved for small values of ϑ, specifically,

$$
|s_m(\varrho,\vartheta)| \le |s_m(\varrho,1)| + \left|\int_1^\vartheta \frac{\partial s_{\mathrm{m}}}{\partial\vartheta}(\varrho,z)\,\mathrm{d}z\right| \lesssim \left(1 + |\log(\varrho)| + [\log(\vartheta)]^+\right). \qquad (12.35)
$$

Consequently, going back to (12.34), we have

$$
\begin{aligned}
&\left|\int_\Omega \varrho s_{\mathrm{m}}(\varrho,\vartheta)\mathbf{u}\cdot\nabla_x\vartheta_B\,\mathrm{d}x\right| \\
&\lesssim \left(\int_\Omega \varrho|\mathbf{u}|^2\,\mathrm{d}x + \int_\Omega \varrho s^2_{\mathrm{m}}(\varrho,\vartheta)\,\mathrm{d}x\right) \\
&\lesssim \left(1 + \int_\Omega \varrho|\mathbf{u}|^2\,\mathrm{d}x + \int_\Omega \varrho^{\frac{5}{3}}\,\mathrm{d}x + \int_\Omega \vartheta^4\,\mathrm{d}x\right) \\
&\lesssim \left(c(\vartheta_B) + \int_\Omega\left(\frac{1}{2}\varrho|\mathbf{u}-\mathbf{u}_B|^2 + \varrho e - \vartheta_B\varrho s\right)\,\mathrm{d}x\right). \qquad (12.36)
\end{aligned}
$$

Finally, we have to control the radiative component of the entropy flux. To this end, we have imposed additional growth conditions on the heat conductivity coefficient $\kappa(\vartheta)$. Specifically, we replace the original hypotheses (4.18)–(4.20) by

$$0 < \underline{\mu}\,(1+\vartheta) \le \mu(\vartheta) \le \overline{\mu}\,(1+\vartheta)\,,\ |\mu'(\vartheta)| \le c \text{ for all } \vartheta \ge 0,$$
$$0 \le \eta(\vartheta) \le \overline{\eta}\,(1+\vartheta)\,,$$
$$0 < \underline{\kappa}\,\left(1+\vartheta^{\beta}\right) \le \kappa(\vartheta) \le \overline{\kappa}\,\left(1+\vartheta^{\beta}\right),\ \beta > 6. \tag{12.37}$$

Consequently, on the one hand,

$$\int_\Omega \frac{\kappa(\vartheta)|\nabla_x\vartheta|^2}{\vartheta^2}\ \mathrm{d}x \gtrsim \int_\Omega \left(\frac{1}{\vartheta^2} + \vartheta^{\beta-2}\right) |\nabla_x\vartheta|^2\ \mathrm{d}x, \tag{12.38}$$

while, on the other hand,

$$\left|\int_\Omega \vartheta^3 \mathbf{u}\cdot\nabla_x\vartheta_B\ \mathrm{d}x\right| \le \varepsilon \|\mathbf{u}\|^2_{W^{1,2}(\Omega;R^d)} + c(\varepsilon)\|\vartheta^3\|^2_{L^2(\Omega)} \tag{12.39}$$

for any $\varepsilon > 0$. Next,

$$\|\vartheta^3\|^2_{L^2(\Omega)} = \int_{\vartheta \le K} \vartheta^6\ \mathrm{d}x + \int_{\vartheta > K} \vartheta^6\ \mathrm{d}x \le |\Omega| K^6 + K^{6-\beta}\int_\Omega \vartheta^\beta\ \mathrm{d}x,$$

while, by Hölder and Poincaré inequalities,

$$\int_\Omega \vartheta^\beta\ \mathrm{d}x \lesssim \|\vartheta^\beta\|_{L^3(\Omega)} = \|\vartheta^{\frac{\beta}{2}}\|^2_{L^6(\Omega)} \lesssim \|\vartheta^{\frac{\beta}{2}}\|^2_{W^{1,2}(\Omega)} \lesssim \|\nabla_x\vartheta^{\frac{\beta}{2}}\|^2_{L^2(\Omega)} + c(\vartheta_B)$$
$$\lesssim \int_\Omega \vartheta^{\beta-2}|\nabla_x\vartheta|^2\ \mathrm{d}x + c(\vartheta_B).$$

As $\beta > 6$, we can first fix $\varepsilon > 0$ small enough and then $K = K(\varepsilon)$ large enough so that the integral (12.39) is dominated by the left-hand side of (12.34).

Thus we conclude

$$\int_\Omega \left(\frac{1}{2}\varrho|\mathbf{u}-\mathbf{u}_B|^2 + \varrho e - \tilde{\vartheta}\varrho s\right)(\tau,\cdot)\ \mathrm{d}x$$
$$+ \int_0^\tau \int_{\partial\Omega} \left[\varrho e(\varrho,\vartheta_B) - \vartheta_B \varrho_B s(\varrho,\vartheta_B)\right]\,[\mathbf{u}_B\cdot\mathbf{n}]^+ \mathrm{d}\sigma_x \mathrm{d}t$$
$$+ \inf_{[0,T]\times\partial\Omega}\{\vartheta_B\}\int_0^\tau\int_\Omega \left(\|\mathbf{u}\|^2_{W^{1,2}(\Omega;R^d)} + \frac{\kappa(\vartheta)|\nabla_x\vartheta|^2}{\vartheta^2}\right)\ \mathrm{d}x\mathrm{d}t$$
$$\le \int_\Omega \left(\frac{1}{2}\varrho_0|\mathbf{u}_0-\mathbf{u}_B|^2 + \varrho_0 e(\varrho_0,\vartheta_0) - \tilde{\vartheta}\varrho_0 s(\varrho_0,\vartheta_0)\right)\ \mathrm{d}x$$

$$+ c(\varrho_B, \vartheta_B, \mathbf{u}_B, \mathbf{g}) \left[c(\vartheta_B) + \int_0^T \int_\Omega \left(\frac{1}{2}\varrho|\mathbf{u} - \mathbf{u}_B|^2 + \varrho e - \tilde{\vartheta}\varrho s \right) \, \mathrm{d}x\mathrm{d}t \right],$$

whence desired *a priori* bounds follow from Gronwall lemma.

12.4.3 Conclusion

The *a priori* bounds obtained in the preceding two sections can be easily implemented in the approximation scheme introduced in Sect. 5.1, with the obvious modification of the boundary conditions for the approximate internal energy balance. The existence proof leading to the results stated in Chap. 9 can be reproduced with obvious modifications. Note that positivity of the temperature discussed in Sect. 8.1.3 follows easily as long as the boundary temperature ϑ_B is strictly positive.

We may state the following existence result:

Theorem 18 (Global Existence, Dirichlet Problem).
Let $\Omega \subset R^d$, $d = 2, 3$, be a bounded domain of class at least C^4. Suppose that the boundary data $\mathbf{u}_B = \mathbf{u}_B(t,x)$, $\varrho_B = \varrho_B(t,x)$, and $\vartheta_B = \vartheta_B(t,x)$ are twice continuously differentiable, and

$$\inf_{[0,T]\times\partial\Omega} \varrho_B > 0, \quad \inf_{[0,T]\times\partial\Omega} \vartheta_B > 0.$$

Let the pressure $p = p(\varrho,\vartheta)$, the internal energy $e = e(\varrho,\vartheta)$, the entropy $s = s(\varrho,\vartheta)$ satisfy the hypotheses (4.2), (4.8)–(4.14), *and let the transport coefficients $\mu = \mu(\vartheta)$, $\eta = \eta(\vartheta)$, and $\kappa = \kappa(\vartheta)$ satisfy* (4.18)–(4.20). *In addition, if ϑ_B depends on the x-variable, let* (12.35) *and* (12.37) *hold.*

Then for any $T > 0$, any initial data

$$\varrho_0, \vartheta_0, \mathbf{u}_0, \ \varrho_0 > 0, \ \vartheta_0 > 0,$$
$$\int_\Omega \left(\frac{1}{2}\varrho_0|\mathbf{u}_0|^2 + \varrho_0 e(\varrho_0,\vartheta_0) - \vartheta_B \varrho_0 s(\varrho_0,\vartheta_0) \right) \, \mathrm{d}x < \infty,$$

and any $\mathbf{g} \in L^\infty((0,T)\times\Omega; R^d)$, there exists a weak solution $(\varrho,\vartheta,\mathbf{u})$ of the Navier–Stokes–Fourier system in $(0,T)\times\Omega$ in the sense of Definition 7.

12.5 Concluding Remarks

The existence of local in time strong solutions for the Navier–Stokes–Fourier system with the temperature prescribed on the boundary was shown by Valli and Zajaczkowski [89], see also Matsumura and Nishida [71, 72], and Valli [86–88] for related results. Global in time weak solutions were obtained in [17] by the method presented in this chapter.

The results concerning the long time behaviour of solutions obtained in Chaps. 10 and 11 can be extended, under appropriate structural restrictions imposed on the constitutive relations, to the case of the Dirichlet boundary conditions for the temperature, cf. [37].

Appendix

A.1 Skorokhod Spaces

Unlike the density ϱ and the momentum $\mathbf{m}$, the total entropy $S = \varrho s$ is not, in general, a weakly continuous function of time, as its weak time derivative satisfies only the entropy inequality. In Chaps. 3, 10, we have identified the instantaneous values of S with its left limits,

$$\int_{\Omega} S\phi(\tau,\cdot)\ \mathrm{d}x \equiv \int_{\Omega} \varrho s(\tau,\cdot)\phi\ \mathrm{d}x = \lim_{\delta\searrow 0}\frac{1}{\delta}\int_{\tau-\delta}^{\tau}\int_{\Omega} \varrho s\phi\ \mathrm{d}x\mathrm{d}t$$

for any $\phi \in C_c^{\infty}(\Omega)$, whence for any $\phi \in C_0(\Omega)$. As observed in Chap. 10, the resulting function

$$t \mapsto \int_{\Omega} S\phi\ \mathrm{d}x$$

is càglàd (left-continuous with right limits) provided we define $\int_{\Omega} S\phi\ \mathrm{d}x$ by its initial value $\int_{\Omega} S_0\phi\ \mathrm{d}x$ for $\tau \leq 0$.

We recall some basic properties of Skorokhod spaces of càglàd functions referring systematically to the monograph by Whitt [91]. As a matter of fact, we only make use of the Skorokhod space $D_{\mathrm{loc}}(R; W^{-k,2}(\Omega))$, where $k > \frac{d}{2}$ so that $W_0^{k,2}(\Omega) \hookrightarrow C_0(\Omega)$.

A.1.1 The Space $D([0,T];R)$

We consider the space $D([0,T];R)$ of scalar-valued càglàd functions in $[0,T]$. For $h \in D([0,T];R)$ we introduce its *graph*,

$$\mathrm{graph}[h] = \left\{ (t,y) \in [0,T] \times R \ \middle|\ t \in [0,T],\ y \in [h(t), h(t+)] \right\}.$$

E. Feireisl, A. Novotný, *Mathematics of Open Fluid Systems*, Nečas Center Series, https://doi.org/10.1007/978-3-030-94793-4

For $(t_1, y_1), (t_2, y_2) \in \text{graph}[h]$ we say

$$(t_1, y_1) \prec (t_2, y_2)$$

if either: (i) $t_1 < t_2$ or (ii) $t_1 = t_2$ and $|y_1 - h(t_1+)| \geq |y_2 - h(t_1+)|$. A mapping

$$[t(s), y(s)] : s \in [0, T] \mapsto [t, y] \in \text{graph}[h]$$

is a parametric representation of h, $[t, y] \in \text{graph}[h]$ if it is non-decreasing with respect to $\prec$ and

$$t(0) = 0,\ y(0) = h(0),\ s(T) = T.$$

Finally, we define the metric d on $D([0,T]; R)$, see Whitt [91, Chapter 12, Theorem 12.3.1],

$$d[h_1; h_2] = \inf_{[t_1, y_1] \in \Pi[h_1],\ [t_2, y_2] \in \Pi[h_2]} \sup_{s \in [0,T]} \max \Big\{ |t_1(s) - t_2(s)|; |y_1(s) - y_2(s)| \Big\}.$$

The space $D([0,T]; R)$ with the metric d is not complete; however, it is possible to find a complete metric that is *topologically equivalent* to d, see Whitt [91, Chapter 12, Theorem 12.8.1].

The following criterion of compactness is useful when dealing with the total entropy, see Whitt [91, Chapter 12, Corollary 12.5.1].

Lemma 12 (Compactness of Monotone Functions).
Let

$$h_n : [0, T] \to R$$

be a sequence of monotone functions.

Then

$$d[h_n; h] \to 0 \text{ as } n \to \infty \text{ for some } h \in D([0,T]; R)$$

if and only if

$$h_n(t) \to h(t) \text{ for all } t \text{ belonging to a dense set in } [0,T] \text{ including } 0 \text{ and } T.$$

A.1.2 Extension to Unbounded Intervals

Denote d_T the Skorokhod metric on the space $D([0,T]; R)$. We may define a metric on the space

$$D([0,\infty); R)$$

of all *bounded* càglàd functions as

$$d[h_1, h_2] = \int_0^\infty \exp(-t) d_t[h_1, h_2] \mathrm{d}t.$$

Finally, we define

the **Skorokhod metric on** $D(R; R)$

$$d[h_1, h_2] = \int_{-\infty}^\infty \exp(-|t|) d_{-t,t}[h_1, h_2],$$

where $d_{-T,T}$ is the Skorokhod metric on $D([-T,T]; R)$.

A.1.3 Extension to Infinite-Dimensional Spaces

Finally, we extend the concept of Skorokhod topology to infinite-dimensional function spaces, in particular to the dual Sobolev spaces $W^{-k,2} \approx (W_0^{k,2})^*$, $k > \frac{d}{2}$. Let $\{\phi_n\}_{n=1}^\infty$ be an orthonormal basis in $W_0^{k,2}$. For

$$h_1, h_2 \text{ such that } t \mapsto \langle h_i; \phi_n \rangle_{W^{-k,2}, W_0^{k,2}} \text{ is càglàd for } i = 1, 2, \text{ and any } \phi_n,$$

we set

$$d_S[h_1, h_2] = \sum_{n=1}^\infty \frac{1}{2^n} \frac{d_{D([0,T];R)} \left[\langle h_1; \phi_n \rangle ; \langle h_2; \phi_n \rangle \right]}{1 + d_{D([0,T];R)} \left[\langle h_1; \phi_n \rangle ; \langle h_2; \phi_n \rangle \right]}.$$

This yields the product metric on bounded subsets of the space

$$D([0,T]; W^{-k,2}).$$

The topology on the spaces $D([0,\infty); W^{-k,2})$ and $D(R; W^{-k,2})$ is defined in a similar way as in Sect. A.1.2.

B.1 Equation of Continuity, Renormalization

We collect some useful properties of the equation of continuity considered in the framework of renormalized solutions introduced by DiPerna and Lions [28]. We consider a smooth (at least C^2) velocity field $\mathbf{u}_B = \mathbf{u}_B(t,x)$, $\mathbf{u}_B \in C^2(R^{d+1}; R^d)$ and a bounded Lipschitz domain $\Omega \subset R^d$, $d = 2, 3$. In addition, we assume that $[0,T] \times \partial\Omega$ admits

a boundary decomposition:

$$
\begin{aligned}
&[0,T] \times \partial\Omega = \Gamma^0_{\rm in} \cup \Gamma^0_{\rm out} \cup \Gamma^0_{\rm wall} \cup \Gamma_{\rm sing},\\
&\Gamma^0_{\rm in},\ \Gamma^0_{\rm out},\ \Gamma^0_{\rm wall} \text{ open in } [0,T]\times\partial\Omega,\\
&\Gamma^0_{\rm in} \subset \left\{ (t,x) \ \Big|\ \mathbf{n}(x) \text{ exists, } \mathbf{u}_B(t,x)\cdot\mathbf{n}(x) < 0 \right\},\\
&\Gamma^0_{\rm out} \subset \left\{ (t,x) \ \Big|\ \mathbf{n}(x) \text{ exists, } \mathbf{u}_B(t,x)\cdot\mathbf{n}(x) > 0 \right\},\\
&\Gamma^0_{\rm wall} \subset \left\{ (t,x) \ \Big|\ \mathbf{n}(x) \text{ exists, } \mathbf{u}_B(t,x)\cdot\mathbf{n}(x) = 0 \right\},\\
&\Gamma_{\rm sing} \subset [0,T]\times\partial\Omega \text{ compact, } |\Gamma_{\rm sing}|_{{\rm d}t\otimes {\rm d}\sigma_x} = 0. \qquad \text{(B.1)}
\end{aligned}
$$

In particular, the "singular" component $\Gamma_{\rm sing}$ contains all singular points of $\partial\Omega$, where the outer normal $\mathbf{n}$ does not exist. The fact that $\Gamma_{\rm sing}$ is compact and of zero surface measure plays a crucial role in the analysis. We point out that Γ^0 does not necessarily coincide with the topological interior of Γ.

B.1.1 Renormalization

Recall that the equation of continuity (1.11), with the initial/boundary conditions

$$
\varrho(0,\cdot) = \varrho_0 \text{ in } \Omega,\ \mathbf{u}|_{\partial\Omega} = \mathbf{u}_B,\ \varrho|_{\Gamma_{\rm in}} = \varrho_B, \qquad \text{(B.2)}
$$

is satisfied in the *renormalized form* if the integral identity

$$
\begin{aligned}
&\int_0^\tau \int_\Omega \left[b(\varrho)\partial_t\varphi + b(\varrho)\mathbf{u}\cdot\nabla_x\varphi + \Big(b(\varrho) - b'(\varrho)\varrho\Big){\rm div}_x\mathbf{u}\varphi \right] \ {\rm d}x{\rm d}t\\
&\quad = \left[\int_\Omega b(\varrho)\varphi \ {\rm d}x \right]_{t=0}^{t=\tau}\\
&\quad + \int_0^\tau \int_{\partial\Omega} \varphi b(\varrho_B)\ [\mathbf{u}_B\cdot\mathbf{n}]^- {\rm d}\sigma_x{\rm d}t + \int_0^\tau \int_{\partial\Omega} \varphi b(\varrho)\ [\mathbf{u}_B\cdot\mathbf{n}]^+ {\rm d}\sigma_x{\rm d}t
\end{aligned}
\qquad \text{(B.3)}
$$

holds for any $0 \le \tau \le T$, any $\varphi \in C^1([0,T]\times\overline{\Omega})$, and any

$$
b \in C^1(R),\ b' \in C_c(R).
$$

As b' is compactly supported, the integral identity (B.3) makes sense whenever the density ϱ is merely measurable. If, in addition, ϱ belongs to some Lebesgue space of integrable functions, the validity of (B.3) can be extended to the class of functions b with appropriate polynomial growth by means of the Lebesgue theorem.

B.1.2 Weak Sequential Stability

Our goal is to find sufficient conditions for the class of renormalized solutions to be stable with respect to weak convergence. Consider a sequence of solutions $\{\varrho_n, \mathbf{u}_n\}_{n=1}^{\infty}$ of the equation of continuity

$$\begin{aligned}\int_0^T \int_\Omega \Big[\varrho_n \partial_t \varphi + \varrho_n \mathbf{u}_n \cdot \nabla_x \varphi\Big] \, \mathrm{d}x\mathrm{d}t &= - \int_\Omega \varrho_0 \varphi \ \mathrm{d}x \\ &+ \int_0^T \int_{\partial\Omega} \varrho_B \varphi \ [\mathbf{u}_B \cdot \mathbf{n}]^- \mathrm{d}\sigma_x \mathrm{d}t + \int_0^T \int_{\partial\Omega} \varrho_n \varphi \ [\mathbf{u}_B \cdot \mathbf{n}]^+ \mathrm{d}\sigma_x \mathrm{d}t\end{aligned} \tag{B.4}$$

for any $\varphi \in C^1_c([0,T) \times \overline{\Omega})$. Suppose, in addition, that the equation is also satisfied in the renormalized sense,

$$\begin{aligned}&\int_0^T \int_\Omega \Big[b(\varrho_n) \partial_t \varphi + b(\varrho_n) \mathbf{u}_n \cdot \nabla_x \varphi + (b(\varrho_n) - b'(\varrho_n)\varrho_n) \mathrm{div}_x \mathbf{u}_n \varphi\Big] \ \mathrm{d}x\mathrm{d}t \\ &= - \int_\Omega b(\varrho_0) \varphi \ \mathrm{d}x \\ &+ \int_0^T \int_{\partial\Omega} b(\varrho_B) \varphi \ [\mathbf{u}_B \cdot \mathbf{n}]^- \mathrm{d}\sigma_x \mathrm{d}t + \int_0^T \int_{\partial\Omega} b(\varrho_n) \varphi \ [\mathbf{u}_B \cdot \mathbf{n}]^+ \ \mathrm{d}\sigma_x \mathrm{d}t\end{aligned} \tag{B.5}$$

for any $b \in C^1(R)$, $b' \in C_c(R)$. Finally, we introduce the oscillation defect measure

$$\begin{aligned}&\mathbf{osc}_q[\varrho_n \to \varrho]((0,T) \times \Omega) \\ &\qquad \equiv \sup_{k \geq 1} \left(\limsup_{n \to \infty} \int_0^T \int_\Omega |T_k(\varrho_n) - T_k(\varrho)|^q \ \mathrm{d}x\mathrm{d}t \right), \\ &T_k(\varrho) = \min\{\varrho, k\}.\end{aligned}$$

We claim the following result.

Theorem 19 (Weak Sequential Stability of Renormalized Solutions).
Let $\mathbf{u}_B \in C^2(R^{d+1}; R^d)$, $\varrho_B \in C^2(R^{d+1})$, $\varrho_B|_{\partial\Omega} > 0$, *be given fields. Suppose* $\Omega \subset R^d$ *is a bounded Lipschitz domain such that the boundary of the space–time cylinder* $(0,T) \times \Omega$ *can be decomposed as in* (B.1). *Let* $\{\varrho_n, \mathbf{u}_n\}_{n=1}^{\infty}$ *be a sequence of renormalized solutions satisfying* (B.4), (B.5),

$$\varrho_n \to \varrho \text{ weakly-(*) in } L^\infty(0,T; L^\gamma(\Omega)) \text{ for some } \gamma > 1,$$
$$\mathbf{u}_n \to \mathbf{u} \text{ weakly in } L^r(0,T; W^{1,r}(\Omega; R^d)) \text{ for some } r > 1.$$

Finally, suppose that

$$\mathbf{osc}_q[\varrho_n \to \varrho]((0,T) \times \Omega) < \infty, \text{ for } \frac{1}{q} + \frac{1}{r} < 1. \tag{B.6}$$

Then the limit $(\varrho, \mathbf{u})$ *is a renormalized solution of the equation of continuity, meaning* (B.3) *holds.*

The rest of this section is devoted to the proof of Theorem 19. In view of the existing theory, see e.g. [43], the limit $(\varrho, \mathbf{u})$ satisfies the renormalized equation *inside* Ω. The main issue is therefore extending its validity up to the boundary.

B.1.3 Extension Outside Γ^0_{in}

Let us denote $\overline{\Omega}^c \equiv R^d \setminus \overline{\Omega}$. As ϱ_B, $\mathbf{u}_B$ are regular, it is easy to see that the functions

$$\tilde{\varrho}_n(t,x) = \begin{cases} \varrho_n(t,x) \text{ if } x \in \Omega, \\ \varrho_B(t,x) \text{ if } x \in \Gamma^0_{\text{in}} \cup \overline{\Omega}^c, \end{cases}$$

$$\tilde{\mathbf{u}}_n(t,x) = \begin{cases} \mathbf{u}_n(t,x) \text{ if } x \in \Omega, \\ \mathbf{u}_B(t,x) \text{ if } x \in \Gamma^0_{\text{in}} \cup \overline{\Omega}^c, \end{cases}$$

represent a weak renormalized solution of the equation

$$\partial_t \tilde{\varrho}_n + \mathrm{div}_x(\tilde{\varrho}_n \tilde{\mathbf{u}}_n) = \mathbb{1}_{\overline{\Omega}^c} f, \ f = \partial_t(\varrho_B) + \mathrm{div}_x(\varrho_B \mathbf{u}_B)$$

in the open set $(0,T) \times (\Omega \cup \Gamma^0_{\text{in}} \cup \overline{\Omega}^c) \subset R^{d+1}$. Thus, by virtue of [41, Chapter 3, Lemma 3.8], the limit

$$\tilde{\varrho}(t,x) = \begin{cases} \varrho(t,x) \text{ if } x \in \Omega, \\ \varrho_B(t,x) \text{ if } x \in \Gamma^0_{\rm in} \cup (R^d \setminus \overline{\Omega}), \end{cases}$$

$$\tilde{\mathbf{u}}(t,x) = \begin{cases} \mathbf{u}(t,x) \text{ if } x \in \Omega, \\ \mathbf{u}_B(t,x) \text{ if } x \in \Gamma^0_{\rm in} \cup (R^d \setminus \overline{\Omega}), \end{cases}$$

is a renormalized solution in $(0,T) \times (\Omega \cup \Gamma^0_{\rm in} \cup \overline{\Omega}^c)$, specifically,

$$\begin{aligned} &\int_0^T \int_{R^d} \Big[b(\tilde{\varrho})\partial_t \varphi + b(\tilde{\varrho})\tilde{\mathbf{u}} \cdot \nabla_x \varphi + (b(\tilde{\varrho}) - b'(\tilde{\varrho})\tilde{\varrho}) \mathrm{div}_x \tilde{\mathbf{u}} \varphi \Big] \mathrm{d}x\mathrm{d}t \\ &\quad + \int_0^T \int_{R^d \setminus \overline{\Omega}} b'(\varrho_B) f \varphi \mathrm{d}x\mathrm{d}t = - \int_{R^d} b(\tilde{\varrho}_0)\varphi \mathrm{d}x \end{aligned} \tag{B.7}$$

for any $\varphi \in C^1_c([0,T) \times (\Omega \cup \Gamma^0_{\rm in} \cup \overline{\Omega}^c))$.

Now, the integral in the left-hand side of (B.7) can be split as

$$\begin{aligned} &\int_0^T \int_{R^d} \Big[b(\tilde{\varrho})\partial_t \varphi + b(\tilde{\varrho})\tilde{\mathbf{u}} \cdot \nabla_x \varphi + (b(\tilde{\varrho}) - b'(\tilde{\varrho})\tilde{\varrho}) \mathrm{div}_x \tilde{\mathbf{u}} \varphi \Big] \mathrm{d}x\mathrm{d}t \\ &\quad = \int_0^T \int_{\Omega} \Big[b(\varrho)\partial_t \varphi + b(\varrho)\mathbf{u} \cdot \nabla_x \varphi + (b(\varrho) - b'(\varrho)\varrho) \mathrm{div}_x \tilde{\mathbf{u}} \varphi \Big] \ \mathrm{d}x\mathrm{d}t \\ &\quad + \int_0^T \int_{R^d \setminus \overline{\Omega}} \Big[b(\varrho_B)\partial_t \varphi + b(\varrho_B)\mathbf{u}_B \cdot \nabla_x \varphi \Big] \mathrm{d}x\mathrm{d}t \\ &\quad + \int_0^T \int_{R^d \setminus \overline{\Omega}} \Big(b(\varrho_B) - b'(\varrho_B)\varrho_B \Big) \mathrm{div}_x \mathbf{u}_B \varphi \mathrm{d}x\mathrm{d}t, \end{aligned} \tag{B.8}$$

where

$$\begin{aligned} &\int_0^T \int_{R^d \setminus \overline{\Omega}} \Big[b(\varrho_B)\partial_t \varphi + b(\varrho_B)\mathbf{u}_B \cdot \nabla_x \varphi + (b(\varrho_B) - b'(\varrho_B)\varrho_B) \mathrm{div}_x \mathbf{u}_B \varphi \Big] \mathrm{d}x\mathrm{d}t \\ &= - \int_0^T \int_{R^d \setminus \overline{\Omega}} b'(\varrho_B) f dx dt - \int_{R^d \setminus \overline{\Omega}} b(\varrho_B)\varphi \mathrm{d}x \\ &- \int_{\Gamma^0_{\rm in}} \varphi b(\varrho_B) \mathbf{u}_B \cdot \mathbf{n} \ \mathrm{d}\sigma_x \ \mathrm{d}t \end{aligned}$$

whenever $\varphi \in C^1_c([0,T) \times (\Omega \cup \Gamma^0_{\rm in} \cup \overline{\Omega}^c)$. Consequently, we conclude

$$\begin{aligned} &\int_0^T \int_{\Omega} \Big[b(\varrho)\partial_t \varphi + b(\varrho)\mathbf{u} \cdot \nabla_x \varphi + (b(\varrho) - b'(\varrho)\varrho) \mathrm{div}_x \mathbf{u} \varphi \Big] \ \mathrm{d}x\mathrm{d}t \\ &\quad = - \int_{\Omega} \varphi b(\varrho_0) \ \mathrm{d}x + \int_0^T \int_{\partial\Omega} \varphi b(\varrho_B) \varphi \ [\mathbf{u}_B \cdot \mathbf{n}]^- \mathrm{d}\sigma_x \mathrm{d}t \end{aligned} \tag{B.9}$$

for any $b \in C^1(R)$, $b' \in C_c(R)$, and any $\varphi \in C^1_c([0,T) \times (\Omega \cup \Gamma^0_{\rm in}))$.

B.1.4 Extension Outside Γ_{out}^0

Extension outside Γ_{out} is more delicate. To facilitate analysis, we suppose, without loss of generality, that $\mathrm{div}_x \mathbf{u}_B = 0$ outside Ω. In view of (B.5), we have

$$\int_0^T \int_\Omega \Big[T_k(\varrho_n)\partial_t\varphi + T_k(\varrho_n)\mathbf{u}_n \cdot \nabla_x\varphi + (T_k(\varrho_n) - T'(\varrho_n)\varrho_n)\mathrm{div}_x \mathbf{u}_n\varphi \Big] \, \mathrm{d}x\mathrm{d}t$$
$$= - \int_\Omega T_k(\varrho_0)\varphi \, \mathrm{d}x + \int_0^T \int_{\partial\Omega} T_k(\varrho_n)\varphi \, [\mathbf{u}_B \cdot \mathbf{n}]^+ \mathrm{d}\sigma_x \mathrm{d}t$$

for any $\varphi \in C_c^1([0,T) \times (\Omega \cup \Gamma_{\mathrm{out}}^0))$. Supposing ϱ_0 has been extended as a compactly supported function outside Ω, we may consider the (unique) weak solution $r_{k,n}$ of the transport equation

$$\partial_t r_{k,n} + \mathbf{u}_B \cdot \nabla_x r_{k,n} = 0 \text{ in } (0,T) \times R^d \setminus \overline{\Omega},$$
$$r_{k,n}|_{\Gamma_{\mathrm{out}}^0} = -T_k(\varrho_n)\mathbf{u}_B \cdot \mathbf{n}, \; r_k(0,\cdot) = T_k(\varrho_0).$$

Note carefully that $\mathbf{n}$ here denotes the normal to $\partial\Omega$ directed out of Ω. The existence of the solution r_k was shown by Crippa et al. [23]. Alternatively, we may set

$$r_{k,n} = T_k(r_n),$$

where r_n is the unique renormalized solution of the problem

$$\partial_t r_n + \mathbf{u}_B \cdot \nabla_x r_n = 0 \text{ in } (0,T) \times R^d \setminus \overline{\Omega}, \; r_n|_{\Gamma_{\mathrm{out}}^0} = -\varrho_n \mathbf{u}_B \cdot \mathbf{n}, \; r(0,\cdot) = \varrho_0$$

constructed by Boyer [11, Theorem 6.1]. Strictly speaking, Boyer supposes certain regularity of $\partial\Omega$ not necessary for the construction by Crippa et al. [23].

Accordingly, setting

$$\tilde{\varrho}_n(t,x) = \begin{cases} \varrho_n(t,x) \text{ if } x \in \Omega, \\ r_n(t,x) \text{ if } x \in \Gamma_{\mathrm{out}}^0 \cup \overline{\Omega}^c, \end{cases}$$

$$\tilde{\mathbf{u}}_n(t,x) = \begin{cases} \mathbf{u}_n(t,x) \text{ if } x \in \Omega, \\ \mathbf{u}_B(t,x) \text{ if } x \in \Gamma_{\mathrm{out}}^0 \cup \overline{\Omega}^c, \end{cases}$$

we deduce

$$\int_0^T \int_{R^d} \Big[T_k(\tilde{\varrho}_n)\partial_t\varphi + T_k(\tilde{\varrho}_n)\mathbf{u}_n \cdot \nabla_x\varphi + (T_k(\tilde{\varrho}_n) - T'(\tilde{\varrho}_n)\tilde{\varrho}_n)\mathrm{div}_x \tilde{\mathbf{u}}_n\varphi \Big] \mathrm{d}x\mathrm{d}t$$
$$= - \int_{R^d} T_k(\varrho_0)\varphi \mathrm{d}x$$

for any $\varphi \in C_c^1([0,T) \times (\Omega \cup \Gamma_{\mathrm{out}}^0 \cup \overline{\Omega}^c))$.

Now, we let $n \to 0$ obtaining

$$\int_0^T \int_{R^d} \left[\overline{T_k(\tilde{\varrho})} \partial_t \varphi + \overline{T_k(\tilde{\varrho})} \tilde{\mathbf{u}} \cdot \nabla_x \varphi + \overline{(T_k(\tilde{\varrho}) - T'(\tilde{\varrho})\tilde{\varrho}) \mathrm{div}_x \tilde{\mathbf{u}} \varphi} \right] \mathrm{d}x \mathrm{d}t \\ = - \int_{R^d} T_k(\varrho_0) \varphi \mathrm{d}x \tag{B.10}$$

for any $\varphi \in C^1_c([0,T) \times (\Omega \cup \Gamma^0_{\rm out} \cup \overline{\Omega}^c))$. Note that

$$\tilde{\mathbf{u}}(t,x) = \begin{cases} \mathbf{u}(t,x) \text{ if } x \in \Omega, \\ \mathbf{u}_B(t,x) \text{ if } x \in \Gamma^0_{\rm out} \cup \overline{\Omega}^c, \end{cases}$$

whereas $\overline{T_k(\tilde{\varrho})}|_{\overline{\Omega}^c}$ is the unique solution of the transport equation

$$\partial_t \overline{T_k(\tilde{\varrho})} + \mathbf{u}_B \cdot \overline{T_k(\tilde{\varrho})} = 0 \text{ in } (0,T) \times \overline{\Omega}^c,$$
$$\overline{T_k(\tilde{\varrho})}|_{\Gamma_{\rm out}} = -\overline{T_k(\varrho)} \mathbf{u}_B \cdot \mathbf{n}, \ r(0, \cdot) = T_k(\varrho_0).$$

Applying the regularization procedure of DiPerna and Lions to (B.10), we deduce, exactly as in the proof of [41, Lemma 3.8],

$$\int_0^T \int_{R^d} \Big[b(\overline{T_k(\tilde{\varrho})}) \partial_t \varphi + b(\overline{T_k(\tilde{\varrho})}) \tilde{\mathbf{u}} \cdot \nabla_x \varphi \mathrm{d}x \mathrm{d}t \\ + \int_0^T \int_{R^d} \Big(b(\overline{T_k(\tilde{\varrho})}) - b'(\overline{T_k(\tilde{\varrho})}) \overline{T_k(\tilde{\varrho})} \Big) \mathrm{div}_x \tilde{\mathbf{u}} \varphi \ \mathrm{d}x \mathrm{d}t \\ + \int_0^T \int_{R^d} b'(\overline{T_k(\tilde{\varrho})}) \overline{(T_k(\tilde{\varrho}) - T'(\tilde{\varrho})\tilde{\varrho}) \mathrm{div}_x \tilde{\mathbf{u}} \varphi} \Big] \mathrm{d}x \mathrm{d}t \\ = - \int_{R^d} T_k(\varrho_0) \varphi \mathrm{d}x$$

for any $\varphi \in C^1_c([0,T) \times (\Omega \cup \Gamma^0_{\rm out} \cup \overline{\Omega}^c))$. This can be rewritten, using the trace theorem for the transport equation (see Boyer [11], Crippa et al. [23]), as

$$\int_0^T \int_{\Omega} \Big[b(\overline{T_k(\varrho)}) \partial_t \varphi + b(\overline{T_k(\varrho)}) \Big] \mathbf{u} \cdot \nabla_x \varphi \ \mathrm{d}x \mathrm{d}t \\ + \int_0^T \int_{\Omega} \Big(b(\overline{T_k(\varrho)}) - b'(\overline{T_k(\varrho)}) \overline{T_k(\varrho)} \Big) \mathrm{div}_x \mathbf{u} \varphi \ \mathrm{d}x \mathrm{d}t \\ + \int_0^T \int_{\Omega} b'(\overline{T_k(\varrho)}) \overline{(T_k(\varrho) - T'(\varrho)\varrho) \mathrm{div}_x \mathbf{u} \varphi} \Big] \ \mathrm{d}x \\ = - \int_{\Omega} T_k(\varrho_0) \varphi \ \mathrm{d}x + \int_0^T \int_{\partial\Omega} \varphi b(\overline{T_k(\varrho)}) \ [\mathbf{u}_B \cdot \mathbf{n}]^+ \mathrm{d}\sigma_x \mathrm{d}t$$

for any $\varphi \in C_c^1([0,T) \times (\Omega \cup \Gamma_{\rm out}^0))$. Now, exactly as in the proof of [41, Lemma 3.8], we use boundedness of the oscillation defect measure (B.6) to let $k \to \infty$ obtaining

$$\begin{aligned}
&\int_0^T \int_\Omega \Big[b(\varrho)\partial_t \varphi + b(\varrho)\mathbf{u} \cdot \nabla_x \varphi \Big] \, {\rm d}x{\rm d}t \\
&\quad + \int_0^T \int_\Omega \Big(b(\varrho) - b'(\varrho)\varrho \Big) {\rm div}_x \mathbf{u} \varphi \, {\rm d}x{\rm d}t \\
&\quad = - \int_\Omega b(\varrho_0)\varphi \, {\rm d}x + \int_0^T \int_{\partial\Omega} \varphi b(\varrho) \, [\mathbf{u}_B \cdot \mathbf{n}]^+ {\rm d}\sigma_x){\rm d}t
\end{aligned} \tag{B.11}$$

for any $\varphi \in C_c^1([0,T) \times (\Omega \cup \Gamma_{\rm out}^0))$.

Finally, combining (B.9) with (B.11), we conclude

$$\begin{aligned}
&\int_0^T \int_\Omega \Big[b(\varrho)\partial_t \varphi + b(\varrho)\mathbf{u} \cdot \nabla_x \varphi \Big] \, {\rm d}x{\rm d}t \\
&\quad + \int_0^T \int_\Omega \Big(b(\varrho) - b'(\varrho)\varrho \Big) {\rm div}_x \mathbf{u} \varphi \, {\rm d}x{\rm d}t \\
&\quad = - \int_\Omega b(\varrho_0)\varphi \, {\rm d}x \\
&\quad + \int_0^T \int_{\partial\Omega} \varphi b(\varrho_B) \, [\mathbf{u}_B \cdot \mathbf{n}]^- {\rm d}\sigma_x + \int_0^T \int_{\partial\Omega} \varphi b(\varrho) \, [\mathbf{u}_B \cdot \mathbf{n}]^+ {\rm d}\sigma_x {\rm d}t
\end{aligned} \tag{B.12}$$

for any $\varphi \in C_c^1([0,T) \times (\Omega \cup \Gamma_{\rm in}^0 \cup \Gamma_{\rm out}^0))$.

B.1.5 Extension of the Class of Test Functions

The final step is to extend validity of (B.12) to the class of test functions $\varphi \in C_c^1([0,T) \times \overline{\Omega})$. To this end, consider

$$\Phi_\varepsilon(t,x) = \min \left\{ 1, \frac{1}{\varepsilon} {\rm dist} \left[(t,x); \Gamma_{\rm wall}^0 \cup \Gamma_{\rm sing} \right] \right\}.$$

By virtue of the hypothesis (B.1), $\Gamma_{\rm wall}^0 \cup \Gamma_{\rm sing}$ is a compact subset of $[0,T] \times R^{d+1}$, in particular, Φ_ε is Lipschitz continuous and vanishes on $\Gamma_{\rm wall}^0 \cup \Gamma_{\rm sing}$.

The first observation is that

$$\varphi = \phi \Phi_\varepsilon, \ \phi \in C_c^1([0,T) \times \overline{\Omega})$$

can be considered as a test function in (B.12). Indeed, we may approximate Φ_ε by

$$\Phi_{\varepsilon,\omega} = \min \left\{ 1, \frac{1}{\varepsilon} \mathrm{dist} \left[(t,x); \mathcal{U}_\omega(\Gamma^0_{\rm wall} \cup \Gamma_{\rm sing}) \right] \right\},$$

where

$$\mathcal{U}_\omega(Q) = \left\{ (t,x) \ \middle| \ \mathrm{dist}[(t,x);Q] < \omega \right\}$$

denotes an ω-neighbourhood of a set Q. Consequently, the standard approximation by regularizing kernels makes $\phi\Phi_{\varepsilon,\omega}$ eligible as test function in (B.12) and letting $\omega \to 0$ yields the desired conclusion.

Thus we start with

$$\begin{aligned}
&\int_0^T \int_\Omega b(\varrho) \Big[\Phi_\varepsilon \partial_t \phi + \phi \partial_t \Phi_\varepsilon + \Phi_\varepsilon \mathbf{u} \cdot \nabla_x \phi + \phi \mathbf{u} \cdot \nabla_x \ \Phi_\varepsilon \Big] \ \mathrm{d}x\mathrm{d}t \\
&\quad + \int_0^T \int_\Omega \Big(b(\varrho) - b'(\varrho)\varrho \Big) \mathrm{div}_x \mathbf{u} \Phi_\varepsilon \phi \ \mathrm{d}x\mathrm{d}t \\
&\quad = - \int_\Omega b(\varrho_0) \phi \Phi_\varepsilon \ \mathrm{d}x \\
&\quad + \int_0^T \int_{\partial\Omega} \phi \Phi_\varepsilon b(\varrho_B) \ [\mathbf{u}_B \cdot \mathbf{n}]^- \mathrm{d}\sigma_x \mathrm{d}t + \int_0^T \int_{\partial\Omega} \phi \Phi_\varepsilon b(\varrho) \ [\mathbf{u}_B \cdot \mathbf{n}]^+ \mathrm{d}\sigma_x \mathrm{d}t.
\end{aligned} \tag{B.13}$$

First, we have

$$\int_0^T \int_\Omega b(\varrho) \Big[\Phi_\varepsilon \partial_t \phi + \Phi_\varepsilon \mathbf{u} \cdot \nabla_x \phi \Big] \ \mathrm{d}x\mathrm{d}t \to \int_0^T \int_\Omega b(\varrho) \Big[\partial_t \phi + \mathbf{u} \cdot \nabla_x \phi \Big] \ \mathrm{d}x\mathrm{d}t,$$

$$\int_0^T \int_\Omega \Big(b(\varrho) - b'(\varrho)\varrho \Big) \mathrm{div}_x \mathbf{u} \Phi_\varepsilon \phi \ \mathrm{d}x\mathrm{d}t, \to \int_0^T \int_\Omega \Big(b(\varrho) - b'(\varrho)\varrho \Big) \mathrm{div}_x \mathbf{u} \phi \ \mathrm{d}x\mathrm{d}t$$

$$\int_\Omega \phi \Phi_\varepsilon b(\varrho_0) \ \mathrm{d}x \to \int_\Omega \phi b(\varrho_0) \ \mathrm{d}x,$$

and

$$\int_0^T \int_{\partial\Omega} \phi \Phi_\varepsilon b(\varrho_B) \ [\mathbf{u}_B \cdot \mathbf{n}]^- \mathrm{d}\sigma_x \mathrm{d}t \to \int_0^T \int_{\partial\Omega} \phi b(\varrho_B) \ [\mathbf{u}_B \cdot \mathbf{n}]^- \mathrm{d}\sigma_x \mathrm{d}t,$$

$$\int_0^T \int_{\partial\Omega} \phi \Phi_\varepsilon b(\varrho) \ [\mathbf{u}_B \cdot \mathbf{n}]^+ \mathrm{d}\sigma_x \mathrm{d}t \to \int_0^T \int_{\partial\Omega} \phi b(\varrho) \ [\mathbf{u}_B \cdot \mathbf{n}]^+ \mathrm{d}\sigma_x \mathrm{d}t.$$

Thus it remains to show

$$\lim_{\varepsilon \to 0} \int_0^T \int_\Omega b(\varrho) \phi \Big[\partial_t \Phi_\varepsilon + \mathbf{u} \cdot \nabla_x \Phi_\varepsilon \Big] \ \mathrm{d}x\mathrm{d}t = 0. \tag{B.14}$$

If

$$\mathrm{dist} \Big[(t,x); \Gamma^0_{\rm wall} \cup \Gamma_{\rm sing} \Big] > \varepsilon,$$

we obviously get

$$\Phi_\varepsilon = 1 \text{ on an open neighbourhood of } (t,x) \Rightarrow \partial_t \Phi_\varepsilon = 0,\ \nabla_x \Phi_\varepsilon = 0.$$

Next, write

$$\begin{aligned}&\int_0^T \int_\Omega b(\varrho)\mathbf{u}\cdot\nabla_x \Phi_\varepsilon \ \mathrm{d}x\mathrm{d}t \\ &= \int_0^T \int_{\mathcal{U}_\varepsilon(\partial\Omega)} b(\varrho)(\mathbf{u}-\mathbf{u}_B)\cdot\nabla_x \Phi_\varepsilon \ \mathrm{d}x\mathrm{d}t + \int_0^T \int_{\mathcal{U}_\varepsilon(\partial\Omega)} b(\varrho)\mathbf{u}_B\cdot\nabla_x \Phi_\varepsilon \ \mathrm{d}x\mathrm{d}t,\end{aligned}$$

where, by Hardy's inequality,

$$\begin{aligned}&\left| \int_0^T \int_{\mathcal{U}_\varepsilon(\partial\Omega)} b(\varrho)(\mathbf{u}-\mathbf{u}_B)\cdot\nabla_x \Phi_\varepsilon \ \mathrm{d}x\mathrm{d}t \right| \\ &\lesssim \frac{1}{\varepsilon}\left| \int_0^T \int_{\mathcal{U}_\varepsilon(\partial\Omega)} \frac{|\mathbf{u}-\mathbf{u}_B|}{\mathrm{dist}[x;\partial\Omega]}\ \mathrm{dist}[x;\partial\Omega]\mathrm{d}x\mathrm{d}t \right| \lesssim \varepsilon^{1-\frac{1}{\alpha}}.\end{aligned}$$

Thus (B.14) reduces to showing

$$\lim_{\varepsilon\to 0} \int_{\overline{\mathcal{U}_\varepsilon(\Gamma^0_{\mathrm{wall}}\cup\Gamma_{\mathrm{sing}})}} b(\varrho)\phi\left[\partial_t\Phi_\varepsilon + \mathbf{u}_B\cdot\nabla_x\Phi_\varepsilon\right]\ \mathrm{d}x\mathrm{d}t = 0. \tag{B.15}$$

As $\partial\Omega$ is Lipschitz, the outer normal vector $\mathbf{n} = \mathbf{n}(x)$ exists σ_x a.a. on $\partial\Omega$. Moreover, for a.a. $x \in \Omega$, there exists a unique nearest point $\Pi(x) \in \partial\Omega$,

$$|\Pi(x) - x| = \mathrm{dist}[x;\partial\Omega].$$

We introduce the sets

$$M_{r,\varepsilon} = \left\{ (t,x) \ \middle|\ \mathrm{dist}[(t,x); G^0_{\mathrm{wall}}\cup\Gamma_{\mathrm{sing}}] < \varepsilon,\ (t,\Pi(x)) \in \Gamma^0_{\mathrm{wall}} \right\},$$

$$M_{s,\varepsilon} = \overline{\mathcal{U}_\varepsilon(\Gamma_{\mathrm{wall}})} \setminus M_{r,\varepsilon}.$$

If $(t,x) \in M_{r,\varepsilon}$, then, as Γ^0_{wall} is open (cf. hypothesis (B.1)),

$$\Phi_\varepsilon = \frac{1}{\varepsilon}\mathrm{dist}\left[(t,x);\Gamma^0_{\mathrm{wall}}\right] \text{ on an open neighbourhood of } (t,x).$$

Moreover, the function

$$(t,x) \mapsto \mathrm{dist}\left[(t,x);\Gamma_{\mathrm{wall}}\right]$$

is differentiable for a.a. (t,x) and

$$\nabla_{t,x} \mathrm{dist}\left[(t,x); \Gamma_{\mathrm{wall}}\right] = -\frac{P(t,x) - (t,x)}{|P(t,x) - (t,x)|},$$

where $P(t,x)$ denotes the nearest point to (t,x) in Γ_{wall}. Consequently,

$$\partial_t \Phi_\varepsilon(t,x) = 0,\ \nabla_x \Phi_\varepsilon(t,x) = -\frac{1}{\varepsilon}\frac{x - \Pi(x)}{|x - \Pi(x)|} = -\frac{1}{\varepsilon}\mathbf{n}(\Pi(x))$$

for a.a. $(t,x) \in M_{r,\varepsilon}$. We infer that

$$\begin{aligned}
&\left| \int_{M_{r,\varepsilon}} b(\varrho)\phi \Big[\partial_t \Phi_\varepsilon + \mathbf{u}_B \cdot \nabla_x \Phi_\varepsilon \Big] \, \mathrm{d}x\mathrm{d}t \right| \\
&= \frac{1}{\varepsilon} \left| \int_{M_{r,\varepsilon}} b(\varrho)\phi \mathbf{u}_B(t,x) \cdot \mathbf{n}(\Pi(x)) \mathrm{d}x\mathrm{d}t \right| \\
&\lesssim \frac{1}{\varepsilon} \left| \int_{M_{r,\varepsilon}} |\mathbf{u}_B(t,x) - \mathbf{u}_B(t,\Pi(x)) \cdot \mathbf{n}(\Pi(x))| \mathrm{d}x\mathrm{d}t \right| \lesssim |M_{r,\varepsilon}| \to 0.
\end{aligned}$$

It remains to show

$$\lim_{\varepsilon \to 0} \int_{M_{s,\varepsilon}} b(\varrho)\phi \Big[\partial_t \Phi_\varepsilon + \mathbf{u}_B \cdot \nabla_x \Phi_\varepsilon \Big] \, \mathrm{d}x\mathrm{d}t = 0. \tag{B.16}$$

We have

$$M_{s,\varepsilon} \subset M^1_{s,\varepsilon} \cup M^2_{s,\varepsilon} \cup M^3_{s,\varepsilon},$$

where

$$M^1_{s,\varepsilon} = \Big\{ (t,x) \ \Big| \ \text{either } |x - \Pi(x)| = \varepsilon \text{ or } \Pi(x) \text{ does not exist} \Big\},$$

$$M^2_{s,\varepsilon} = \Big\{ (t,x) \ \Big| \ \mathrm{dist}[(t,x); \Gamma^0_{\mathrm{wall}} \cup \Gamma_{\mathrm{sing}}] \le \varepsilon,\ [t, \Pi(x)] \in \Gamma_{\mathrm{in}} \cup \Gamma_{\mathrm{out}} \Big\},$$

$$M^3_{s,\varepsilon} = \Big\{ (t,x) \ \Big| \ x \in \mathcal{U}_\varepsilon(\partial\Omega),\ [t, \Pi(x)] \in \Gamma_{\mathrm{sing}} \Big\}.$$

Seeing that $|M^1_{s,\varepsilon}| = 0$, the desired conclusion (B.16) follows as soon as we show

$$\lim_{\varepsilon \to 0} \frac{1}{\varepsilon} \left(|M^2_{s,\varepsilon}| + |M^3_{s,\varepsilon}| \right) = 0 \tag{B.17}$$

As $\Gamma_{\mathrm{in}} \cup \Gamma_{\mathrm{out}}$ is open, we have

$$M^2_{s,\varepsilon} = \cup_{h>0} M^2_{s,\varepsilon,h},$$

where

$$M^2_{s,\varepsilon,h} = \Big\{ (t,x) \ \Big| \ \mathrm{dist}[(t,x); \Gamma_{\mathrm{wall}}] \le \varepsilon, \mathrm{dist}[(t, \Pi(x)); \Gamma^0_{\mathrm{wall}} \cup \Gamma_{\mathrm{sing}}] > h \Big\}.$$

Now, for any $h > 0$, there is $\varepsilon(h) > 0$ such that

$$\Big\{(t,x) \ \Big|\ x \in \mathcal{U}_\varepsilon(\partial\Omega), \mathrm{dist}[(t,\Pi(x)); \Gamma^0_{\mathrm{wall}} \cup \Gamma_{\mathrm{sing}}] > h\Big\}$$
$$\subset \left\{(t,x) \ \middle|\ x \in \mathcal{U}_\varepsilon(\partial\Omega), \mathrm{dist}[(t,x); \Gamma^0_{\mathrm{wall}} \cup \Gamma_{\mathrm{sing}}] > \frac{h}{2}\right\} \text{ for any } 0 < \varepsilon < \varepsilon(h).$$

Consequently,

$$M^2_{s,\varepsilon,h} = \emptyset \text{ for all } 0 < \varepsilon < \min\left\{\varepsilon(h), \frac{h}{2}\right\},$$

and

$$\lim_{\varepsilon\to 0}\frac{1}{\varepsilon}|M^2_{s,\varepsilon}|$$
$$\leq \limsup_{\varepsilon\to 0}\frac{1}{\varepsilon}\left|\Big\{(t,x) \ \Big|\ \mathrm{dist}[(t,x); \Gamma^0_{\mathrm{wall}} \cup \Gamma_{\mathrm{sing}}] \leq \varepsilon,\right.$$
$$\left. 0 < \mathrm{dist}[(t,\Pi(x)); \Gamma^0_{\mathrm{wall}} \cup \Gamma_{\mathrm{sing}}] \leq h\Big\}\right|$$

for any $h > 0$. As Γ_{in}, Γ_{out}, and Γ^0_{wall} are open in $(0,T) \times \partial\Omega$, we have $\partial\Gamma^0_{\mathrm{wall}} \subset \Gamma_{\mathrm{sing}}$, and, consequently,

$$\Big\{(t,x) \ \Big|\ t \in (0,T),\ x \in \partial\Omega,\ 0 < \mathrm{dist}[(t,x); \Gamma^0_{\mathrm{wall}} \cup \Gamma_{\mathrm{sing}}] \leq h\Big\}$$
$$\subset \Big\{(t,x) \ \Big|\ t \in (0,T),\ x \in \partial\Omega,\ \mathrm{dist}[(t,x); \Gamma_{\mathrm{sing}}] \leq h\Big\}.$$

Thus the proof of (B.17) reduces to showing

$$\limsup_{\varepsilon\to 0}\frac{1}{\varepsilon}\left|\Big\{(t,x)\Big|\ x \in \mathcal{U}_\varepsilon(\partial\Omega),\ \mathrm{dist}\,[(t,\Pi(x)); \Gamma_{\mathrm{sing}}] \leq h\Big\}\right| = \mathcal{O}[h].$$

Since

$$x \in \mathcal{U}_\varepsilon(\partial\Omega),\ \mathrm{dist}\,[(t,\Pi(x)); \Gamma_{\mathrm{sing}}] \leq h \ \Rightarrow\ \mathrm{dist}\,[(t,x); \Gamma_{\mathrm{sing}}] \leq (h+\varepsilon),$$

it is enough to observe

$$\limsup_{\varepsilon\to 0}\frac{1}{\varepsilon}\left|\Big\{(t,x)\Big|\ x \in \mathcal{U}_\varepsilon(\partial\Omega),\ \mathrm{dist}\,[(t,x); \Gamma_{\mathrm{sing}}] \leq h\Big\}\right| = \mathcal{O}[h]. \tag{B.18}$$

On the one hand, as Γ_{sing} is compact, we have

$$0 = |\Gamma_{\mathrm{sing}}|_{\mathrm{d}t\otimes\mathrm{d}\sigma_x} = \lim_{h\to 0}|\mathcal{U}_h(\Gamma_{\mathrm{sing}})|_{\mathrm{d}t\otimes\mathrm{d}\sigma_x}. \tag{B.19}$$

On the other hand, for any $h > 0$ there is a continuous function $\chi \in C(R^{d+1})$ such that

$$0 \leq \chi \leq 1, \ \chi|_{\mathcal{U}_h(\Gamma_{\text{sing}})} = 1, \ \chi|_{R^{d+1} \setminus \mathcal{U}_{2h}(\Gamma_{\text{sing}})} = 0.$$

Consequently,

$$|\mathcal{U}_h(\Gamma_{\text{sing}})|_{\mathrm{d}t \otimes \mathrm{d}\sigma_x} \leq \int_0^T \int_{\partial\Omega} \chi \ \mathrm{d}\sigma_x \mathrm{d}t \leq |\mathcal{U}_{2h}(\Gamma_{\text{sing}})|_{\mathrm{d}t \otimes \mathrm{d}\sigma_x}.$$

By trace theorem

$$\int_0^T \int_{\partial\Omega} \chi \ \mathrm{d}\sigma_x \mathrm{d}t = \lim_{\varepsilon \to 0} \frac{1}{\varepsilon} \int_0^T \int_{\mathcal{U}_\varepsilon(\partial\Omega)} \chi \ \mathrm{d}x \mathrm{d}t.$$

Consequently,

$$\limsup_{\varepsilon \to 0} \frac{1}{\varepsilon} \left| \left\{ (t,x) \middle| \ x \in \mathcal{U}_\varepsilon(\partial\Omega), \ \text{dist}\left[(t,x); \Gamma_{\text{sing}}\right] \leq h \right\} \right|$$
$$\leq \lim_{\varepsilon \to 0} \frac{1}{\varepsilon} \int_0^T \int_{\mathcal{U}_\varepsilon(\partial\Omega)} \chi \ \mathrm{d}x \mathrm{d}t \leq |\mathcal{U}_{2h}(\Gamma_{\text{sing}})|_{\mathrm{d}t \otimes \mathrm{d}\sigma_x}, \tag{B.20}$$

whence (B.19) yields (B.18).

We have proved Theorem 19.

Remark 21. Let us point out that hypothesis (B.6) is trivially satisfied if the densities/velocities are integrable with sufficiently high exponents,

$$\varrho_n \to \varrho \text{ weakly-(*) in } L^\infty(0,T;L^\gamma(\Omega)), \ \gamma > 1,$$
$$\mathbf{u}_n \to \mathbf{u} \text{ weakly in } L^r(0,T;W^{1,r}(\Omega;R^d)), \ r > 1,$$
$$\frac{1}{\gamma} + \frac{1}{r} < 1.$$

If this is the case, we can simply assume that the limit $(\varrho, \mathbf{u})$ is a weak solution of the equation of continuity (B.4) belonging to the class

$$\varrho \in L^\infty(0,T;L^\gamma(\Omega)), \ \mathbf{u} \in L^r(0,T;W^{1,r}(\Omega;R^d)), \ \frac{1}{\gamma} + \frac{1}{r} \leq 1.$$

Indeed, the validity of the renormalized equation for the limit $(\varrho, \mathbf{u})$ in the extended domain established in Sect. B.1.3 can be shown by means of the regularization method of DiPerna and Lions [28] based on standard mollifiers, see [43, Chapter 11, Section 11.19].

References

1. R.A. Adams, *Sobolev Spaces* (Academic Press, New York, 1975)
2. J.M. Ball, G.-Q.G. Chen, Entropy and convexity for nonlinear partial differential equations. Philos. Trans. R. Soc. Lond. Ser. A Math. Phys. Eng. Sci., **371**(2005), 20120340 (2013)
3. J.M. Ball, A version of the fundamental theorem for Young measures. Lect. Notes in Physics, vol. 344 (Springer, 1989), pp. 207–215
4. G.K. Batchelor, *An Introduction to Fluid Dynamics* (Cambridge University Press, Cambridge, 1967)
5. A. Battaner, *Astrophysical Fluid Dynamics* (Cambridge University Press, Cambridge, 1996)
6. S.E. Bechtel, F.J. Rooney, M.G. Forest, Connection between stability, convexity of internal energy, and the second law for compressible Newtonian fluids. J. Appl. Mech. **72**, 299–300 (2005)
7. E. Becker, *Gasdynamik* (Teubner-Verlag, Stuttgart, 1966)
8. F. Belgiorno, Notes on the third law of thermodynamics, I. J. Phys. A **36**, 8165–8193 (2003)
9. F. Belgiorno, Notes on the third law of thermodynamics, II. J. Phys. A **36**, 8195–8221 (2003)
10. M.E. Bogovskii, Solution of some vector analysis problems connected with operators div and grad (in Russian). Trudy Sem. S.L. Sobolev **80**(1), 5–40 (1980)
11. F. Boyer, Trace theorems and spatial continuity properties for the solutions of the transport equation. Differ. Integral Equ. **18**(8), 891–934 (2005)
12. D. Bresch, B. Desjardins, Stabilité de solutions faibles globales pour les équations de Navier-Stokes compressibles avec température. C.R. Acad. Sci. Paris **343**, 219–224 (2006)
13. D. Bresch, B. Desjardins, On the existence of global weak solutions to the Navier-Stokes equations for viscous compressible and heat conducting fluids. J. Math. Pures Appl. **87**, 57–90 (2007)

E. Feireisl, A. Novotný, *Mathematics of Open Fluid Systems*, Nečas Center Series, https://doi.org/10.1007/978-3-030-94793-4

14. D. Bresch, P.-E. Jabin, Global existence of weak solutions for compressible Navier-Stokes equations: thermodynamically unstable pressure and anisotropic viscous stress tensor. Ann. Math. (2) **188**(2), 577–684 (2018)
15. M. Bulíček, J. Málek, K.R. Rajagopal, Navier's slip and evolutionary Navier-Stokes-like systems with pressure and shear-rate dependent viscosity. Indiana Univ. Math. J. **56**, 51–86 (2007)
16. T. Chang, B.J. Jin, A. Novotný, Compressible Navier-Stokes system with general inflow-outflow boundary data. SIAM J. Math. Anal. **51**(2), 1238–1278 (2019)
17. N. Chaudhuri, E. Feireisl, Navier–Stokes–Fourier system with Dirichlet boundary conditions. Preprint (2021). arxiv:2106.05315
18. G.-Q. Chen, M. Torres, W.P. Ziemer, Gauss-Green theorem for weakly differentiable vector fields, sets of finite perimeter, and balance laws. Comm. Pure Appl. Math. **62**(2), 242–304 (2009)
19. H.J. Choe, A. Novotný, M. Yang, Compressible Navier-Stokes system with hard sphere pressure law and general inflow-outflow boundary conditions. J. Differ. Equ. **266**(6), 3066–3099 (2019)
20. A.J. Chorin, J.E. Marsden, *A Mathematical Introduction to Fluid Mechanics* (Springer, New York, 1979)
21. R. Coifman, Y. Meyer, On commutators of singular integrals and bilinear singular integrals. Trans. Am. Math. Soc. **212**, 315–331 (1975)
22. P. Constantin, J. Wu, Statistical solutions of the Navier-Stokes equations on the phase space of vorticity and the inviscid limits. J. Math. Phys. **38**(6), 3031–3045 (1997)
23. G. Crippa, C. Donadello, L.V. Spinolo, A note on the initial–boundary value problem for continuity equations with rough coefficients. HYP 2012 Confer. Proc. AIMS Ser. Appl. Math. **8**, 957–966 (2013)
24. C.M. Dafermos, The entropy rate admissibility criterion for solutions of hyperbolic conservation laws. J. Differ. Equ. **14**, 202–212 (1973)
25. C.M. Dafermos, The second law of thermodynamics and stability. Arch. Ration. Mech. Anal. **70**, 167–179 (1979)
26. P.A. Davidson, *Turbulence: An Introduction for Scientists and Engineers* (Oxford University Press, Oxford, 2004)
27. F. Demengel, R. Temam, Convex functions of a measure and applications. Indiana Univ. Math. J. **33**(5), 673–709 (1984)
28. R.J. DiPerna, P.-L. Lions, Ordinary differential equations, transport theory and Sobolev spaces. Invent. Math. **98**, 511–547 (1989)
29. I. Ekeland, R. Temam, *Convex Analysis and Variational Problems* (North-Holland, Amsterdam, 1976)
30. S. Eliezer, A. Ghatak, H. Hora, *An Introduction to Equations of States, Theory and Applications* (Cambridge University Press, Cambridge, 1986)
31. L.C. Evans, A survey of entropy methods for partial differential equations. Bull. Am. Math. Soc. (N.S.) **41**(4), 409–438 (2004)

32. L.C. Evans, R.F. Gariepy, *Measure Theory and Fine Properties of Functions* (CRC Press, Boca Raton, 1992)
33. F. Fanelli, E. Feireisl, Statistical solutions to the barotropic Navier-Stokes system. J. Stat. Phys. **181**(1), 212–245 (2020)
34. F. Fanelli, E. Feireisl, M. Hofmanová, Ergodic theory for energetically open compressible fluid flows. Phys. D **423**, Paper No. 132914, 25 (2021)
35. E. Feireisl, On compactness of solutions to the compressible isentropic Navier-Stokes equations when the density is not square integrable. Comment. Math. Univ. Carolinae **42**(1), 83–98 (2001)
36. E. Feireisl, *Dynamics of Viscous Compressible Fluids* (Oxford University Press, Oxford, 2004)
37. E. Feireisl, Y.-S. Kwon, Asymptotic stability of solutions to the Navier–Stokes–Fourier system driven by inhomogeneous Dirichlet boundary conditions. Arxiv Preprint Series. Preprint (2021). arxiv:2109.00980
38. E. Feireisl, Y. Lu, J. Málek, On PDE analysis of flows of quasi-incompressible fluids. ZAMM Z. Angew. Math. Mech. **96**(4), 491–508 (2016)
39. E. Feireisl, Y. Lu, A. Novotný, Weak-strong uniqueness for the compressible Navier-Stokes equations with a hard-sphere pressure law. Sci. China Math. **61**(11), 2003–2016 (2018)
40. E. Feireisl, P. Mucha, A. Novotný, M. Pokorný, Time periodic solutions to the full Navier-Stokes-Fourier system. Arch. Rational. Mech. Anal. **204**, 745–786 (2012)
41. E. Feireisl, A. Novotný, *Singular Limits in Thermodynamics of Viscous Fluids* (Birkhäuser-Verlag, Basel, 2009)
42. E. Feireisl, A. Novotný, Weak-strong uniqueness property for the full Navier-Stokes-Fourier system. Arch. Rational Mech. Anal. **204**, 683–706 (2012)
43. E. Feireisl, A. Novotný, *Singular Limits in Thermodynamics of Viscous Fluids*, 2nd edn. Advances in Mathematical Fluid Mechanics (Birkhäuser/Springer, Cham, 2017)
44. E. Feireisl, A. Novotný, Navier-Stokes-Fourier system with general boundary conditions. Comm. Math. Phys. **386**(2), 975–1010 (2021)
45. E. Feireisl, A. Novotný, H. Petzeltová, On the existence of globally defined weak solutions to the Navier-Stokes equations of compressible isentropic fluids. J. Math. Fluid Mech. **3**, 358–392 (2001)
46. E. Feireisl, H. Petzeltová, On the long-time behaviour of solutions to the Navier-Stokes-Fourier system with a time-dependent driving force. J. Dynam. Differ. Equ. **19**(3), 685–707 (2007)
47. E. Feireisl, D. Pražák, *Asymptotic Behavior of Dynamical Systems in Fluid Mechanics* (AIMS, Springfield, 2010)
48. C. Foias, Statistical study of Navier-Stokes equations. I, II. Rend. Sem. Mat. Univ. Padova **48**, 219–348 (1973); ibid. 49 (1973), 9–123 (1972)

49. C. Foias, R.M.S. Rosa, R. Temam, Properties of time-dependent statistical solutions of the three-dimensional Navier-Stokes equations. Ann. Inst. Fourier (Grenoble) **63**(6), 2515–2573 (2013)
50. C. Foias, R.M.S. Rosa, R.M. Temam, Convergence of time averages of weak solutions of the three-dimensional Navier-Stokes equations. J. Stat. Phys. **160**(3), 519–531 (2015)
51. C. Foias, R.M.S. Rosa, R.M. Temam, Properties of stationary statistical solutions of the three-dimensional Navier-Stokes equations. J. Dynam. Differ. Equ. **31**(3), 1689–1741 (2019)
52. R.L. Foote, Regularity of the distance function. Proc. Am. Math. Soc. **92**, 153–155 (1984)
53. H. Gajewski, K. Gröger, K. Zacharias, *Nichtlineare Operatorgleichungen und Operatordifferentialgleichungen* (Akademie Verlag, Berlin, 1974)
54. G.P. Galdi, *An Introduction to the Mathematical Theory of the Navier-Stokes Equations, I* (Springer, New York, 1994)
55. G. Gallavotti, *Foundations of Fluid Dynamics* (Springer, New York, 2002)
56. M. Geißert, H. Heck, M. Hieber, On the equation $\operatorname{div} u = g$ and Bogovskiĭ's operator in Sobolev spaces of negative order, in *Partial Differential Equations and Functional Analysis*, vol. 168 of *Oper. Theory Adv. Appl.* (Birkhäuser, Basel, 2006), pp. 113–121
57. P. Germain, Weak-strong uniqueness for the isentropic compressible Navier-Stokes system. J. Math. Fluid Mech. **13**(1), 137–146 (2011)
58. V. Girinon, Navier-Stokes equations with nonhomogeneous boundary conditions in a bounded three-dimensional domain. J. Math. Fluid Mech. **13**(3), 309–339 (2011)
59. K. Itô, M. Nisio, On stationary solutions of a stochastic differential equation. J. Math. Kyoto Univ. **4**, 1–75 (1964)
60. A. Kastler, R. Vichnievsky, G. Bruhat, *Cours de physique générale à l'usage de l'enseignement supérieur scientifique et technique: Thermodynamique* (1962)
61. J. Kolafa, S. Labik, A. Malijevsky, Accurate equation of state of the hard sphere fluid in stable and metastable regions. Phys. Chem. Chem. Phys. **6**, 2335–2340 (2004)
62. A.N. Kolmogorov, *Selected Works of A. N. Kolmogorov. Vol. I*, vol. 25 of *Mathematics and Its Applications (Soviet Series)* (Kluwer Academic Publishers Group, Dordrecht, 1991). Mathematics and mechanics, With commentaries by V. I. Arnol′d, V. A. Skvortsov, P. L. Ul′yanov et al, Translated from the Russian original by V. M. Volosov, Edited and with a preface, foreword and brief biography by V. M. Tikhomirov
63. N.V. Krylov, *Introduction to the Theory of Random Processes*, vol. 43 of *Graduate Studies in Mathematics* (American Mathematical Society, Providence, RI, 2002)
64. Y.-S. Kwon, A. Novotny, Dissipative solutions to compressible Navier-Stokes equations with general inflow-outflow data: existence, stability

and weak strong uniqueness. J. Math. Fluid Mech. **23**(1), Paper No. 23, 27 (2021)
65. O.A. Ladyzhenskaya, V.A. Solonnikov, N.N. Uraltseva, *Linear and Quasilinear Equations of Parabolic Type.* Trans. Math. Monograph, vol. 23 (Amer. Math. Soc., Providence, 1968)
66. M.W. Licht, Smoothed projections over weakly Lipschitz domains. Math. Comp. **88**(315), 179–210 (2019)
67. P.-L. Lions, *Mathematical Topics in Fluid Dynamics, Vol.1, Incompressible Models* (Oxford Science Publication, Oxford, 1996)
68. P.-L. Lions, *Mathematical Topics in Fluid Dynamics, Vol.2, Compressible Models* (Oxford Science Publication, Oxford, 1998)
69. A. Lunardi, *Analytic Semigroups and Optimal Regularity in Parabolic Problems* (Birkhäuser, Berlin, 1995)
70. J. Málek, J. Nečas, A finite-dimensional attractor for the three dimensional flow of incompressible fluid. J. Differ. Equ. **127**, 498–518 (1996)
71. A. Matsumura, T. Nishida, The initial value problem for the equations of motion of viscous and heat-conductive gases. J. Math. Kyoto Univ. **20**, 67–104 (1980)
72. A. Matsumura, T. Nishida, The initial value problem for the equations of motion of compressible and heat conductive fluids. Comm. Math. Phys. **89**, 445–464 (1983)
73. A. Mellet, A. Vasseur, Existence and uniqueness of global strong solutions for one-dimensional compressible Navier-Stokes equations. SIAM J. Math. Anal. **39**(4), 1344–1365 (2007/08)
74. F Merle, P. Raphael, I. Rodnianski, J. Szeftel, On the implosion of a three dimensional compressible fluid. Arxiv Preprint Series. Preprint (2019). arxiv:1912.11009
75. I. Müller, T. Ruggeri, *Rational Extended Thermodynamics.* Springer Tracts in Natural Philosophy, vol. 37 (Springer, Heidelberg, 1998)
76. F. Murat, Compacité par compensation. Ann. Sc. Norm. Sup. Pisa, Cl. Sci. Ser. 5 **IV**, 489–507 (1978)
77. D.E. Norman, Chemically reacting fluid flows: weak solutions and global attractors. J. Differ. Equ. **152**(1), 75–135 (1999)
78. J. Oxenius, *Kinetic Theory of Particles and Photons* (Springer, Berlin, 1986)
79. P. Pedregal, Optimization, relaxation and Young measures. Bull. Am. Math. Soc. **36**, 27–58 (1999)
80. P. Plotnikov, J. Sokoł owski, *Compressible Navier-Stokes Equations*, vol. 73 of *Instytut Matematyczny Polskiej Akademii Nauk. Monografie Matematyczne (New Series) [Mathematics Institute of the Polish Academy of Sciences. Mathematical Monographs (New Series)].* (Birkhäuser/Springer Basel AG, Basel, 2012). Theory and shape optimization.
81. P.I. Plotnikov, W. Weigant, Isothermal Navier-Stokes equations and Radon transform. SIAM J. Math. Anal. **47**(1), 626–653 (2015)

82. L. Saint-Raymond, Hydrodynamic limits: some improvements of the relative entropy method. Annal. I.H.Poincaré - AN **26**, 705–744 (2009)
83. G.R. Sell, Global attractors for the three-dimensional Navier-Stokes equations. J. Dynam. Differ. Equ. **8**(1), 1–33 (1996)
84. B. Sprung, Upper and lower bounds for the Bregman divergence. J. Inequal. Appl., Paper No. 4, 12 (2019)
85. L. Tartar, Compensated compactness and applications to partial differential equations, in *Nonlinear Anal. and Mech., Heriot-Watt Sympos., L.J. Knopps editor, Research Notes in Math 39, Pitman, Boston*, pp. 136–211 (1975)
86. A. Valli, A correction to the paper: "An existence theorem for compressible viscous fluids" [Ann. Mat. Pura Appl. (4) **130** (1982), 197–213; MR 83h:35112]. Ann. Mat. Pura Appl. (4) **132**, 399–400 (1983)
87. A. Valli, An existence theorem for compressible viscous fluids. Ann. Mat. Pura Appl. (4) **130**, 197–213 (1982)
88. A. Valli, Periodic and stationary solutions for compressible Navier-Stokes equations via a stability method. Ann. Scuola Normale Sup. Pisa **10**(1), 607–646 (1983)
89. A. Valli, M. Zajaczkowski, Navier-Stokes equations for compressible fluids: Global existence and qualitative properties of the solutions in the general case. Commun. Math. Phys. **103**, 259–296 (1986)
90. M.J. Vishik, A.V. Fursikov, *Mathematical Problems of Statistical Hydromechanics*, vol. 9 of *Mathematics and Its Applications (Soviet Series)* (Kluwer Academic Publishers Group, Dordrecht, 1988). Translated from the 1980 Russian original [MR0591678] by D. A. Leites
91. W. Whitt, *Stochastic-Process Limits*. Springer Series in Operations Research (Springer, New York, 2002). An introduction to stochastic-process limits and their application to queues
92. W.P. Ziemer, *Weakly Differentiable Functions* (Springer, New York, 1989)

Index

E. Feireisl, A. Novotný, *Mathematics of Open Fluid Systems*, Nečas Center Series, https://doi.org/10.1007/978-3-030-94793-4

Printed in the United States
by Baker & Taylor Publisher Services